BRITISH
PARASITIC FUNGI

BRITISH PARASITIC FUNGI

A HOST-PARASITE INDEX AND A GUIDE TO BRITISH LITERATURE ON THE FUNGUS DISEASES OF CULTIVATED PLANTS

BY

W. C. MOORE
C.B.E., M.A.

Director of the Plant Pathology Laboratory, Ministry of Agriculture, Fisheries and Food, Harpenden

CAMBRIDGE
AT THE UNIVERSITY PRESS
1959

CAMBRIDGE UNIVERSITY PRESS
Cambridge, New York, Melbourne, Madrid, Cape Town,
Singapore, São Paulo, Delhi, Tokyo, Mexico City

Cambridge University Press
The Edinburgh Building, Cambridge CB2 8RU, UK

Published in the United States of America by
Cambridge University Press, New York

www.cambridge.org
Information on this title: www.cambridge.org/9780521279239

First published 1959
First paperback edition 2011

A catalogue record for this publication is available from the British Library

ISBN 978-0-521-05758-5 Hardback
ISBN 978-0-521-27923-9 Paperback

TO MY WIFE

CONTENTS

Preface   *page* ix

Abbreviations of geographical names   xi

Map showing the regions named in indicating the distri-
    bution of many plant diseases occurring in Scotland   xii

Books and periodicals   xiii

Abbreviations of selected references   xiv

PART I. HOSTS   1

PART II. PARASITES   49

PREFACE

My aim has been to prepare a reference book from which the plant pathologist can readily trace the parasitic fungi which have been reported on cultivated plants in Great Britain, the diseases caused by these fungi, and British literature on the subject.

To this end the book is divided into two parts, dealing respectively with hosts and parasites. The first part consists of a single alphabetical list of the scientific and common names of cultivated host plants. Under the scientific name of each host there is an alphabetical list of the fungus parasites recorded on it, including (in brackets) the imperfect stages and a few well-known synonyms.

In the second part there is an alphabetical list of the names of the parasites. Along with the accepted name, references are given to the place where the fungus was first described, to its Latin diagnosis in Saccardo's *Sylloge Fungorum*, to important British papers in which it is compiled, and, where appropriate, to well-known synonyms and imperfect stages. Under each parasite the disease or diseases which it causes are enumerated, with asterisks (*) to facilitate finding the host plants, and notes are added about the distribution and prevalence of the diseases. Annotated references to British literature on the subject are also given. The fungi that invade mushroom beds, and certain troublesome myxomycetes, though not parasites in the true sense, are purposely included, as also are a few plant pathogenic Actinomycetes, which are more often treated as allied to bacteria than to fungi.

The scientific names of the parasites and the common names of the diseases used are, where possible, those given in the third edition (Cambridge University Press, 1944) of the British Mycological Society's *List of Common British Plant Diseases*, but some have been changed to conform with names likely to be adopted in the next edition of that list, a draft of which I have been privileged to see. For the rest, a conservative attitude has been taken where there is still controversy about valid names of fungi.

ix

For the names of the host plants I have followed L. H. Bailey's *Manual of Cultivated Plants* (1949) and Alfred Rehder's *Manual of Cultivated Trees and Shrubs* (1947).

A selection of the books and journals consulted and searched is given on pp. xiii–xiv, but the *Reports on Fungus, Bacterial and other Diseases of Crops in England and Wales*, issued periodically since 1917 by the Ministry of Agriculture, Fisheries and Food (see pp. xiv–xv), provided the main source of information about the distribution and prevalence of the diseases. The references to British research have been extracted from a personal card-index of the world literature on plant diseases built up during the past forty years, and owing much to the compilers of the *Review of Applied Mycology*. There seemed little point in citing all the publications on widespread diseases such as potato blight, club root and apple scab, which have been the subject of extensive research in this country for many years, but I trust nothing really substantial has been overlooked.

I am deeply indebted to Dr F. Joan Moore for reading the manuscript and the proofs, and for her generous help in many ways. My warmest thanks are also due to Miss G. M. Waterhouse for tracing the original descriptions of a number of the parasites, to Mr T. R. Peace for assistance and guidance in dealing with diseases of forest trees, to Dr C. E. Foister for generously allowing me to draw freely on his own records of the distribution of plant diseases in Scotland, and to Mr E. C. Large for reading and criticizing the section on potato blight. My task was made much easier by the care with which Mrs J. O. Durden typed a difficult manuscript and checked the journal abbreviations with the *World List of Scientific Periodicals*, 3rd ed., 1952. The map on p. xii is reproduced by kind permission of Dr Foister and the editors of the British Mycological Society.

W.C.M.

ABBREVIATIONS OF GEOGRAPHICAL NAMES

The names of most English and Welsh counties have been abbreviated as shown below, and where possible these abbreviations are those recommended in the *Authors' and Printers' Dictionary* by F. H. Collins, 8th ed., 1938. Many of the Scottish records are listed according to the main drainage areas as laid down for *Insecta Scotica* and employed by Dennis and Foister in their list of diseases recorded in Scotland (*Trans. Brit. mycol. Soc.* **25**, 1942, 266–306)—see map on p. xii. Otherwise the names of Scottish counties are spelt out.

Anglesey	Anglesey	Kent	Kent
Beds	Bedfordshire	Lancs	Lancashire
Berks	Berkshire	Leics	Leicestershire
Brecon	Breconshire	Lincs	Lincolnshire
Bucks	Buckinghamshire	Merioneth	Merionethshire
Caerns	Caernarvonshire	Middx	Middlesex
Cambs	Cambridgeshire	Mon	Monmouthshire
Cards	Cardiganshire	Montgomery	Montgomeryshire
Carms	Carmarthenshire	Norfolk	Norfolk
Ches	Cheshire	Northants	Northamptonshire
Corn	Cornwall	Northumb	Northumberland
Cumb	Cumberland	Notts	Nottinghamshire
Denbigh	Denbighshire	Oxon	Oxfordshire
Derby	Derbyshire	Pemb	Pembrokeshire
Devon	Devonshire	Radnor	Radnorshire
Dorset	Dorset	Rutland	Rutland
Dur	County Durham	Salop	Shropshire
Essex	Essex	Som	Somerset
Flints	Flintshire	Staffs	Staffordshire
Glam	Glamorganshire	Suffolk	Suffolk
Glos	Gloucestershire	Surrey	Surrey
Hants	Hampshire	Sussex	Sussex
Herefs	Herefordshire	War	Warwickshire
Herts	Hertfordshire	Westmorland	Westmorland
Hunts	Huntingdonshire	Wilts	Wiltshire
I. of Ely	Isle of Ely	Worcs	Worcestershire
I.W.	Isle of Wight	Yorks	Yorkshire

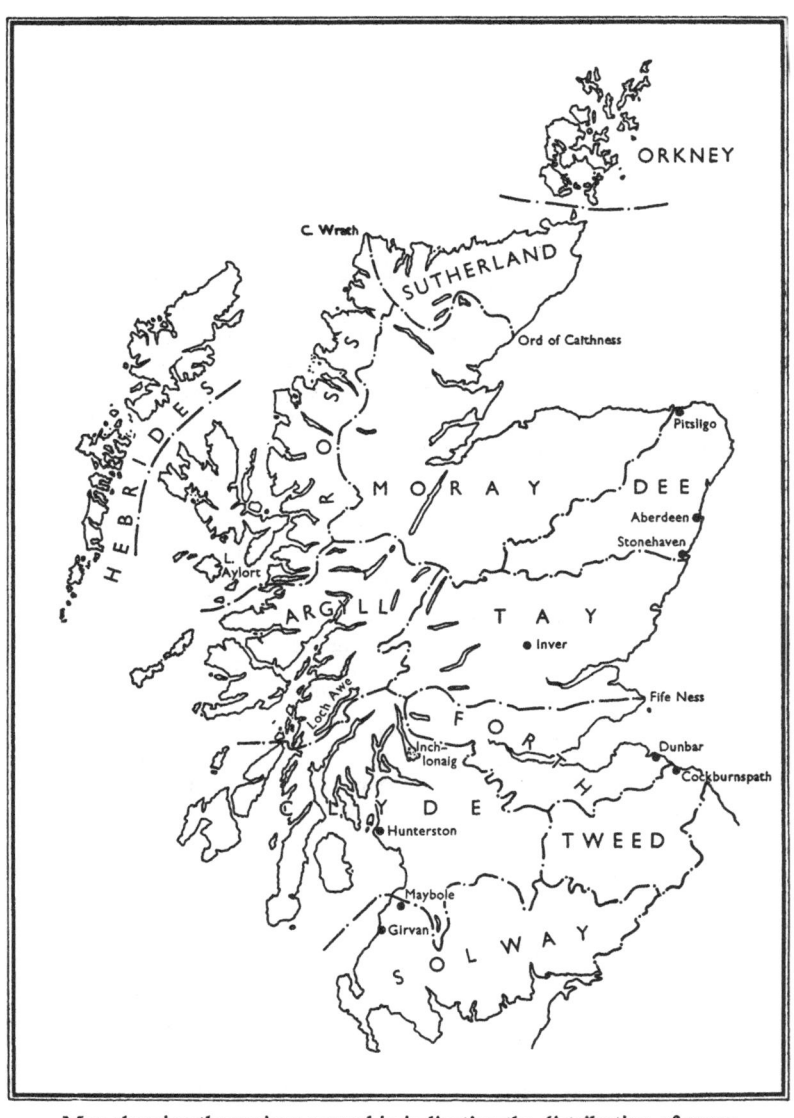

Map showing the regions named in indicating the distribution of many
plant diseases occurring in Scotland

BOOKS AND PERIODICALS

The following books and periodicals were among those consulted and searched for British records of fungus diseases of cultivated plants.

Ainsworth, G. C. *The Plant Diseases of Great Britain* (1937).
Ainsworth, G. C. & Sampson, K. *The British Smut Fungi* (1950).
Beaumont, A. *Diseases of Garden Plants* (1956).
Bewley, W. F. *Diseases of Glasshouse Plants* (1923).
British Mycological Society. *List of Common British Plant Diseases* (1944).
Brooks, F. T. *Plant Diseases* (1953).
Cooke, M. C. *Fungoid Pests of Cultivated Plants* (1906).
Green, D. E. *Diseases of Vegetables* (1943).
Grove, W. B. *The British Rust Fungi* (1913).
Grove, W. B. *British Stem- and Leaf-Fungi*, vols. i (1935) and ii (1937).
Massee, G. *Diseases of Cultivated Plants and Trees* (1915).
Moore, W. C. Diseases of Bulbs (*Bull. Minist. agric. Lond.* no. 117, 1939).
Muskett, A. E. & Colhoun, J. *The Diseases of the Flax Plant* (1947).
Plowright, C. B. *British Uredineae and Ustilagineae* (1889).
Sampson, K. & Western, J. H. *Diseases of British Grasses and Herbage Legumes* (1954).
Wilson, M. *The Distribution of the Uredineae of Scotland* (1934).
Wormald, H. *Diseases of Fruit and Hops* (1955).

Annals of Applied Biology (1914–).
Gardeners' Chronicle (1841–).
Journal of Pomology and Horticultural Science (1919–).
Journal of the Royal Agricultural Society (1840–).
Journal of the Royal Horticultural Society (1893–).
Journal of the South-Eastern Agricultural College, Wye (1902–).

Ministry of Agriculture, Fisheries and Food publications, notably the periodic Reports on Crop Diseases (see pp. xiv–xv); *Bulletins* nos. 25 (Celery Diseases), 117 (Diseases of Bulbs), 123 (Vegetable Diseases), 129 (Cereal Diseases), 142 (Sugar Beet Diseases); and *Technical Bull.* no. 3 (Brown Rot Diseases of Fruit Trees).

BOOKS AND PERIODICALS

Plant Pathology (1952–).
Review of Applied Mycology (1922–).
Transactions of the British Mycological Society (1898–).

The Annual Reports of the Agricultural and Horticultural Research Stations at Cheshunt, East Malling, Long Ashton, Rothamsted, and Wellesbourne.

ABBREVIATIONS OF SELECTED REFERENCES

Many references are made throughout the text to the *Transactions of the British Mycological Society* and the *Annals of Applied Biology*, as well as to the periodic Reports on Diseases of Crop Plants issued by the Ministry of Agriculture, Fisheries and Food. For brevity's sake the special abbreviations given below are used for these. Otherwise the *World List of Scientific Periodicals* has been followed wherever possible.

TBMS: *Transactions of the British Mycological Society.*
AAB: *Annals of Applied Biology.*
Report i: Report on the Occurrence of Insect and Fungus Pests on Plants in England and Wales in the year 1917. *Misc. Publ. Bd Agric., Lond.*, no. 21, 1918, 32 pp.
Report ii: Report on the Occurrence of Insect and Fungus Pests on Plants in England and Wales in the year 1918. *Misc. Publ. Bd Agric., Lond.*, no. 23, 1920, 65 pp.
Report iii: Report on the Occurrence of Insect and Fungus Pests on Plants in England and Wales for the year 1919. *Misc. Publ. Minist. Agric., Lond.*, no. 33, 1921, 68 pp.
Report iv: Fungus Diseases of Crops 1920–1921. *Misc. Publ. Minist. Agric., Lond.*, no. 38, 1922, 104 pp.
Report v: Report on the Occurrence of Fungus, Bacterial and Allied Diseases of Crops in England and Wales for the years 1922–1924. *Misc. Publ. Minist. Agric., Lond.*, no. 52, 1926, 95 pp.
Report vi: Report on the Occurrence of Fungus, Bacterial and Allied Diseases of Crops in England and Wales for the years 1925, 1926 and 1927. *Misc. Publ. Minist. Agric., Lond.*, no. 70, 1929, 75 pp.
Report vii: Fungus and other Diseases of Crops 1928–1932. *Bull. Minist. Agric., Lond.*, no. 79, 1934, 117 pp.
Report viii: Diseases of Crop Plants. A ten years' review (1933–1942). *Bull. Minist. Agric., Lond.*, no. 126, 1943, 101 pp.

ABBREVIATIONS OF SELECTED REFERENCES

Report ix: Diseases of Crop Plants, 1943–1946. *Bull. Minist. Agric., Lond.*, no. 139, 1948, 90 pp.

The following books and pamphlets are referred to in an abbreviated form as shown below:

Ainsworth & Sampson, *Brit. Smut Fungi*: Ainsworth, G. C. & Sampson, K., *The British Smut Fungi (Ustilaginales)*, Kew, 1950.

de Bary, *Morphol.*: Bary, A. de, *Vergleichende Morphologie und Biologie der Pilze*, Leipzig, 1st ed., 1866; 2nd ed., 1884.

Berk., *Outl.* 1860: Berkeley, M. J., *Outlines of British Fungology*, London, 1860.

Bon., *Handb.*: Bonorden, H. F., *Handbuch der allgemeinen Mykologie*, Stuttgart, 1851.

Bri. & Cav., *Fungh. Parass.*: Briosi, G. & Cavara, F., *I funghi parassiti delle piante coltivate de utili essicati, delineati e descritti*, Pavia, Fascs. i–xvi, 1888–1905.

Ces. & de Not., *Schema Sfer.*: Cesati, V. de & de Notaris, G., 'Schema di classificazione degli Sferiacei italici', *Comment. Soc. Crittogam. Ital.* v, Pt. iv, 1863, 177–240.

Cooke, *Fung. Pests Cult. Pl.*: Cooke, M. C., *Fungoid Pests of Cultivated Plants*, London, 1906.

Cooke, *Handb.*: Cooke, M. C., *Handbook of British Fungi*, 2 vols., London, 1871.

Corda, *Ic. Fung.*: Corda, A. C. J., *Icones Fungorum hucusque cognitorum*, Parts i–vi, Prague, 1837–54.

DC., *Flor. Fr.*: De Candolle, A. P. & de Lamarck, J., *Flore Française*, vols. ii and vi, Paris (dated 1815 but published 1805).

Fr., *Epicr.*: Fries, E. M., *Epicrisis systematis mycologici*, Upsaliae et Lundae, 1836–8.

Fr., *Summ. Veg. Scand.*: Fries, E. M., *Summa vegetabilium Scandinaviae*, Holmiae et Lipsiae, 1846–9.

Fr., *Syst. Myc.*: Fries, E. M., *Systema Mycologicum*, Gryphiswaldiae, vol. i, 520 pp., 1821; ii, pp. 1–274, 1822; pp. 275–620, 1823; iii, pp. 1–260, 1829; pp. 261–524, 1832.

Fuckel, *Symb. Myc.* [*Nachtr.*]: Fuckel, L., *Symbolae Mycologicae, Beiträge zur Kenntniss der Rheinischen Pilze*, Wiesbaden, 1869. *Nachtrag* (Suppl.), i, 1871; ii, 1873; iii, 1877.

Grove, *Brit. Rust Fungi*: Grove, W. B., *The British Rust Fungi (Uredinales)*, Cambridge, 1913.

Grove, *Coelomycetes*: Grove, W. B., *British Stem- and Leaf-Fungi (Coelomycetes)*, Cambridge, vol. i, 1935; vol. ii, 1937.

Karst., *Myc. fenn.*: Karsten, P. A., *Mycologia Fennica*, Helsingfors, part i, 1871; ii, 1873; iii, 1876; iv, 1878.

ABBREVIATIONS OF SELECTED REFERENCES

Kickx, *Fl. Cr. Fl.*: Kickx, J., *Flore cryptogamique des Flandres*, Bruxelles, 2 vols., 1867.

Massee, *Brit. Fung. Fl.*: Massee, G., *British Fungus-Flora*, London, 4 vols., 1892–5.

Massee, *Dis. Cult. Pl. Trees*: Massee, G., *Diseases of Cultivated Plants and Trees*, London, 1910; 2nd ed., 1915.

Pers., *Syn. Fung.*: Persoon, C. H., *Synopsis methodica fungorum*, Göttingen, 1801.

Plowr., *Brit. Ured. Ustil.*: Plowright, C. B., *A Monograph of the British Uredineae and Ustilagineae*, London, 1889.

Pringsh., *Jahrb.*: *Pringsheim's Jährbucher für wissenschaftliche Botanik*, Berlin, Themen i–xiv, 1858–84.

Rabenh., *Fung. Eur.*: Rabenhorst, G. L., *Fungi Europaei Exsiccati*, Dresden, Cent. i–xxxvi, 1859–86.

Rabenh., *Krypt. Fl.*: Rabenhorst, G. L., *Deutschlands Kryptogamen-Flora*, Leipzig, 1844–53; 2nd ed. as *Kryptogamen-Flora von Deutschland, Oesterreich und der Schweiz*, 1884–.

Rea, *Brit. Basid.*: Rea, Carleton, *British Basidiomycetae*, Cambridge, 1922.

Sacc., *Mich.*: Saccardo, P. A., *Fungi italici autographice delineati. Michelia*, Patavia, vols. i and ii, 1877–82.

Sacc., *Syll.*: Saccardo, P. A., *Sylloge Fungorum omnium hucusque cognitorum*, Patavia, 25 vols., 1882–1931.

Salmon, *Monogr. Erysiph.*: Salmon, E. S., A Monograph of the Erysiphaceae. *Mem. Torrey bot. Cl. 9*, 1900, 292 pp.

Schröt., *Kr. Fl. Schles.*: Schröter, J., *Kryptogamen-Flora von Schlesien. Pilze*, i, Breslau, 1885–9.

Speg., *Fung. Arg.*: Spegazzini, C., *Fungi Argentini. Pugillus* i–iv Buenos-Aires, 1880–2.

Sydow, *Monogr. Uredin.*: Sydow, H. & Sydow, P., *Monographia Uredinearum*, Lipsiae, 1902–24.

Tul., *Sel. Fung. Carp.*: Tulasne, L. R. & Tulasne, C., *Selecta Fungorum Carpologia*, Paris, vol. i, 1861; ii, 1863; iii, 1865.

Westend., *Exs.*: Westendorp, G. D., *Herbier cryptogamique belge*, Fasc. 1–28, 1841–59.

Wint., *Pilze Deutschl.*: Winter, G. in Rabenhorst's *Kryptogamen-Flora*, 1881–7.

PART I

HOSTS

HOSTS

ABIES
Acanthostigma parasiticum
Armillaria mellea
Botrytis *sp.*
Calyptospora goeppertiana
(Camarosporium abietis)
Fomes annosus
Melampsorella caryophylla-
cearum
Milesia blechni
Milesia kriegeriana
Milesia polypodii
Milesia scolopendrii
Myxocyclus cenangioides
Peridermium coruscans
Phomopsis pseudotsugae
Pucciniastrum epilobii
Rehmiellopsis bohemica
Rhizosphaera kalkhoffii
Acacia, False, see Robinia
ACER CAMPESTRE L.
(maple)
Cryptostroma corticale
(Melasmia acerina)
Phleospora aceris
Phyllosticta aceris
Rhytisma acerinum
ACER PSEUDO-PLATA-
NUS L. *(sycamore)*
Armillaria mellea
Cristulariella depraedens
Cryptostroma corticale
Leptothyrium platanoidis
(Melasmia acerina)
Nectria cinnabarina
Phleospora aceris
Phyllosticta platanoidis
Phytophthora cactorum
Polyporus squamosus
Rhytisma acerinum
Rhytisma pseudoplatani
Rhytisma punctatum

Stereum purpureum
Uncinula aceris
ACER *spp.*
Armillaria mellea
(Melasmia acerina)
Phleospora pseudoplatani
Rhytisma acerinum
ACHILLEA
Schizothyrium ptarmicae
ACIDANTHERA
Botrytis gladiolorum
Sclerotinia gladioli
Septoria gladioli
Aconite, see Eranthis
AESCHYNANTHUS
Gloeosporium affine
AESCULUS
Botrytis cinerea
Guignardia aesculi
Phyllosticta paviae
Septoria hippocastani
Stereum purpureum
AGARICUS *(mushroom)*
Parasites on mushrooms
Acremonium *sp.*
Badhamia utricularis
Cephalosporium costantinii
Cephalosporium lamellicola
Dactylium dendroides
Fusarium oxysporum
Fusarium solani
Fusarium *sp.*
Hormiactis *sp.*
Mycogone perniciosa
Mycogone rosea
Verticillium malthousei
Verticillium psalliotae
Weed fungi of mushroom beds
Aleuria vesiculosa
Ascobolus leveillei
Botrytis gemella
Chaetomium globosum

Chaetomium olivaceum
Clitocybe dealbata
Clitopilus cretatus
Dialonectria peziza
Heleococcum aurantiacum
Lilliputia insigne
Myceliophthora lutea
Oedocephalum pallidum
Paneolus sub-balteatus
Papulaspora byssina
Plicaria fulva
Pseudobalsamea microspora
Scopulariopsis fimicola
Sporendonema purpurascens
Trichoderma viride
Volvaria *sp.*
Xylaria vaporaria
AGAVE
Coniothyrium concentricum
var. agaves
AGROPYRON REPENS L.
(*couch grass*)
Cladochytrium caespitis
Claviceps purpurea
Epichloe typhina
Erysiphe graminis
Helminthosporium tritici-
repentis
Ophiobolus herpotrichus
Phyllachora graminis
Physoderma graminis
Puccinia agropyrina
Puccinia graminis
Rhynchosporium secalis
Septoria affinis
Urocystis agropyri
Ustilago bullata
Ustilago hypodytes
Ustilago macrospora
Ustilago striiformis
AGROSTIS
Cercosporella herpotrichoides
Cladochytrium caespitis
Corticium fuciforme
Dilophospora alopecuri

Epichloe typhina
Fusarium culmorum
(Fusarium nivale)
Griphosphaeria nivalis
Hadrotrichum virescens
Helminthosporium stenacrum
Mastigosporium rubricosum
var. agrostidis
Olpidium agrostidis
Ophiobolus graminis
Ophiobolus graminis *var.* avenae
Phyllachora graminis
Puccinia agrostidis
Puccinia graminis
(Rhizoctonia monteithianum)
Sclerotinia homoeocarpa
Sclerotium rhizodes
Selenophoma donacis
(Septoria oxyspora)
Tilletia decipiens
Alder, see Alnus glutinosa (L.)
Gaertn.
Alfalfa, see Medicago sativa L.
ALLIUM ASCALONICUM
L. (*shallot*)
Botrytis allii
Fusarium avenaceum
Fusarium *sp.*
Peronospora destructor
Phytophthora *sp.*
Sclerotium cepivorum
ALLIUM CEPA L. (*onion*)
Alternaria porri
Armillaria mellea
Botrytis allii
Botrytis byssoidea
Botrytis cinerea
Botrytis squamosa
Colletotrichum circinans
Corticium solani
Fusarium *sp.*
Heterosporium allii *var.* cepi-
vorum
Melampsora allii-fragilis
Peronospora destructor

4

Phytophthora *sp.*
Pleospora herbarum
Puccinia porri
Pythium *sp.*
(Rhizoctonia solani)
Sclerotinia porri
Sclerotium cepivorum
(Stemphylium botryosum)
Urocystis cepulae
ALLIUM PORRUM L.
(*leek*)
Botrytis cinerea
Corticium solani
Fusarium culmorum
Heterosporium allii
Phytophthora porri
Puccinia porri
(Rhizoctonia solani)
Sclerotium cepivorum
Urocystis cepulae
ALLIUM SATIVUM L.
(*garlic*)
Helminthosporium allii
Puccinia allii
Sclerotium cepivorum
**ALLIUM SCHOENO-
PRASUM L.** (*chives*)
Heterosporium allii
Puccinia porri
Uromyces ambiguus
ALLIUM *spp.*
Melampsora allii-fragilis
Melampsora allii-populina
Melampsora allii-salicis-albae
Puccinia porri
Sclerotium cepivorum
Urocystis cepulae
Uromyces ambiguus
Almond, see Prunus amygdalus
Batsch
ALNUS GLUTINOSA (L.)
Gaertn. (*alder*)
Armillaria mellea
Fomes annosus
Polyporus radiatus

Taphrina alni-incanae
Taphrina sadebeckii
Taphrina tosquinetii
ALOE
Uromyces aloes
**ALOPECURUS PRATEN-
SIS L.** (*meadow foxtail*) *and*
ALOPECURUS *spp.*
Ascochyta graminicola
Cladochytrium caespitis
Claviceps purpurea
Dilophospora alopecuri
Epichloe typhina
Mastigosporium album
Passalora graminis
Puccinia coronata
Puccinia perplexans
Selenophoma donacis
(Septoria oxyspora)
ALTHAEA ROSEA Cav.
(*hollyhock*)
Puccinia malvacearum
Sclerotinia sclerotiorum
ALYSSUM
Cystopus candidus
Fusarium *sp.*
Peronospora galligena
Pythium *sp.*
Sclerotinia sclerotiorum
AMARYLLIS
Corticium rolfsii
(Sclerotium rolfsii)
Stagonospora curtisii
Thielaviopsis basicola
ANCHUSA
Sclerotinia sclerotiorum
ANDROMEDA
Armillaria mellea
ANEMONE, *see also* **HEPA-
TICA**
Ascochyta *sp.*
Botrytis cinerea
Itersonilia *sp.*
Leptothyrium anemones
Ochropsora ariae

Oidium *sp.*
Peronospora ficariae
Phyllosticta *sp.*
Phytophthora *sp.*
Plasmopara pygmaea
(Puccinia pruni-spinosae)
Rhizoctonia *sp.*
Sclerotinia tuberosa
Septoria anemones *var.* coro-
   nariae
Thielaviopsis basicola
Tranzschelia pruni-spinosae
Urocystis anemones
ANTHEMIS
Verticillium dahliae
ANTHOCERCIS
Phytophthora infestans
**ANTHOXANTHUM ODO-
   RATUM L.** (*sweet vernal
   grass*)
Helminthosporium dematioi-
   deum
Puccinia anthoxanthi
Puccinia poae-nemoralis
ANTIRRHINUM MAJUS L.
   (*snapdragon*)
Botrytis cinerea
(Cercosporella antirrhini)
Corticium solani
Diplodina passerinii
Fusarium *sp.*
Helicobasidium purpureum
Heteropatella antirrhini
Myrothecium roridum
Oidium *sp.*
Peronospora antirrhini
Phyllosticta antirrhini
Phytophthora citricola
Phytophthora cryptogea
(Pseudodiscosia antirrhini)
Puccinia antirrhini
Pythium *sp.*
(Rhizoctonia crocorum)
(Rhizoctonia solani)
Sclerotinia sclerotiorum

Septoria antirrhini
Thielaviopsis basicola
Verticillium albo-atrum
Verticillium dahliae
Verticillium nigrescens
APIUM GRAVEOLENS L.
   var. **DULCE** Pers. (*celery*)
(Ansatospora acerina)
Centrospora acerina
Cercospora apii
Corticium solani
Helicobasidium purpureum
Phoma apiicola
Phyllosticta apii
Phytophthora cryptogea
Puccinia apii
Pythium *sp.*
(Rhizoctonia crocorum)
(Rhizoctonia solani)
Sclerotinia sclerotiorum
Septoria apii
Septoria apii-graveolentis
Stemphylium radicinum
Thielaviopsis basicola
Verticillium dahliae
Apple, see Malus sylvestris Mill.
Apricot, see Prunus armeniaca L.
AQUILEGIA
Actinonema aquilegiae
Erysiphe polygoni
Haplobasidium pavoninum
Puccinia agrostidis
ARABIS
Cystopus candidus
ARAUCARIA
Phomopsis araucariae
ARBUTUS
Septoria unedonis
ARCTOSTAPHYLOS
Pleospora herbarum
(Stemphylium botryosum)
ARISAEMA
Uromyces ari-triphylli
ARISTOLOCHIA
Septoria aristolochiae

6

ARISTOTELIA
Gloeosporium aristoteliae
ARMERIA
Botrytis cinerea
Septoria armeriae
Uromyces armeriae
ARMORACIA LAPATHI-
FOLIA Gilib. (horse-
radish)
Corticium solani
Cystopus candidus
Mycosphaerella brassicicola
Peronospora parasitica
(Phyllosticta brassicicola)
Ramularia armoraciae
(Rhizoctonia solani)
Septoria armoraciae
ARRHENATHERUM ELA-
TIUS Mert. & Koch (tall
oat grass)
Claviceps purpurea
Puccinia arrhenatheri
Puccinia coronata
Puccinia graminis
Selenophoma donacis
(Septoria oxyspora)
Urocystis agropyri
Ustilago avenae
Ustilago striiformis
ARTEMISIA
Erysiphe cichoracearum
Artichoke (globe), see Cynara
scolymus L.
Artichoke (Jerusalem), see
Helianthus tuberosus L.
Arum lily, see Zantedeschia
Ash, see Fraxinus excelsior L.
ASPARAGUS
Botrytis cinerea
Fusarium sp.
Helicobasidium purpureum
Phytophthora sp.
Puccinia asparagi
Pythium sp.
(Rhizoctonia crocorum)

Zopfia rhizophila
ASPIDISTRA
Ascochyta aspidistrae
ASPLENIUM, see FERNS
ASTER, including Michaelmas
daisy
Botrytis cinerea
Erysiphe cichoracearum
Sclerotinia sclerotiorum
Thielaviopsis basicola
Verticillium vilmorinii
Aster (China), see Callistephus
chinensis Nees
ATHROTAXIS
Armillaria mellea
ATHYRIUM, see FERNS
ATROPA BELLADONNA L.
Phytophthora erythroseptica
var. atropae
AUBRETIA
Cystopus candidus
Peronospora parasitica
AUCUBA
Botrytis cinerea
Diplodia aucubae
Phomopsis aucubicola
AURICULA, see PRIMULA
AVENA SATIVA L. (oats)
Cercosporella herpotrichoides
Claviceps purpurea
Corticium solani
Dilophospora alopecuri
Erysiphe graminis
Fusarium avenaceum
Fusarium culmorum
Fusarium equiseti
(Fusarium graminearum)
(Fusarium nivale)
Fusarium tricinctum f. poae
Gibberella pulicaris
Gibberella zeae
Griphosphaeria nivalis
(Helminthosporium avenae)
Leptosphaeria avenaria
Ophiobolus graminis var. avenae

7

Puccinia coronata
Puccinia graminis
Pyrenophora avenae
Pythium *sp.*
(Rhizoctonia solani)
(Septoria avenae)
Thielaviopsis basicola
Typhula incarnata
Ustilago avenae
Ustilago hordei
AZALEA, *see* **RHODODEN-
DRON**
AZARA
Stereum purpureum

Barley, see Hordeum
Bean (broad and *field), see* Vicia
faba L.
Bean (dwarf and *runner), see*
Phaseolus
Beech, see Fagus sylvatica L.
Beet, see Beta vulgaris L.
BEGONIA
Botrytis cinerea
Corticium solani
Gloeosporium begoniae
Oidium begoniae
Phyllosticta begoniae
Phytophthora parasitica
Pythium *sp.*
(Rhizoctonia solani)
Sclerotinia sclerotiorum
Thielaviopsis basicola
BERBERIS
Armillaria mellea
Microsphaera berberidis
Ovularia berberidis
Phoma berberidicola
Puccinia arrhenatheri
Puccinia graminis
Septoria berberidis
BETA VULGARIS L. *and vars.*
*(beet, fodder beet, mangold,
spinach beet, sugar beet)*
(Actinomyces scabies)

(Actinomyces tumuli)
Aphanomyces cochlioides
Aphanomyces laevis
Ascochyta betae
Botrytis cinerea
Cercospora beticola
Corticium solani
Erysiphe *sp.*
Fusarium culmorum
Helicobasidium purpureum
Peronospora schachtii
(Phoma betae)
Phytophthora megasperma
Phytophthora *sp.*
Pleospora betae
Pleospora herbarum
Pythium intermedium
Pythium ultimum
Pythium *sp.*
Ramularia beticola
(Rhizoctonia crocorum)
(Rhizoctonia solani)
Rosellinia necatrix
Sclerotinia sclerotiorum
Sporodesmium putrefaciens
(Stemphylium botryosum)
Streptomyces scabies
Streptomyces tumuli
Thielaviopsis basicola
Uromyces betae
BETULA *(birch)*
Armillaria mellea
Fomes annosus
Fomes fomentarius
Melampsoridium betulinum
Phyllosticta betulina
Plowrightia virgultorum
Polyporus betulinus
Poria obliqua
Stereum purpureum
Taphrina betulina
Birch, see Betula
Blackberry, see Rubus *spp.*
Black currant, see Ribes
nigrum L.

8

BLECHNUM, *see* FERNS
Box, see Buxus
BRACHYPODIUM
  Ustilago olida
BRASSICA NAPOBRAS-
  SICA Mill., *see* B. RAPA
BRASSICA NAPUS L. (*rape*)
  Alternaria brassicae
  Corticium solani
  Erysiphe polygoni
  Peronospora parasitica
  Phytophthora cryptogea
  Plasmodiophora brassicae
  Pythium de baryanum
  (Rhizoctonia solani)
BRASSICA NIGRA Koch
  (*mustard*)
  Corticium solani
  Cystopus candidus
  Fusarium *sp.*
  Plasmodiophora brassicae
  Pythium mamillatum
  (Rhizoctonia solani)
  Sclerotinia sclerotiorum
BRASSICA OLERACEA L.
  and vars. (*broccoli, Brussels
  sprouts, cabbage, cauli-
  flower, kale, kohl-rabi,
  marrow-stem kale*)
  Alternaria brassicae
  Alternaria brassicicola
  Ascochyta brassicae
  Botrytis cinerea
  Corticium solani
  Cystopus candidus
  Erysiphe polygoni
  Fuligo septica
  Fusarium *sp.*
  Gloeosporium concentricum
  Mycosphaerella brassicicola
  Olpidium brassicae
  Peronospora parasitica
  Phoma lingam
  (Phyllosticta brassicicola)
  Phytophthora cryptogea

  Phytophthora megasperma
  Plasmodiophora brassicae
  Pythium *sp.*
  (Rhizoctonia solani)
  Sclerotinia sclerotiorum
  Thielaviopsis basicola
  Verticillium dahliae
BRASSICA RAPA L. (*tur-
  nip*) *and* B. NAPOBRAS-
  SICA Mill. (*swede*)
  (Actinomyces scabies)
  Alternaria brassicae
  Ascochyta brassicae
  Botrytis cinerea
  Cercosporella brassicae
  Corticium solani
  Cystopus candidus
  Erysiphe polygoni
  Gloeosporium concentricum
  Helicobasidium purpureum
  Olpidium brassicae
  Peronospora parasitica
  Phoma lingam
  Plasmodiophora brassicae
  Pythium *sp.*
  (Rhizoctonia crocorum)
  (Rhizoctonia solani)
  Sclerotinia sclerotiorum
  Streptomyces scabies
  Typhula gyrans
Broad bean, see Vicia faba L.
Broccoli, see Brassica oleracea
  L.
BROMUS
  Claviceps purpurea
  Erysiphe graminis
  Phyllachora graminis
  Puccinia bromina
  Puccinia graminis
  Rhynchosporium secalis
  Stagonospora bromi
  Ustilago bullata
  Ustilago hypodytes
  Ustilago macrospora
Broom, see Cytisus

Brussels sprouts, see Brassica
oleracea L.
BUDDLEIA
Phyllosticta *sp.*
BULBOCODIUM, *see* **COL-
CHICUM**
BUXUS (*box*)
Phyllosticta buxina
Phyllosticta limbalis
Puccinia buxi
Trochila buxi
Volutella buxi

Cabbage, see Brassica oleracea
L.
CACALIA, *see* **EMILIA**
CACTUS—*including* Cereus,
Echinocereus, Myrtillo-
cactus, Phyllocactus
Botrytis cinerea (*on* Myrtillo-
cactus)
Corticium solani
Diplodia opuntiae (*on* Phyllo-
cactus)
Phytophthora *sp.*
(Rhizoctonia solani)
Sclerotinia sclerotiorum (*on*
Echinocereus)
CALAMAGROSTIS
Ustilago macrospora
CALCEOLARIA
Corticium solani
Phytophthora *sp.*
(Rhizoctonia solani)
Thielaviopsis basicola
CALENDULA (*marigold*)
Botrytis cinerea
Cercospora calendulae
Coleosporium senecionis
Entyloma calendulae
Phytophthora *sp.*
Sclerotinia *sp.*
Sphaerotheca fuliginea
**CALLISTEPHUS CHINEN-
SIS** Nees (*China aster*)

Ascochyta asteris
Botrytis cinerea
Corticium solani
Erysiphe *sp.*
Fusarium oxysporum *f.* calli-
stephi
Phytophthora cryptogea
Phytophthora parasitica
(Rhizoctonia solani)
Sclerotinia sclerotiorum
Thielaviopsis basicola
CALLUNA (*heather*)
Armillaria mellea
Fomes annosus
Marasmius androsaceus
Phytophthora cinnamomi
Sporonema obturatum
CAMELLIA
Armillaria mellea
Exobasidium camelliae
Pestalotia guepini
Phyllosticta camelliae
Phyllosticta camelliaecola *var.*
meranensis
Rhizoctonia *sp.*
CAMPANULA
Ascochyta bohemica
Botrytis cinerea
Coleosporium campanulae
Erysiphe cichoracearum
Peronospora corollae
Phyllosticta carpathica
Phytophthora megasperma
Phytophthora porri
Puccinia campanulae
Ramularia campanulae-lati-
foliae
Ramularia macrospora
Sclerotinia sclerotiorum
Septoria obscura
Verticillium albo-atrum
Canary grass, see Phalaris
canariensis L.
CANNA
Armillaria mellea

Carnation, see Dianthus caryophyllus L.
CARPINUS BETULUS L.
(hornbeam)
Gloeosporium carpini
Carrot, see Daucus carota L.
var. sativa DC.
CASTANEA SATIVA Mill.
(chestnut)
Armillaria mellea
Cryptodiaporthe castanea
Diplodina castaneae
Fistulina hepatica
(Fusicoccum castaneum)
Phytophthora cambivora
Phytophthora cinnamomi
Polyporus sulphureus
Septoria castanicola
CATTLEYA, see also
ORCHIDS
Hemileia americana
Pythium *sp.*
Cauliflower, see Brassica
oleracea L.
CEANOTHUS
Armillaria mellea
Cedar, Western Red, see
Thuja
CEDRUS
Armillaria mellea
Phomopsis pseudotsugae
Stereum purpureum
Celery, see Apium graveolens L.
var. dulce Pers.
CELOSIA
Phytophthora cryptogea
Pythium *sp.*
CENTAUREA, including corn-
flower, sweet sultan
Bremia lactucae
Diplodina passerinii
Oidium *sp.*
Phytophthora cryptogea
Puccinia cyani
Sclerotinia sclerotiorum

CEREUS, see CACTUS
CHAENOMELES LAGE-
NARIA (Loisel.) Koidz.
Botrytis cinerea
(Monilia cinerea)
(Monilia fructigena)
Sclerotinia fructigena
Sclerotinia laxa
CHAMAECYPARIS
Fomes annosus
CHEIRANTHUS, see also
ERYSIMUM
Alternaria cheiranthi
Botrytis cinerea
Corticium solani
Cystopus candidus
Erysiphe polygoni
Peronospora parasitica
Phoma lingam
Plasmodiophora brassicae
(Rhizoctonia solani)
Sclerotinia sclerotiorum
Cherry, see Prunus cerasus L.
and P. avium L.
Cherry laurel, see Prunus
laurocerasus L.
Cherry, Winter, see Solanum
capsicastrum Link
Chestnut, see Castanea sativa
Mill.
Chicory, see Cichorium intybus
L.
China aster, see Callistephus
chinensis Nees
CHIONODOXA
Ustilago vaillantii
Chives, see Allium schoeno-
prasum L.
CHOISYA
Armillaria mellea
Botryodiplodia *sp.*
Christmas rose, see Helleborus
niger L.
CHRYSANTHEMUM
Botrytis cinerea

Coleosporium senecionis
Corticium sambuci
Corticium solani
Cylindrocarpon radicicola
Didymella *sp.*
Fusarium *sp.*
Itersonilia perplexans
Oidium chrysanthemi
Phyllosticta *sp.*
Phytophthora cryptogea
Puccinia chrysanthemi
Puccinia leucanthemi
Pythium megalacanthum
Ramularia bellunensis
(Rhizoctonia solani)
Sclerotinia sclerotiorum
Septoria cercosporioides
Septoria chrysanthemella
Septoria leucanthemi
Thielaviopsis basicola
Verticillium albo-atrum
Verticillium dahliae
CICHORIUM ENDIVIA L.
    (*endive*)
Marssonina panattoniana
Pleospora herbarum
Puccinia endiviae
(Stemphylium botryosum)
CICHORIUM INTYBUS L.
    (*chicory*)
Helicobasidium purpureum
Puccinia cichorii
(Rhizoctonia crocorum)
Sclerotinia sclerotiorum
Cineraria, see Senecio cruentis
    DC.
CLARKIA
Diplodina passerinii
Peronospora arthuri
Phytophthora cactorum
CLEMATIS
Ascochyta *sp.*
Erysiphe polygoni
Septoria clematidis
Clover, see Trifolium

Cobnut, see Corylus
Cocksfoot, see Dactylis glo-
    merata L.
COELOGYNE, *see also*
    ORCHIDS
Cladosporium orchidearum
Colletotrichum orchidearum
COLCHICUM
Pythium ultimum
Sclerotium tuliparum
Urocystis colchici
Uromyces colchici
COLEUS
Botrytis cinerea
Verticillium *sp.*
COLUTEA
Oidium *sp.*
Comfrey, see Symphytum
CONVALLARIA (*lily-of-
    the-valley*)
Botrytis *sp.*
Puccinia sessilis
Septoria brunneola
COPROSMA
Rhizoctonia *sp.*
Rosellinia necatrix
CORDYLINE
Phyllosticta draconis
CORIANDRUM
Fusarium *sp.*
Cornflower, see Centaurea
Corsican pine, see Pinus spp.
CORYLUS
Armillaria mellea
Botrytis cinerea
Fomes annosus
(Fusarium lateritium)
Gibberella baccata
(Monilia fructigena)
Phyllactinia corylea
Sclerotinia fructigena
Stereum purpureum
COTONEASTER
Armillaria mellea
Phyllosticta sanguinea

Stereum purpureum
Couch grass, see Agropyron
repens L.
CRAMBE MARITIMA L.
(*seakale*)
Botrytis cinerea
Corticium solani
Helicobasidium purpureum
Peronospora parasitica
Plasmodiophora brassicae
(Rhizoctonia crocorum)
(Rhizoctonia solani)
**CRATAEGUS OXYA-
CANTHA L.** (*hawthorn*)
Armillaria mellea
Fomes annosus
Gymnosporangium clavarii-
forme
Gymnosporangium confu-
sum
Phleospora oxyacanthae
Physalospora obtusa
Podosphaera oxyacanthae
Sclerotinia crataegi
(Sphaeropsis malorum)
Stereum purpureum
CRATAEGUS spp.
Fomes annosus
Fusicladium pyracanthae
Gymnosporangium clavarii-
forme
Gymnosporangium confusum
Cress, see Lepidium sativum
L.
CRINODENDRON, *see*
TRICUSPIDARIA
CROCUS
Botrytis croci
Fusarium oxysporum *f.* narcissi
Helicobasidium purpureum
Penicillium gladioli
Pythium ultimum
(Rhizoctonia crocorum)
Sclerotinia gladioli
Sclerotium tuliparum

Septoria gladioli
CROTALARIA
Gloeosporium crotalariae
Cucumber, see Cucumis sativus
L.
CUCUMIS MELO L. (*melon*)
Botrytis cinerea
Cercospora melonis
Cladosporium cucumerinum
Colletotrichum lagenarium
Erysiphe cichoracearum
Fusarium *sp.*
Gloeosporium orbiculare
Mycosphaerella citrullina
Phytophthora parasitica
Pseudoperonospora cubensis
Sclerotinia sclerotiorum
Verticillium albo-atrum
Verticillium dahliae
CUCUMIS SATIVUS L.
(*cucumber*)
Alternaria cucumerina
Botrytis cinerea
Cercospora melonis
Cladosporium cucumerinum
Colletotrichum lagenarium
Corticium solani
Didymella lycopersici
(Diplodina lycopersici)
Erysiphe cichoracearum
Fuligo septica
Fusarium *sp.*
Helicobasidium purpureum
Mycosphaerella citrullina
Olpidium majus
Penicillium *sp.*
Phyllosticta cucurbitacearum
Phytophthora *sp.*
Pseudoperonospora cubensis
Pythium aphanidermatum
Pythium ultimum
(Rhizoctonia crocorum)
(Rhizoctonia solani)
Sclerotinia sclerotiorum
Scolecotrichum melopthorum

Thielaviopsis basicola
Trichothecium roseum
Verticillium albo-atrum
Verticillium dahliae
CUCURBITA PEPO L.
    (*vegetable marrow*)
Ascochyta cucumeris
Botrytis cinerea
Cladosporium cucumerinum
Colletotrichum lagenarium
Erysiphe cichoracearum
Gloeosporium orbiculare
Phoma *sp.*
Pythium *sp.*
Sclerotinia sclerotiorum
CUPRESSUS
Armillaria mellea
Pestalotia hartigii
Currant, black, *see* Ribes
    nigrum L.
Currant, red, *see* Ribes sativum
    Syme
CYCLAMEN
Botrytis cinerea
Cercosporella *sp.*
Corticium solani
Fusarium oxysporum
Oidium *sp.*
Phyllosticta *sp.*
Ramularia *sp.*
(Rhizoctonia solani)
Rosellinia necatrix
Thielaviopsis basicola
CYDONIA JAPONICA
    Pers., *see* CHAENO-
    MELES LAGENARIA
CYDONIA OBLONGA Mill.
    (*quince*)
Armillaria mellea
Botrytis cinerea
Diaporthe perniciosa
(Entomosporium maculatum)
Fabraea maculata
Gloeosporium album
(Gloeosporium fructigenum)

Glomerella cingulata
Gymnosporangium confusum
(Monilia cinerea)
(Monilia fructigena)
(Phomopsis perniciosa)
Podosphaera leucotricha
Sclerotinia cydoniae
Sclerotinia fructigena
Sclerotinia laxa
Stereum purpureum
Verticillium albo-atrum
Verticillium dahliae
CYMBIDIUM, *see also*
    ORCHIDS
Gloeosporium bidgoodi
Hypodermium orchidearum
CYNARA SCOLYMUS L.
    (*globe artichoke*)
Botrytis cinerea
Bremia lactucae
CYNOSURUS
Dilophospora alopecuri
Mastigosporium album
Passalora graminis
CYPRIPEDIUM, *see also*
    ORCHIDS
Cladosporium orchidearum
Penicillium thomii
Rhizoctonia *sp.*
Thielaviopsis basicola
CYSTOPTERIS, *see* FERNS
CYTISUS
Armillaria mellea
(Ceratophorum setosum)
Pleiochaeta setosa
Stereum purpureum
Uromyces laburni

DACTYLIS GLOMERATA
    L. (*cocksfoot grass*)
Ascochyta graminicola
Cladochytrium caespitis
Claviceps purpurea
Dilophospora alopecuri
Epichloe typhina

Erysiphe graminis
Fusarium tricinctum *f.* poae
Mastigosporium rubricosum
Ophiobolus graminis
Phyllachora graminis
Puccinia glumarum
Puccinia graminis
Puccinia graminis *var.* phlei-
  pratensis
Rhynchosporium orthosporum
Selenophoma donacis
(Septoria oxyspora)
Uromyces dactylidis
Ustilago striiformis
Daffodil, see Narcissus
DAHLIA
Ascochyta dahliicola
Botrytis cinerea
Corticium solani
Entyloma calendulae *f.* dahliae
Oidium *sp.*
(Phyllosticta dahliicola)
Phytophthora cryptogea
(Rhizoctonia solani)
Sclerotinia sclerotiorum
Thielaviopsis basicola
Verticillium albo-atrum
Verticillium *sp.*
DAPHNE
Armillaria mellea
Botrytis cinerea
Gloeosporium mezerei
Marssonina daphnes
Thielaviopsis basicola
Verticillium dahliae
DATURA
Phytophthora infestans
DAUCUS CAROTA L. *var.*
  SATIVA DC. (*carrot*)
(Alternaria radicina)
(Ansatospora acerina)
Armillaria mellea
Botrytis cinerea
Centrospora acerina
Chalaropsis thielavioides

Corticium solani
Helicobasidium purpureum
Phoma rostrupii
Plasmopara nivea
Pythium *sp.*
(Rhizoctonia crocorum)
(Rhizoctonia solani)
Rosellinia necatrix
Sclerotinia sclerotiorum
Stemphylium radicinum
DAVIDIA
Armillaria mellea
DELPHINIUM, *including*
  annual larkspur
Ascochyta *sp.*
Botrytis cinerea
Corticium solani
Diplodina delphinii
Diplodina passerinii
Erysiphe polygoni
Myrothecium roridum
Phyllosticta ajacis
Pythium *sp.*
(Rhizoctonia solani)
Sclerotinia sclerotiorum
Sclerotium delphinii
Thielaviopsis basicola
DENDROBIUM, *see also*
  ORCHIDS
Cladosporium orchidearum
Colletotrichum orchidearum
Hypodermium orchidearum
DESCHAMPSIA
Mastigosporium album
Mastigosporium deschampsiae
Passalora graminis
Puccinia airae
Puccinia graminis
Uromyces airae-flexuosae
Ustilago striiformis
DEUTZIA
Ascochyta deutziae
Dewberry, see Rubus *spp.*
DIANTHUS BARBATUS L.
  (*sweet william*)

15

Ascochyta dianthi
Didymellina dianthi
Fusarium *sp.*
Gloeosporium dianthi
(Heterosporium echinulatum)
Phomopsis caryophylli
Puccinia arenariae
Septoria sinarum
**DIANTHUS CARYOPHYL-
LUS L.** (*carnation, pink*)
Alternaria dianthi
Alternaria dianthicola
Ascochyta dianthi
Botrytis cinerea
Corticium solani
Didymellina dianthi
Fusarium culmorum
Fusarium oxysporum *f.* dianthi
Fusarium tricinctum *f.* poae
Fusarium *sp.*
Gloeosporium dianthi
Helicobasidium purpureum
Heteropatella valtellinensis
(Heterosporium echinulatum)
Oidium *sp.*
Phomopsis caryophylli
Phytophthora cryptogea
Pythium *sp.*
(Rhizoctonia crocorum)
(Rhizoctonia solani)
Septoria dianthi
Uromyces dianthi
Ustilago violacea
Vermicularia herbarum *f.*
 dianthi
Verticillium cinerescens
DIANTHUS *spp., excluding
 carnation, pink* and *sweet
 william*
Fusarium culmorum
Puccinia arenariae
Septoria dianthi
Sorosporium saponariae
DICENTRA
Ascochyta papaveris

DIELYTRA, *see* **DICENTRA**
DIGITALIS
Colletotrichum fuscum
Phyllosticta digitalis
Septoria digitalis $_s$
Verticillium *sp.*
DIMORPHOTHECA
(Fusarium lateritium)
Gibberella baccata
Verticillium *sp.*
DIPSACUS FULLONUM L.
 (*teazle*)
Helicobasidium purpureum
Peronospora dipsaci
(Rhizoctonia crocorum)
DORONICUM
Septoria doronici
Sphaerotheca fuliginea
Douglas fir, see Pseudotsuga
 taxifolia (Poir.) Britt.
DRACAENA
Macrophoma draconis
Phyllosticta draconis
DRYOPTERIS, *see* **FERNS**
Dwarf bean, see Phaseolus

ECHINOCEREUS, *see*
 CACTUS
Eggplant, see Solanum melon-
 gena L.
ELAEAGNUS
Phyllosticta argyrea
Elder, see Sambucus nigra L.
Elm, see Ulmus
EMILIA
Coleosporium cacaliae
Endive, see Cichorium endivia L.
ERANTHIS (*aconite*)
Urocystis eranthidis
ERICA
Dematium pullulans
Oidium ericinum
Phyllosticta *sp.*
Phytophthora cactorum
Phytophthora cinnamomi

ERIOBOTRYA (*loquat*)
(Entomosporium maculatum)
Fabraea maculata
ERYSIMUM, *see also*
CHEIRANTHUS
Fusarium *sp.*
Sclerotinia sclerotiorum
Septoria cheiranthi
ERYTHRONIUM
Botrytis cinerea
Botrytis *sp.*
Sclerotium delphinii
Uromyces erythronii
ESCALLONIA
Stereum purpureum
ESCHSCHOLZIA
Phytophthora *sp.*
EUCALYPTUS
Stereum purpureum
EUONYMUS
Armillaria mellea
Melampsora euonymi-
caprearum
Oidium euonymi-japonicae
Phyllosticta bolleana
Phyllosticta euonymi
Phyllosticta subnervisequa
Septoria euonymi
EUPHORBIA, *including*
Poinsettia
Botrytis cinerea
Rhizoctonia *sp.*
Sphaerotheca euphorbiae
Thielaviopsis basicola
Uromyces pisi
EXOCHORDA
Stereum purpureum

FAGUS SYLVATICA L.
(*beech*)
Armillaria mellea
Auricularia auricula-judae
Chalaropsis *sp.*
Fomes annosus
Ganoderma applanatum

Gloeosporium fagi
Hydnum diversidens
Microsphaera alphitoides
Nectria cinnabarina
Nectria coccinea
(Oidium quercinum)
Phaeobulgaria inquinans
Phytophthora cambivora
Phytophthora syringae
Polyporus adustus
Polyporus giganteus
Polyporus quercinus
Stereum purpureum
Stereum rugosum
Ustulina deusta
FERNS, *includes* Asplenium,
Athyrium, Blechnum,
Cystopteris, Dryopteris,
Phyllitis, Polypodium,
Polystichum, Pteris
Herpobasidium filicinum
Hyalopsora polypodii
Milesia blechni
Milesia carpatica
Milesia kriegeriana
Milesia murariae
Milesia polypodii
Milesia scolopendrii
Milesia whitei
Pythium *sp.*
Taphrina athyrii
Fescue, see Festuca
FESTUCA (*fescue*)
Ascochyta graminicola
Cladochytrium caespitis
Claviceps purpurea
Corticium fuciforme
Epichloe typhina
Erysiphe graminis
Fusarium culmorum
(Fusarium nivale)
Griphosphaeria nivalis
Helicobasidium purpureum
(Helminthosporium sativum)
Helminthosporium siccans

Ophiobolus graminis *var.*
  avenae
Ophiobolus sativus
Phyllachora sylvatica
Puccinia festucae
Puccinia graminis *var.* phlei-
  pratensis
(Rhizoctonia crocorum)
(Rhizoctonia monteithianum)
Sclerotinia homoeocarpa
Uredo festucae
Uromyces dactylidis
Uromyces festucae
Uromyces poae
Ustilago hypodytes
Ustilago striiformis
FICUS CARICA L. (*fig*)
  Armillaria mellea
  Botrytis cinerea
  Cercospora bolleana
  (Fusarium lateritium *var.* mori)
  Gibberella moricola
  (Gloeosporium fructigenum)
  Glomerella cingulata
  Nectria cinnabarina
  Phomopsis cinerascens
Field bean, see Vicia faba L.
Fig, see Ficus carica L.
Filbert nut, see Corylus
Fir, Douglas, see Pseudotsuga
  taxifolia (Poir.) Britt.
Fir, silver, see Abies.
Flax, see Linum usitatissimum
  L.
Flax, New Zealand, see Phor-
  mium tenax Forst.
FORSYTHIA
  Armillaria mellea
  Phyllosticta forsythiae
  Sclerotinia sclerotiorum
FRAGARIA (*strawberry*)
  Armillaria mellea
  Botrytis cinerea
  Byssochlamys fulva
  (Coniothyrium fuckelii)

Corticium solani
Cylindrocarpon radicicola
Didymella lycopersici
Diplocarpon earlianum
(Diplodina lycopersici)
Fusarium oxysporum
Fusarium solani
Gnomonia fructicola
Hainesia lythri
Helicobasidium purpureum
Leptosphaeria coniothyrium
(Marssonina fragariae)
Mycosphaerella fragariae
Pachybasium hamatum *var.*
  candidum
(Phyllosticta grandimaculans)
Physarum cinereum
Phytophthora cactorum
Phytophthora fragariae
Phytophthora megasperma
Plasmodiophora brassicae
(Ramularia tulasnei)
(Rhizoctonia crocorum)
(Rhizoctonia solani)
Rhizopus *sp.*
Rosellinia necatrix
Sclerotinia sclerotiorum
Sphaerotheca humuli
Spumaria alba
Stagonospora fragariae
Verticillium albo-atrum
Verticillium dahliae
(Zythia fragariae)
FRAXINUS EXCELSIOR
  L. (*ash*)
  Armillaria mellea
  Daldinia concentrica
  Diaporthe eres
  Fomes fraxineus
  Helicobasidium purpureum
  Nectria galligena
  Phomopsis controversa
  Phyllactinia corylea
  Phyllosticta fraxinicola
  Physalospora mutila

Phytophthora cactorum
Polyporus hispidus
(Rhizoctonia crocorum)
Septoria fraxini
FREESIA
Armillaria mellea
Botrytis cinerea
Botrytis gladiolorum
Cladosporium *sp.*
Corticium solani
Curvularia trifolii
Fusarium oxysporum
(Rhizoctonia solani)
Sclerotinia gladioli
Septoria gladioli
FUCHSIA
Phyllosticta fuchsiicola
Uredo fuchsiae

GAILLARDIA
Bremia lactucae
GALANTHUS (*snowdrop*)
Botrytis galanthina
(Botrytis narcissicola)
Penicillium corymbiferum
Sclerotinia narcissicola
Stagonospora curtisii
GALTONIA
Fusarium culmorum
Penicillium corymbiferum
GARDENIA
Botrytis cinerea
Phomopsis gardeniae
Garlic, see Allium sativum L.
GARRYA
Armillaria mellea
Phyllosticta garryae
GENISTA
Armillaria mellea
Stereum purpureum
Uromyces laburni
GENTIANA
Macrophoma *sp.*
Puccinia gentianae
Pycnothyrium gentianicola

Rhizoctonia *sp.*
Geranium, see Pelargonium
GERBERA
Ascochyta gerberae
Corticium solani
Phytophthora *sp.*
(Rhizoctonia solani)
GEUM
Gnomonia fructicola
Peronospora gei
(Phyllosticta gei)
(Zythia fragariae)
GLADIOLUS
Botrytis cinerea
Botrytis gladioli
Botrytis gladiolorum
Corticium solani
Didymellina macrospora
Fusarium oxysporum *f.* gladioli
(Heterosporium gracile)
Penicillium gladioli
Phytophthora cactorum
Puccinia gladioli
Pythium *sp.*
(Rhizoctonia solani)
Sclerotinia draytoni
Sclerotinia gladioli
Septoria gladioli
Urocystis gladiolicola
Globe artichoke, see Cynara
scolymus L.
GLOXINIA, *see* SIN-
NINGIA
GLYCINE MAX Merr. (*soya
bean*)
Sclerotinia sclerotiorum
GODETIA
Alternaria godetiae
Botrytis cinerea
Corticium solani
Diplodina passerinii
Phytophthora cactorum
(Rhizoctonia solani)
Gooseberry, see Ribes grossu-
laria L.

19   2-2

Grape vine, see Vitis vinifera L.

GRASSES, *see under*
Agropyron repens
Agrostis
Alopecurus pratensis
Anthoxanthum odoratum
Arrhenatherum elatius
Bromus
Cynosurus
Dactylis glomerata
Deschampsia
Festuca
Holcus
Phalaris canariensis
Phleum pratense
Poa
Turf

GYPSOPHILA
Botrytis cinerea
Fusarium *sp.*
Puccinia arenariae
Sclerotinia serica

Hawthorn, see Crataegus oxyacantha L.
Heather, see Calluna
HEBE, *see* **VERONICA**
HEDERA HELIX L. (*ivy*)
Colletotrichum trichellum
Gloeosporium paradoxum
Mycosphaerella hedericola
Phyllosticta hedericola
(Septoria hederae)
HELENIUM
Septoria helenii
Verticillium dahliae
HELIANTHEMUM (*rock rose*)
Leveillula taurica
Oidium *sp.*
Peronospora leptoclada
Septoria chamaecisti
HELIANTHUS ANNUUS L.
(*sunflower*)

Botrytis cinerea
Sclerotinia sclerotiorum
HELIANTHUS TUBEROSUS L. (*Jerusalem artichoke*)
Botrytis cinerea
Sclerotinia minor
Sclerotinia sclerotiorum
HELICHRYSUM
Bremia lactucae
Verticillium dahliae
HELIOTROPIUM
Corticium solani
(Rhizoctonia solani)
HELLEBORUS NIGER L. (*Christmas rose*)
Coniothyrium hellebori
Peronospora pulveracea
Ramularia hellebori
Septoria helleborina
Urocystis floccosa
Henbane, see Hyoscyamus niger L.
HEPATICA AMERICANA Ker. (syn. Anemone hepatica L.)
Phyllosticta hepaticae
Septoria hepaticae
Urocystis hepaticae-trilobae
HESPERIS
Plasmodiophora brassicae
HEUCHERA
Phyllosticta heucherae *f.* sanguinae
HIPPEASTRUM, *see* **AMARYLLIS**
HOLCUS
Ascochyta graminicola
Colletotrichum holci
Dilophospora alopecuri
Helminthosporium triseptatum
Plasmodiophora brassicae
Puccinia holcina
Ramulaspera holci-lanati
Sclerotium rhizodes

Tilletia holci
Ustilago striiformis
Holly, see Ilex aquifolium L.
Hollyhock, see Althaea rosea
   Cav.
Honesty, see Lunaria
Honeysuckle, see Lonicera
Hop, see Humulus lupulus L.
HORDEUM (*barley*)
  Botrytis cinerea
  Cercosporella herpotrichoides
  Claviceps purpurea
  Corticium solani
  Erysiphe graminis
  Fusarium avenaceum
  Fusarium culmorum
  Fusarium equiseti
  (Fusarium graminearum)
  (Fusarium nivale)
  (Fusarium sambucinum)
  Gibberella pulicaris
  Gibberella zeae
  Griphosphaeria nivalis
  (Helminthosporium gramineum)
  (Helminthosporium sativum)
  (Helminthosporium teres)
  Ophiobolus graminis
  Ophiobolus graminis *var.* avenae
  Ophiobolus sativus
  Puccinia glumarum
  Puccinia graminis
  Puccinia hordei
  Pyrenophora graminea
  Pyrenophora teres
  (Rhizoctonia solani)
  Rhynchosporium secalis
  Selenophoma donacis
  (Septoria oxyspora)
  Septoria passerinii
  Typhula incarnata
  Ustilago hordei
  Ustilago nuda
  Verticillium *sp.*
Hornbeam, see Carpinus
   betulus L.

Horse-chestnut, see Aesculus
Horse-radish, see Armoracia
   lapathifolia Gilib.
HOYA
  Gloeosporium affine
  Phyllosticta hoyae
HUMULUS LUPULUS L.
   (*hop*)
  Armillaria mellea
  Ascochyta humuli
  Botrytis cinerea
  Cercospora cantuariensis
  Cladosporium *sp.*
  (Fusarium sambucinum)
  Gibberella pulicaris
  Macrosporium *sp.*
  Marasmius rotula
  Phoma herbarum
  Phytophthora cactorum
  Phytophthora citricola
  Pseudoperonospora humuli
  Pythium *sp.*
  Sclerotinia sclerotiorum
  Septoria humuli
  Sphaerotheca humuli
  Verticillium albo-atrum
  Verticillium dahliae
HYACINTHUS
  Botrytis hyacinthi
  Penicillium corymbiferum
  Penicillium cyclopium
  Penicillium hirsutum
  Penicillium *sp.*
  Pythium *sp.*
  Sclerotinia bulborum
  Sclerotium tuliparum
  Stemphylium *sp.*
HYDRANGEA
  Botrytis cinerea
  Corticium solani
  Microsphaera polonica
  (Oidium hortensiae)
  Pythium *sp.*
  (Rhizoctonia solani)
  Septoria hydrangeae

HYOSCYAMUS NIGER L.
(*henbane*)
Peronospora hyoscyami
HYPERICUM
Melampsora hypericorum

ILEX AQUIFOLIUM L.
(*holly*)
Armillaria mellea
Coniothyrium ilicis
Helicobasidium purpureum
Phyllosticta aquifolia
(Rhizoctonia crocorum)
Vialaea insculpta
IMPATIENS
Ascochyta impatientis
Phyllosticta impatientis
Pythium debaryanum
IRIS
Armillaria mellea
Botrytis cinerea
(Botrytis convoluta)
Botrytis *sp.*
Corticium rolfsii
Corticium solani
Didymellina macrospora
Fusarium *sp.*
Helicobasidium purpureum
(Heterosporium gracile)
Leptosphaeria heterospora
Mystrosporium adustum
Penicillium corymbiferum
Penicillium *sp.*
Phytophthora *sp.*
Puccinia iridis
Pythium *sp.*
(Rhizoctonia crocorum)
(Rhizoctonia solani)
Rosellinia necatrix
Sclerotinia convoluta
Sclerotium delphinii
(Sclerotium rolfsii)
Sclerotium tuliparum
Ivy, see Hedera helix.
  L.

IXIA
Botrytis cinerea
Rosellinia necatrix
Stemphylium *sp.*

JASMINUM
Botrytis cinerea
Rosellinia necatrix
Jerusalem artichoke, see Helianthus tuberosus L.
JUGLANS REGIA L. (*walnut*)
Armillaria mellea
Botrytis cinerea
Chalaropsis thielavioides
Cytospora juglandina
Fusarium *sp.*
Gnomonia leptostyla
(Marssonina juglandis)
Polyporus sulphureus
Septoria nigro-maculans
Stereum purpureum
JUNIPERUS
Armillaria mellea
Didymascella tetraspora
Fomes annosus
Gymnosporangium clavariiforme
Gymnosporangium confusum
Gymnosporangium fuscum
Gymnosporangium juniperi
(Keithia tetraspora)
Lophodermium juniperinum

KALANCHOE
Erysiphe polyphaga
Kale, see Brassica oleracea L.
KOCHIA
Pythium *sp.*
Kohl-rabi, see Brassica oleracea
  L.

LABURNUM
Armillaria mellea
Ascochyta kabatiana
Botrytis cinerea

(Ceratophorum setosum)
Cucurbitaria laburni
Fomes fraxineus
Gloeosporium cytisi
Oidium *sp.*
Peronospora cytisi
(Phyllosticta cytisi)
Pleiochaeta setosa
Stereum purpureum
Uromyces laburni
LACHENALIA
Fusarium *sp.*
Mystrosporium adustum
LACTUCA SATIVA L.
(lettuce)
Botrytis cinerea
Bremia lactucae
Corticium praticola
Corticium solani
Marssonina panattoniana
Pleospora herbarum
Puccinia opizii
Pythium *sp.*
(Rhizoctonia solani)
Sclerotinia minor
Sclerotinia sclerotiorum
Septoria lactucae
(Stemphylium botryosum)
Thielaviopsis basicola
LAMPRANTHUS
Fusarium culmorum
Larch, see Larix
LARIX *(larch)*
Armillaria mellea
Botrytis cinerea
(Dasyscypha willkommii)
Fomes annosus
Hypholoma fasciculare
Melampsora epitea
Melampsora larici-caprearum
Melampsora larici-populina
Melampsora tremulae
Melampsoridium betulinum
Meria laricis
Phomopsis pseudotsugae

Phytophthora cactorum
Phytophthora *sp.*
Polyporus schweinitzii
Rhizina inflata
Stereum purpureum
Stereum sanguinolentum
Thielaviopsis basicola
Trichoscyphella resinaria
Trichoscyphella willkommii
Larkspur, see Delphinium
LATHYRUS ODORATUS
L. *(sweet pea)*
Aphanomyces euteiches
Ascochyta pinodella
Ascochyta pisi
Botrytis cinerea
Botrytis fabae
Corticium solani
Erostrotheca multiformis
Erysiphe polygoni
Fusarium avenaceum
Fusarium culmorum
(Fusarium sambucinum)
Gibberella pulicaris
(Hyalodendron album)
Peronospora viciae
Pythium *sp.*
Ramularia deusta
(Rhizoctonia solani)
Thielaviopsis basicola
Verticillium albo-atrum
Verticillium dahliae
LAURUS
Armillaria mellea
Phyllosticta lauri
**LAVANDULA OFFICI-
NALIS** Chaix *(lavender)*
Armillaria mellea
Botrytis cinerea
Phoma lavandulae
Rosellinia necatrix
Septoria lavandulae
LAVATERA
Colletotrichum malvarum
Puccinia malvacearum

Lavender, see Lavandula
    officinalis Chaix
Leek, see Allium porrum L.
LEPIDIUM SATIVUM L.
    (*cress*)
  Alternaria brassicae
  Corticium solani
  Cystopus candidus
  Mucor *sp.*
  Phycomyces nitens
  Pythium *sp.*
  (Rhizoctonia solani)
  Sclerotinia sclerotiorum
Lettuce, see Lactuca sativa L.
LIGUSTRUM VULGARE L.
    (*privet*)
  Armillaria mellea
  (Gloeosporium fructigenum)
  Glomerella cingulata
  Mycosphaerella ligustri
  (Phyllosticta ligustri)
  Phytophthora *sp.*
  Polystictus velutinus
  Rosellinia necatrix
  Septoria ligustri
  Verticillium dahliae
Lilac, see Syringa vulgaris L.
LILIUM
  Botrytis cinerea
  Botrytis elliptica
  Corticium solani
  Cylindrocarpon radicicola
  Fusarium oxysporum *f.* narcissi
  Kabatiella microsticta
  Penicillium cyclopium
  Phytophthora cactorum
  Phytophthora parasitica
  Pythium *sp.*
  (Rhizoctonia solani)
  Rhizopus necans
  Sclerotium delphinii
  Thielaviopsis basicola
Lily-of-the-valley, see Conval-
    laria
Lime, see Tilia

LIMONIUM
  Armillaria mellea
  Botrytis cinerea
  Cercosporella *sp.*
  Erysiphe polygoni
  Phoma *sp.*
  Phomopsis *sp.*
  Uromyces limonii
  Verticillium *sp.*
LINARIA
  Verticillium *sp.*
LINUM USITATISSIMUM
    L. (*flax*)
  Alternaria linicola
  Ascochyta linicola
  Botrytis cinerea
  Colletotrichum linicola
  Corticium solani
  Fusarium oxysporum *f.* lini
  Fusicladium lini
  Melampsora lini
  Oidium lini
  Phoma *sp.*
  Pleospora herbarum
  Polyspora lini
  (Rhizoctonia solani)
  (Septoria linicola)
  Sphaerella linorum
  (Stemphylium botryosum)
LOBELIA
  Corticium solani
  Fusarium culmorum
  Phoma devastatrix
  Pythium *sp.*
  (Rhizoctonia solani)
  Sclerotinia sclerotiorum
  Septoria lobeliae
Loganberry, see Rubus *spp.*
LOLIUM (*ryegrass*)
  Ascochyta graminicola
  Claviceps purpurea
  Corticium fuciforme
  Corticium solani
  Erysiphe graminis
  Fusarium avenaceum

(Fusarium graminearum)
(Fusarium nivale)
Gibberella zeae
Gloeotinia temulenta
Griphosphaeria nivalis
Helminthosporium dictyoides
(Helminthosporium sativum)
Helminthosporium siccans
Ophiobolus graminis
Ophiobolus sativus
(Phialea temulenta)
Phytophthora sp.
Puccinia coronata
Puccinia graminis
Pyrenophora lolii
Pythium sp.
(Rhizoctonia solani)
Tilletia lolii
Ustilago striiformis

LONICERA (honeysuckle)
Ascochyta vulgaris var. lonicerae
Lasiobotrys lonicerae
Leptothyrium periclymeni
Phyllosticta lonicerae
Puccinia festucae
Stereum purpureum
Verticillium sp.

Loquat, see Eriobotrya

LOTUS
Uromyces loti
Urophlyctis potteri

Lowberry, see Rubus spp.

Lucerne, see Medicago sativa L.

LUNARIA (honesty)
Cystopus candidus
Plasmodiophora brassicae
Sclerotinia sclerotiorum

LUPINUS
Aphanomyces euteiches
Armillaria mellea
(Ceratophorum setosum)
Cylindrocladium scoparium
Erysiphe polygoni
Fusarium culmorum

Fusarium sp.
Myrothecium roridum
Phyllosticta sp.
Phytophthora cryptogea
Pleiochaeta setosa
Pleospora herbarum
Sclerotinia sclerotiorum
(Stemphylium botryosum)
Stereum purpureum
Thielaviopsis basicola
Uromyces anthyllidis
Verticillium dahliae

LYCHNIS
Botrytis sp.

LYCIUM
Microsphaera mougeotii
Phytophthora infestans

LYCOPERSICON ESCU- LENTUM Mill. (tomato)
Alternaria solani
Armillaria mellea
Botrytis cinerea
Chaetomium cochlioides
Cladosporium fulvum
Colletotrichum atramentarium
Colletotrichum phomoides
Corticium solani
Didymella lycopersici
(Diplodina lycopersici)
Fusarium oxysporum f. lyco- persici
Fusarium sp.
Myrothecium roridum
Olpidium brassicae
Penicillium sp.
Petriella asymmetrica
Phoma alternariacearum
Phoma destructiva
Phomopsis sp.
Phytophthora cryptogea
Phytophthora infestans
Phytophthora parasitica
Phytophthora verrucosa
Pleospora herbarum
Pythium ultimum

(Rhizoctonia solani)
Rhizopus stolonifera
Sclerotinia minor
Sclerotinia sclerotiorum
Septoria lycopersici
Spongospora subterranea
(Stemphylium botryosum)
Thielaviopsis basicola
Trichothecium roseum
Verticillium albo-atrum
Verticillium dahliae
Verticillium nigrescens
Verticillium nubilum
Verticillium tricorpus

MAGNOLIA
Armillaria mellea
Botrytis cinerea
Phyllosticta magnoliae
MAHONIA
Cumminsiella mirabilissima
Microsphaera berberidis
Phyllosticta mahoniae
Phyllosticta mahoniana
Puccinia graminis
(Puccinia mirabilissima)
Maize, see Zea mays L.
MALUS SYLVESTRIS Mill.
(*apple*)
Alternaria *sp.*
Armillaria mellea
Botrytis cinerea
Camarosporium karstenii
Cladosporium herbarum
Colletotrichum gloeosporioides
Coniothecium chomatosporum
(Coniothyrium fuckelii)
Coniothyrium pirinum
Corticium centrifugum
(Coryneopsis microsticta)
Cylindrocarpon album
(Cylindrocarpon mali)
(Cytospora ambiens)
Cytospora *sp.*
Diaporthe perniciosa

Diplodia griffoni
Fomes annosus
Fusarium avenaceum
Fusarium culmorum
(Fusarium lateritium)
Fusarium solani
(Fusicladium dendriticum)
Gibberella baccata
Gloeodes pomigena
Gloeosporium album
(Gloeosporium fructigenum)
(Gloeosporium perennans)
Glomerella cingulata
Griphosphaeria corticola
Gymnosporangium juniperi-
virginianae
Helicobasidium purpureum
Leptosphaeria coniothyrium
Leptothyrium pomi
(Monilia cinerea)
(Monilia fructigena)
(Myxosporium corticola)
Nectria cinnabarina
Nectria galligena
Neofabraea perennans
Nummularia discreta
Ochropsora ariae
Penicillium expansum
Pezicula corticola
Phacidiella discolor
(Phacidiopycnis malorum)
Pholiota squarrosa
Phoma mali
(Phomopsis perniciosa)
Phyllosticta angulata
Phyllosticta mali
Physalospora mutila
Physalospora obtusa
Phytophthora cactorum
Phytophthora syringae
Pleospora herbarum
Pleospora pomorum
Podosphaera leucotricha
Polyopeus purpureus
Polyporus adustus

Polyporus hispidus
Polyporus sulphureus
(Rhizoctonia crocorum)
Rhizopus stolonifera
Rosellinia necatrix
Sclerotinia fructigena
Sclerotinia laxa
Sclerotinia laxa f. mali
(Sphaeropsis malorum)
(Stemphylium botryosum)
Stemphylium graminis
Stereum purpureum
Strasseria carpophila
Trichothecium roseum
Valsa ambiens
Venturia inaequalis
MALUS *spp.*
(Monilia cinerea)
(Monilia fructigena)
Podosphaera leucotricha
Sclerotinia fructigena
Sclerotinia laxa
Sclerotinia laxa f. mali
MALVA
Phyllosticta destructiva
Puccinia malvacearum
Mangold, see Beta vulgaris
L.
Maple, see Acer campestre L.
Marguerite, see Chrysanthemum
Marigold, see Calendula
Marigold, African, see Tagetes
Marrow-stem kale, see Brassica
oleracea L.
MATTHIOLA (*stock*)
Alternaria raphani
Botrytis cinerea
Corticium solani
Peronospora parasitica
Phoma lingam
Phytophthora cryptogea
Plasmodiophora brassicae
(Rhizoctonia solani)
Sclerotinia sclerotiorum
Thielaviopsis basicola

Meadow foxtail, see Alopecurus
pratensis L.
MECONOPSIS
Peronospora arborescens
Phytophthora cactorum
Phytophthora parasitica
Phytophthora verrucosa
Sclerotinia sclerotiorum
Spumaria alba
MEDICAGO LUPULINA L.
(*trefoil*)
Ascochyta imperfecta
Helicobasidium purpureum
Peronospora trifoliorum
Phytophthora *sp.*
Pleospora herbarum
Pseudopeziza medicaginis
(Rhizoctonia crocorum)
Sclerotinia trifoliorum
(Stemphylium botryosum)
Thielaviopsis basicola
Uromyces striatus
MEDICAGO SATIVA L.
(*lucerne, alfalfa*)
Ascochyta imperfecta
Colletotrichum trifolii
Corticium solani
Helicobasidium purpureum
Peronospora trifoliorum
Phomopsis *sp.*
Pleospora herbarum
Pseudopeziza medicaginis
(Rhizoctonia crocorum)
(Rhizoctonia solani)
Sclerotinia trifoliorum
Sphaerulina trifolii
(Stemphylium botryosum)
Urophlyctis alfalfae
Verticillium albo-atrum
Verticillium dahliae
Medlar, see Mespilus germanica
L.
MELICA
Claviceps purpurea
Melon, see Cucumis melo L.

27

MENTHA (*mint*)
Armillaria mellea
Fusarium *sp.*
Oidium *sp.*
Phoma *sp.*
Phyllosticta *sp.*
Puccinia menthae
Thielaviopsis basicola
Verticillium albo-atrum
Verticillium dahliae
MESEMBRYANTHEMUM,
see **LAMPRANTHUS**
MESPILUS GERMANICA
L. (*medlar*)
Armillaria mellea
Diaporthe perniciosa
(Entomosporium maculatum)
Fabraea maculata
Gymnosporangium confusum
(Monilia cinerea)
(Monilia fructigena)
(Phomopsis perniciosa)
Podosphaera leucotricha
Sclerotinia fructigena
Sclerotinia laxa
Sclerotinia mespili
Stereum purpureum
METASEQUOIA
Armillaria mellea
Michaelmas daisy, *see* Aster
Mignonette, *see* Reseda
Millet, *see* Panicum miliaceum L.
Mint, *see* Mentha
MONTBRETIA, *see* **TRI-
TONIA**
MORUS (*mulberry*)
Cercospora moricola
(Fusarium lateritium *var.* mori)
Gibberella moricola
Phoma mororum
Septogloeum mori
Mountain ash, *see* Sorbus aucu-
paria L.
Mulberry, *see* Morus
MUSCARI

Sclerotinia bulborum
Uromyces scillarum
Ustilago vaillantii
Mushroom, *see* Agaricus
Mustard, *see* Brassica nigra Koch
MYOSOTIS
Botrytis cinerea
Corticium solani
Entyloma fergussoni
Erysiphe cichoracearum
Peronospora myosotidis
(Rhizoctonia solani)
MYRTILLOCACTUS, *see*
CACTUS
Myrtle, *see* Myrtus
MYRTUS (*myrtle*)
Cercospora myrticola
Pestalotia decolorata
Phyllosticta nuptialis

NARCISSUS
Armillaria mellea
Botrytis cinerea
(Botrytis narcissicola)
(Botrytis polyblastis)
Coleosporium narcissi
Cylindrocarpon radicicola
Fusarium oxysporum *f.* narcissi
Ophiostoma narcissi
Penicillium *sp.*
Phytophthora megasperma
Puccinia schroeteri
Pythium intermedium
Ramularia vallisumbrosae
Rosellinia necatrix
Sclerotinia narcissicola
Sclerotinia polyblastis
Sclerotium tuliparum
Stagonospora curtisii
Trichoderma viride
**NASTURTIUM OFFICI-
NALE** R.Br. (*watercress*)
Peronospora parasitica
Pythium megalacanthum
Pythium *sp.*

28

Spongospora *sp.*
Nectarine, see Prunus persica
(L.) Batsch.
NEMESIA
Phytophthora cryptogea
Phytophthora parasitica
Pythium *sp.*
Sclerotinia sclerotiorum
Thielaviopsis basicola
NEPETA
Oidium *sp.*
NEVIUSIA
Stereum purpureum
New Zealand flax, see Phormium tenax Forst.
NICOTIANA (*including tobacco*)
Oidium *sp.*
Thielaviopsis basicola
Norway spruce, see Picea abies
(L.) Karst.
NOTONIA
Coleosporium senecionis
NYMPHAEA (*water-lily*)
Ovularia nymphaearum
Pythium undulatum

Oak, see Quercus petraea (Matt.)
Liebl. *and* Q. robur L.
Oats, see Avena sativa L.
OBERONIA, *see also*
ORCHIDS
Colletotrichum orchidearum
ODONTOGLOSSUM, *see
also* **ORCHIDS**
Cercospora angraeci
Cercospora odontoglossi
Colletotrichum orchidearum
Gloeosporium bidgoodi
OENOTHERA
Septoria oenotherae
OLEANDER
Botrytis cinerea
OLEARIA
Armillaria mellea

ONCIDIUM, *see also*
ORCHIDS
Cladosporium orchidearum
Hemileia americana
Hemileia oncidii
Uredo oncidii
Onion, see Allium cepa L.
**ONOBRYCHIS VICIAE-
FOLIA** Scop. (*sainfoin*)
Ascochyta orobi
Botrytis cinerea
Erysiphe polygoni
Pleospora herbarum
Ramularia onobrychidis
Sclerotinia trifoliorum
Septoria orobina
(Stemphylium botryosum)
Thielaviopsis basicola
Uromyces onobrychidis
Verticillium dahliae
ORCHIDS, *see* Cattleya,
Coelogyne, Cymbidium,
Cypripedium, Dendrobium,
Oberonia, Odontoglossum,
Oncidium, Phalaenopsis,
Satyrium, Thunia *and* Vanda
(*for convenience the names
given under the different orchid
genera are collected below*)
Botrytis cinerea
Cercospora angraeci
Cercospora odontoglossi
Cladosporium orchidearum
(Colletotrichum cinctum)
Colletotrichum orchidearum
Corticium solani
Gloeosporium bidgoodi
Glomerella cincta
Hemileia americana
Hemileia oncidii
Hypodermium orchidearum
Penicillium thomii
Puccinia satyrii
Pythium *sp.*
(Rhizoctonia solani)

Rhizoctonia *sp.*
Thielaviopsis basicola
Uredo oncidii
ORNITHOGALUM
Heterosporium ornithogali
Penicillium *sp.*
Puccinia hordei
Puccinia liliacearum
OXALIS
Oidium oxalidis

PAEONIA (*peony*)
Armillaria mellea
Botrytis cinerea
Botrytis paeoniae
Cladosporium paeoniae
Cronartium flaccidum
Rosellinia necatrix
Septoria paeoniae *var.* bero-
linensis
PALMAE
Graphiola phoenicis
PANDANUS
Melanconium pandani
PANICUM MILIACEUM
L. (*millet*)
Sphacelotheca destruens
Pansy, see Viola *spp.*
PAPAVER (*poppy*)
Botrytis cinerea
Helicobasidium purpureum
(Helminthosporium papaveris)
Oidium *sp.*
Peronospora arborescens
Pleospora calvescens
(Rhizoctonia crocorum)
Sclerotinia sclerotiorum
Verticillium dahliae
Parsley, see Petroselinum
crispum Nym.
Parsnip, see Pastinaca sativa L.
PASTINACA SATIVA L.
(*parsnip*)
(Actinomyces scabies)
Botrytis cinerea

Cercosporella pastinacae
(Cylindrosporium pastinacae)
Erysiphe polygoni
Helicobasidium purpureum
Itersonilia perplexans
Phoma *sp.*
Phomopsis astericus
Phyllachora pastinacae
Plasmopara nivea
Ramularia pastinacae
(Rhizoctonia crocorum)
Sclerotinia sclerotiorum
Streptomyces scabies
PAULOWNIA
Armillaria mellea
Ascochyta paulowniae
Phyllosticta paulowniae
Pea, see Pisum sativum L.
Peach, see Prunus persica (L.)
Batsch
Pear, see Pyrus communis L.
PELARGONIUM (*geranium*)
Botrytis cinerea
Colletotrichum *sp.*
Fusarium pelargonii
Phytophthora *sp.*
Pythium *sp.*
Thielaviopsis basicola
Uromyces geranii
Verticillium dahliae
PELTANDRA
Aecidium importatum
PENSTEMON
Botrytis *sp.*
Phyllosticta pentstemonis
Peony, see Paeonia
Peppermint, see Mentha
Periwinkle, see Vinca
PERNETTYA
Stereum purpureum
PETASITES
Coleosporium petasitis
PETROSELINUM CRIS-
PUM Nym. (*parsley*)
Erysiphe polygoni

Helicobasidium purpureum
Plasmopara nivea
Puccinia aethusae
(Rhizoctonia crocorum)
Septoria petroselini
PETUNIA
Corticium solani
Phytophthora cryptogea
Phytophthora infestans
Ramularia petuniae
(Rhizoctonia solani)
PHAIUS
Uredo phaji
PHALAENOPSIS, see also
ORCHIDS
Colletotrichum orchidearum
PHALARIS CANARIENSIS
L. (canary grass)
Claviceps purpurea
PHASEOLUS (dwarf and
runner bean)
Aphanomyces euteiches
Ascochyta boltshauseri
Ascochyta phaseolorum
Botrytis cinerea
Colletotrichum lindemuthianum
Corticium solani
Fusarium oxysporum
Fusarium solani f. phaseoli
Pythium sp.
(Rhizoctonia solani)
Sclerotinia sclerotiorum
Sclerotinia trifoliorum var. fabae
(Stagonospora hortensis)
Thielaviopsis basicola
Uromyces appendiculatus
Verticillium dahliae
Phenomenal berry, see Rubus
spp.
PHILADELPHUS
Ascochyta philadelphi
Stereum purpureum
PHILLYREA
Phyllosticta phillyreae
Zaghouania phillyreae

PHLEUM PRATENSE L.
(Timothy grass)
Ascochyta graminicola
Epichloe typhina
Fusarium tricinctum f. poae
Helminthosporium dictyoides
var. phlei
Heterosporium phlei
Mastigosporium cylindricum
Puccinia graminis var. phlei-
pratensis
Rhynchosporium secalis
Selenophoma donacis
Septogloeum oxysporum
(Septoria oxyspora)
Ustilago striiformis
PHLOX
Erysiphe cichoracearum
Helicobasidium purpureum
Phoma phlogis
Phytophthora parasitica
Pyrenochaeta phlogis
(Rhizoctonia crocorum)
Septoria drummondii
Verticillium dahliae
PHOENIX, see PALMAE
PHORMIUM TENAX Forst.
(New Zealand flax)
(Colletotrichum rhodocyclum)
Coniothyrium phormii
Glomerella phacidiomorpha
PHYLLITIS, see FERNS
PICEA ABIES (L.) Karst.
(Norway spruce) and P.
SITCHENSIS (Bong.)
Carr. (Sitka spruce)
Armillaria mellea
Ascochyta piniperda
Botrytis cinerea
Chrysomyxa abietis
Chrysomyxa rhododendri
Corticium solani
Cucurbitaria piceae
Dasyscypha calyciformis
Fomes annosus

31

Helicobasidium purpureum
Herpotrichia nigra
Hypholoma fasciculare
Lophodermium macrosporum
Nectria cucurbitula
Pestalotia hartigii
Pholiota squarrosa
Phomopsis occulta
Phytophthora cactorum
Polyporus schweinitzii
Pythium ultimum
Rhizina inflata
(Rhizoctonia crocorum)
(Rhizoctonia solani)
Rhizosphaera kalkhoffii
Rosellinia aquila
Stereum sanguinolentum
Thekopsora areolata
Trichoscyphella resinaria
PICEA *spp.*
Cucurbitaria piceae
Rhizosphaera kalkhoffii
Pink, see Dianthus caryophyllus
L.
PINUS STROBUS L. (*Weymouth pine*)
Armillaria mellea
(Brunchorstia pinea)
Cronartium ribicola
Crumenula pinea
Hypoderma brachysporum
PINUS SYLVESTRIS L.
(*Scots pine*)
Armillaria mellea
Botrytis cinerea
(Brunchorstia pinea)
Coleosporium campanulae
Coleosporium euphrasiae
Coleosporium melampyri
Coleosporium petasites
Coleosporium senecionis
Coleosporium sonchi
Coleosporium tussilaginis
Cronartium flaccidum
Crumenula pinea

Diplodia pinea
Fomes annosus
Hendersonia acicola
Hypodermella conjuncta
Hypodermella sulcigena
Lophodermium pinastri
Melampsora tremulae
Paxillus giganteus
Peridermium pini
Phytophthora cactorum
Polyporus schweinitzii
Rhizina inflata
Sclerophoma pithyophila
Trametes pini
PINUS *spp.*
Armillaria mellea
(Brunchorstia pinea)
Coleosporium petasites
Coleosporium senecionis
Cronartium ribicola
Crumenula pinea
Diplodia pinea
Fomes annosus
Hendersonia acicola
Hypoderma brachysporum
Hypodermella conjuncta
Hypodermella sulcigena
Melampsora tremulae
Rhizina inflata
Rhizosphaera kalkhoffii
Stereum purpureum
Trichoscyphella resinaria
PISUM SATIVUM L. (*pea*)
Aphanomyces euteiches
Ascochyta pinodella
Ascochyta pisi
Botrytis cinerea
Cladosporium herbarum
Cladosporium pisicola
Corticium solani
Erysiphe polygoni
Fusarium culmorum
Fusarium oxysporum *f.* pisi
Fusarium oxysporum *var.*
redolens

32

(Fusarium sambucinum)
Fusarium solani *f.* pisi
Gibberella pulicaris
Mycosphaerella pinodes
Myrothecium roridum
Peronospora viciae
Phytophthora *sp.*
Pythium de baryanum
Pythium ultimum
(Rhizoctonia solani)
Thielaviopsis basicola
Uromyces pisi
PITTOSPORUM
Rosellinia necatrix
Plane, see Platanus acerifolia
(Ait.) Willd.
PLATANUS ACERIFOLIA
(Ait.) Willd. (*plane*)
Gloeosporium nervisequum
Stereum purpureum
Plum, see Prunus domestica L.
POA
Ascochyta graminicola
Colletotrichum graminicola
Corticium fuciforme
Dilophospora alopecuri
Epichloe typhina
Erysiphe graminis
(Fusarium nivale)
Griphosphaeria nivalis
Helicobasidium purpureum
Helminthosporium vagans
Ophiobolus graminis
Ophiobolus graminis *var.* avenae
Passalora graminis
Puccinia graminis
Puccinia poae-nemoralis
Puccinia poarum
(Rhizoctonia crocorum)
Selenophoma donacis
(Septoria oxyspora)
Uromyces dactylidis
Ustilago striiformis
Poinsettia, see Euphorbia
Polyanthus, see Primula

POLYGONATUM (*Solomon's seal*)
Helicobasidium purpureum
(Rhizoctonia crocorum)
POLYPODIUM, *see* **FERNS**
POLYSTICHUM, *see*
FERNS
Poplar, see Populus
Poppy, see Papaver
POPULUS (*poplar*)
Armillaria mellea
Cytospora chrysosperma
Dothichiza populea
Fomes annosus
Ganoderma applanatum
Marssonina populi
Melampsora allii-populina
Melampsora larici-populina
Melampsora tremulae
Nectria coccinea
Phyllosticta populina
Septoria populi
Stereum purpureum
Taphrina populina
Portugal laurel, see Prunus *spp.*
Potato, see Solanum tuberosum
L.
POTENTILLA
Phragmidium potentillae
PRIMULA
Ascochyta primulae
Botrytis cinerea
Cercosporella primulae
Corticium solani
Fusarium *sp.*
Helicobasidium purpureum
Heterosporium auriculae
Oidium *sp.*
Peronospora oerteliana
Phyllosticta primulicola
Phytophthora cactorum
Phytophthora citricola
Phytophthora parasitica
Phytophthora primulae
Phytophthora verrucosa

HOSTS

Puccinia primulae
Pythium spinosum
Ramularia primulae
(Rhizoctonia crocorum)
(Rhizoctonia solani)
Thielaviopsis basicola
Tuburcinia primulicola
Privet, see Ligustrum vulgare L.
PRUNUS AMYGDALUS
  Batsch (*almond*)
Armillaria mellea
Clasterosporium carpophilum
(Monilia cinerea)
(Monilia fructigena)
Rosellinia necatrix
Sclerotinia fructigena
Sclerotinia laxa
Stereum purpureum
Taphrina deformans
PRUNUS ARMENIACA L.
  (*apricot*)
Botrytis cinerea
Cylindrocladium scoparium
Cytospora *sp.*
(Monilia fructigena)
Nectria cinnabarina
(Puccinia pruni-spinosae)
Sclerotinia fructigena
Sclerotinia laxa
Stereum purpureum
Tranzschelia pruni-spinosae
PRUNUS CERASUS L. *and*
  P. AVIUM L. (*cherry*)
Armillaria mellea
Botrytis cinerea
Clasterosporium carpophilum
Cylindrocladium scoparium
(Cytospora leucostoma)
Fomes pomaceus
(Fusicladium cerasi)
(Gloeosporium fructigenum)
Glomerella cingulata
Gnomonia erythrostoma
(Monilia cinerea)
(Monilia fructigena)

Pholiota squarrosa
Podosphaera oxyacanthae
Polyporus squamosus
Polyporus sulphureus
Rosellinia necatrix
Sclerotinia fructigena
Sclerotinia laxa
Stereum purpureum
Taphrina cerasi
Valsa leucostoma
Venturia cerasi
Verticillium dahliae
PRUNUS DOMESTICA L.
  (*plum*)
Armillaria mellea
Botrytis cinerea
Byssochlamys fulva
Cercospora circumcissa
Cylindrocladium scoparium
Cylindrosporium padi
(Cytospora leucostoma)
Dermatella prunastri
Dermea prunastri
Diaporthe perniciosa
Fomes igniarius
Fomes pomaceus
Fusicladium carpophilum
Gloeodes pomigena
Leptothyrium pomi
(Monilia cinerea)
(Monilia fructigena)
(Phomopsis perniciosa)
Podosphaera oxyacanthae *var.*
  tridactyla
Polyporus hispidus
Polyporus sulphureus
Polystigma rubrum
(Puccinia pruni-spinosae)
Roesleria pallida
Rosellinia necatrix
Sclerotinia fructigena
Sclerotinia laxa
Stereum purpureum
Taphrina insititiae
Taphrina pruni

34

Tranzschelia pruni-spinosae
Valsa leucostoma
Verticillium albo-atrum
**PRUNUS LAUROCERA-
SUS L.** (*cherry laurel*)
Armillaria mellea
Coryneum laurocerasi
(Gloeosporium phacidiellum)
Podosphaera oxyacanthae *var.*
tridactyla
Stereum purpureum
Trochila laurocerasi
PRUNUS PERSICA (L.)
Batsch *and vars.* (*peach* and
nectarine)
Armillaria mellea
Botrytis cinerea
Cercospora circumscissa
Clasterosporium carpophilum
Cylindrocladium scoparium
Cytospora *sp.*
Diaporthe perniciosa
Fomes annosus
Fusicladium carpophilum
(Gloeosporium fructigenum)
(Gloeosporium laeticolor)
Glomerella cingulata
(Monilia cinerea)
(Monilia fructigena)
Naemospora crocea
(Phomopsis perniciosa)
(Puccinia pruni-spinosae)
Rhizopus stolonifera
Roesleria pallida
Rosellinia necatrix
Sclerotinia fructigena
Sclerotinia laxa
Sphaerotheca pannosa *var.* per-
sicae
Stereum purpureum
Taphrina deformans
Tranzschelia pruni-spinosae
Verticillium dahliae
PRUNUS TRILOBA Lindl.
Armillaria mellea

Botrytis cinerea
(Monilia cinerea)
Sclerotinia laxa
Stereum purpureum
PRUNUS *spp.*
(Actinonema padi)
Armillaria mellea
Botrytis cinerea
Fomes annosus
Fusicladium carpophilum
Gnomonia padicola
(Monilia cinerea)
Physalospora obtusa
Podosphaera oxyacanthae *var.*
tridactyla
Polystigma rubrum
Sclerotinia laxa
(Sphaeropsis malorum)
Stereum purpureum
Taphrina cerasi
Taphrina insititiae
Taphrina pruni
Thekopsora areolata
Thielaviopsis basicola
PSALLIOTA, *see* **AGARI-
CUS**
**PSEUDOTSUGA TAXI-
FOLIA (Poir.) Britt.**
(*Douglas fir*)
Armillaria mellea
Botrytis cinerea
Corticium solani
Fomes annosus
Phaeocryptopus gaumannii
Phomopsis conorum
Phomopsis pseudotsugae
Polyporus schweinitzii
Rhabdocline pseudotsugae
(Rhizoctonia solani)
Rhizosphaera kalkhoffii
PTERIS, *see* **FERNS**
PYRACANTHUS
(Fusicladium pirinum *var.*
pyracanthae)
Fusicladium pyracanthae

PYRETHRUM
Botrytis *sp.*
Corticium solani
Erysiphe *sp.*
Fusarium *sp.*
(Rhizoctonia solani)
PYRUS AUCUPARIA
Gaertn., *see* **SORBUS
AUCUPARIA L.**
PYRUS COMMUNIS L.
(*pear*)
Armillaria mellea
Botrytis cinerea
(Cylindrocarpon mali)
Diaporthe perniciosa
(Entomosporium maculatum)
Fabraea maculata
(Fusarium lateritium)
(Fusicladium pirinum)
Gibberella baccata
Gloeodes pomigena
Gloeosporium album
(Gloeosporium fructigenum)
(Gloeosporium perennans)
Glomerella cingulata
Gymnosporangium clavariae-
forme
Gymnosporangium fuscum
Leptothyrium pomi
(Monilia cinerea)
(Monilia fructigena)
Mycosphaerella sentina
Nectria cinnabarina
Nectria galligena
Neofabraea perennans
Phacidiella discolor
(Phacidiopycnis malorum)
(Phomopsis perniciosa)
Physalospora obtusa
Phytophthora cactorum
Phytophthora syringae
Podosphaera leucotricha
Polyporus sulphureus
Roesleria pallida
Rosellinia necatrix

Sclerotinia fructigena
Sclerotinia laxa
(Septoria piricola)
(Sphaeropsis malorum)
Stereum purpureum
Taphrina bullata
Venturia pirina
PYRUS MALUS L., *see*
MALUS SYLVESTRIS
Mill.
PYRUS *spp.*
Fomes annosus
(Monilia cinerea)
(Monilia fructigena)
Sclerotinia fructigena
Sclerotinia laxa

QUERCUS PETRAEA
(Matt.) Liebl. *and* **Q.
ROBUR L.** (*oak*)
Armillaria mellea
Colpoma quercina
Diaporthe leiphaemia
Diaporthe taleola
Fistulina hepatica
Fomes robustus
Ganoderma applanatum
Gloeosporium quercinum
Microsphaera alphitoides
(Oidium quercinum)
Phyllosticta maculiformis
Polyporus dryadeus
Polyporus sulphureus
Rosellinia quercina
Sclerotinia candolleana
Stereum frustulosum
Stereum gausapatum
Stereum hirsutum
Stereum rugosum
Uredo quercus
Quince, see Cydonia oblonga
Mill.

Radish, see Raphanus sativus L.
RANUNCULUS

36

Corticium solani
Erysiphe polygoni
(Rhizoctonia solani)
Sphaerella *sp.*
Rape, see Brassica napus L.
RAPHANUS SATIVUS L.
  (*radish*)
(Actinomyces scabies)
Alternaria brassicae
Alternaria brassicicola
Alternaria raphani
Corticium solani
Peronospora parasitica
Plasmodiophora brassicae
(Rhizoctonia solani)
Streptomyces scabies
Raspberry, see Rubus idaeus L.
Red currant, see Ribes sativum
  Syme
RESEDA, *including mignonette*
Cercospora resedae
Corticium solani
(Rhizoctonia solani)
Verticillium *sp.*
RHEUM RHAPONTICUM
  L. (*rhubarb*)
Armillaria mellea
Botrytis cinerea
Helicobasidium purpureum
Hypholoma fasciculare
Puccinia phragmitis
Ramularia rhei
(Rhizoctonia crocorum)
Verticillium *sp.*
RHODODENDRON (*includ-
  ing* Azalea)
Armillaria mellea
Ascochyta *sp.*
Botrytis cinerea
Cercospora *sp.*
Chrysomyxa rhododendri
Corticium solani
Cylindrocarpon radicicola
Diplodina eurhododendri
Echidnodes aulographoides

Exobasidium vaccinii
Fomes annosus
Gloeosporium rhododendri
Lophodermium melaleucum
Lophodermium vagulum
Ovulinia azaleae
Pestalotia guepini
Pestalotia macrotricha
Phoma rhodorae
Phyllosticta rhododendri
Pycnostysanus azaleae
(Rhizoctonia solani)
Septoria azaleae
Septoria rhododendri
Stereum purpureum
Xylaria vaporaria
Rhubarb, see Rheum rhaponti-
  cum L.
RHUS (*sumach*)
Phyllosticta *sp.*
Verticillium dahliae
RIBES GROSSULARIA L.
  (*gooseberry*)
Armillaria mellea
Ascochyta ribesia
Botrytis cinerea
Coniothyrium vagabundum
Cronartium ribicola
Cytosporina ribis
Fomes ribis
(Gloeosporium ribis)
Helicobasidium purpureum
Microsphaera grossulariae
Nectria cinnabarina
Phyllosticta grossulariae
Plowrightia ribesia
Pseudopeziza ribis
Puccinia pringsheimiana
(Rhizoctonia crocorum)
Rosellinia necatrix
Sphaerotheca mors-uvae
Stereum purpureum
RIBES NIGRUM L. (*black
  currant*)
Armillaria mellea

37

Botrytis cinerea
Cronartium ribicola
Cytosporina ribis
Diaporthe perniciosa
Fomes ribis
(Fusarium lateritium)
Gibberella baccata
(Gloeosporium ribis)
Helicobasidium purpureum
Microsphaera grossulariae
Mycosphaerella ribis
(Phomopsis perniciosa)
Phyllosticta sp.
Plowrightia ribesia
Pseudopeziza ribis
Puccinia pringsheimiana
(Rhizoctonia crocorum)
Rosellinia necatrix
(Septoria ribis)
Sphaerotheca mors-uvae
Stereum purpureum
Verticillium dahliae
RIBES SATIVUM Syme (red
    currant)
Armillaria mellea
Botrytis cinerea
Collybia velutipes
Cronartium ribicola
Cytosporina ribis
Fomes ribis
(Gloeosporium ribis)
Microsphaera grossulariae
Mycosphaerella ribis
Nectria cinnabarina
Plowrightia ribesia
Pseudopeziza ribis
Puccinia pringsheimiana
Puccinia ribis
(Septoria ribis)
Sphaerotheca mors-uvae
Stereum purpureum
Verticillium dahliae
RIBES spp.
Armillaria mellea
Botrytis cinerea

Cronartium ribicola
Phytophthora cactorum
Puccinia pringsheimiana
Sphaerotheca mors-uvae
RICHARDIA, see ZANTE-
    DESCHIA
RIVEA
Oidium erumpens
ROBINIA
Armillaria mellea
Fomes fraxineus
Phleospora robiniae
Rock rose, see Helianthemum
ROMNEYA
Phytophthora sp.
Sclerotinia sclerotiorum
Verticillium albo-atrum
ROSA
(Actinonema rosae)
Armillaria mellea
Botryosphaeria dothidea
Botrytis cinerea
(Coniothyrium fuckelii)
Coniothyrium rosarum
Coniothyrium wernsdorffiae
(Coryneopsis microsticta)
Cryptosporella umbrina
Cryptosporium minimum
Cylindrocladium scoparium
(Cytospora rhodophila)
Didymella sepincoliformis
Diplocarpon rosae
Fusarium sp.
Gnomonia rubi
Griphosphaeria corticola
Leptosphaeria coniothyrium
Myxosporium rosae
Peronospora sparsa
Pestalotia sp.
Phragmidium mucronatum
Phytophthora sp.
Roesleria pallida
(Septoria rosae)
Sphaceloma rosarum
Sphaerotheca pannosa

Sphaerulina rehmiana
Stereum purpureum
Thielaviopsis basicola
Valsa rhodophila
Verticillium albo-atrum
Verticillium dahliae
Rowan, see Sorbus aucuparia L.
RUBUS IDAEUS L. *(raspberry)*
Armillaria mellea
Ascochyta idaei
Botrytis cinerea
(Coniothyrium fuckelii)
Cryptosporium minimum
Diaporthe perniciosa
Didymella applanata
Elsinoe veneta
Fusarium avenaceum
Fusarium culmorum
Fusarium *sp.*
Hapalosphaeria deformans
Helicobasidium purpureum
Hendersonia rubi
Leptosphaeria coniothyrium
Microthyriella rubi
Nectria mammoidea *var.* rubi
(Phomopsis perniciosa)
Phragmidium rubi-idaei
Phytophthora citricola
(Plectodiscella veneta)
(Rhizoctonia crocorum)
Rosellinia necatrix
Sclerotinia sclerotiorum
Sphaerotheca humuli
Stereum purpureum
Thielaviopsis basicola
Verticillium dahliae
RUBUS *spp., including blackberry, dewberry, loganberry, lowberry, wineberry*
Armillaria mellea
Botrytis cinerea
Cladosporium herbarum
Coniothyrium tumefaciens
Diaporthe perniciosa

Didymella applanata
Elsinoe veneta
Gnomonia rubi
Hapalosphaeria deformans
Hendersonia rubi
Kuehneola uredinis
Kunkelia nitens
(Monilia fructigena)
Oidium *sp.*
(Phomopsis perniciosa)
Phragmidium violaceum
(Plectodiscella veneta)
Rhabdospora ramealis
Sclerotinia fructigena
Sclerotinia sclerotiorum
Septoria rubi
Verticillium dahliae
Runner bean, see Phaseolus
Rye, see Secale cereale L.
Ryegrass, see Lolium

Sage, see Salvia
Sainfoin, see Onobrychis viciae-folia Scop.
SALIX *(willow)*
Armillaria mellea
Cryptodiaporthe salicella
Cryptomyces maximum
Cytospora chrysosperma
Diplodina salicis
Discella carbonacea
Fomes igniarius
(Fusicladium saliciperdum)
Ganoderma applanatum
Gloeosporium salicis
Marssonina salicicola
Melampsora allii-fragilis
Melampsora allii-salicis-albae
Melampsora amygdalinae
Melampsora epitea
Melampsora euonymi-caprearum
Melampsora larici-caprearum
Melampsora ribesii-purpureae
Melampsora ribesii-viminalis

39

(Melasmia salicina)
Physalospora miyabeana
Polyporus sulphureus
Rhytisma salicinum
Rhytisma symmetricum
Scleroderris fuliginosa
Septoria salicicola
Stereum purpureum
Trametes suaveolens
Venturia chlorospora
SALPIGLOSSIS
Phytophthora *sp.*
Salsify, see Tragopogon porri-
folius L.
SALVIA *(sage)*
Corticium solani
Oidium *sp.*
Pythium *sp.*
(Rhizoctonia solani)
Spumaria alba
Verticillium dahliae
SAMBUCUS NIGRA L.
(elder)
Armillaria mellea
Auricularia auricula-judae
Cercospora depazeoides
SAPONARIA
Alternaria dianthi
Corticium solani
(Rhizoctonia solani)
SATUREJA *(savory)*
Puccinia menthae
SATYRIUM, *see also*
ORCHIDS
Puccinia satyrii
Savory, see Satureja
SAXIFRAGA
Pestalotia gracilis
Puccinia pazschkei
Puccinia saxifragae
SCABIOSA
Armillaria mellea
Ascochyta scabiosae
Cercospora *sp.*
Erysiphe polygoni

Phoma *sp.*
Pythium *sp.*
Ramularia knautiae
Rosellinia necatrix
Thielaviopsis basicola
SCHIZANTHUS
Corticium solani
Oidium *sp.*
Phytophthora cryptogea
Pythium *sp.*
(Rhizoctonia solani)
Sclerotinia sclerotiorum
Verticillium *sp.*
SCHIZOSTYLIS
Didymellina macrospora
(Heterosporium gracile)
SCILLA
Penicillium cyclopium
Penicillium hirsutum
Sclerotinia bulborum
Septoria scillae
Uromyces scillarum
Ustilago vaillantii
SCOLOPENDRIUM, *see*
PHYLLITIS
SCORZONERA
Cystopus cubicus
Erysiphe cichoracearum
Scots pine, see Pinus sylvestris L.
Seakale, see Crambe maritima L.
SECALE CEREALE L. *(rye)*
Botrytis cinerea
Cercosporella herpotrichoides
Claviceps purpurea
Corticium solani
Erysiphe graminis
Fusarium avenaceum
Fusarium culmorum
(Fusarium graminearum)
(Fusarium nivale)
(Fusarium sambucinum)
Gibberella pulicaris
Gibberella zeae
Griphosphaeria nivalis
(Helminthosporium sativum)

Ophiobolus graminis
Ophiobolus sativus
Puccinia dispersa
Puccinia glumarum
Puccinia graminis
(Rhizoctonia solani)
Rhynchosporium secalis
Tilletia caries
Urocystis occulta
SEMPERVIVUM
Endophyllum sempervivi
SENECIO CRUENTIS DC.
(cineraria)
Alternaria senecionis
Ascochyta cinerariae
Botrytis cinerea
Bremia lactucae
Coleosporium senecionis
Corticium solani
Fusarium sp.
Oidium sp.
Phytophthora cambivora
Phytophthora cinnamomi
Phytophthora cryptogea
Phytophthora parasitica
Pythium sp.
(Rhizoctonia solani)
SENECIO spp.
Ascochyta senecionis
Bremia sp.
Coleosporium senecionis
SEQUOIADENDRON
(Wellingtonia)
Armillaria mellea
Botrytis cinerea
Shallot, see Allium ascalonicum
L.
SIDALCEA
Puccinia malvacearum
Verticillium sp.
Silver fir, see Abies
SINNINGIA (gloxinia)
Cylindrocarpon radicicola
Didymella lycopersici
(Diplodina lycopersici)

Oidium sp.
Phytophthora cryptogea
Phytophthora parasitica
Pythium debaryanum
Thielaviopsis basicola
Sitka spruce, see Picea sitchen-
sis (Bong.) Carr.
Snake gourd, see Trichosanthes
Snapdragon, see Antirrhinum
majus L.
Snowberry, see Symphoricarpus
Snowdrop, see Galanthus
SOLANUM CAPSI-
CASTRUM Link (winter
cherry)
Botrytis cinerea
Colletotrichum atramentarium
Didymella lycopersici
(Diplodina lycopersici)
Phytophthora parasitica
Rhizoctonia sp.
Sclerotinia sclerotiorum
Thielaviopsis basicola
Verticillium sp.
SOLANUM MELONGENA
L. (eggplant)
Alternaria solani
Botrytis cinerea
Phoma sp.
SOLANUM TUBEROSUM
L. (potato)
(Actinomyces scabies)
Alternaria solani
Armillaria mellea
Ascochyta sp.
Botrytis cinerea
Colletotrichum atramentarium
Corticium solani
Fusarium arthrosporioides
Fusarium avenaceum
Fusarium caeruleum
Fusarium culmorum
Fusarium oxysporum
Fusarium solani
Fusarium tricinctum

Helicobasidium purpureum
Oidium *sp.*
Oospora lactis
Oospora pustulans
Penicillium *sp.*
Phoma eupyrena
Phoma foveata
Phoma tuberosa
Phomopsis tuberivora
Phytophthora erythroseptica
Phytophthora infestans
Phytophthora megasperma
Pyrenochaeta ferox
Pythium intermedium
Pythium ultimum
(Rhizoctonia crocorum)
(Rhizoctonia solani)
Rosellinia necatrix
Sclerotinia sclerotiorum
Spondylocladium atrovirens
Spongospora subterranea
Streptomyces scabies
Synchytrium endobioticum
Verticillium albo-atrum
Verticillium nigrescens
Verticillium nubilum
SOLANUM *spp.*
　Didymella lycopersici
　(Diplodina lycopersici)
　Phytophthora infestans
　Spongospora subterranea
SOLDANELLA
　Puccinia soldanellae
Solomon's seal, see
　　Polygonatum
SORBUS AUCUPARIA L.
　(*mountain ash, rowan*)
　Armillaria mellea
　Fomes annosus
　Gymnosporangium juniperi
　Stereum purpureum
SORBUS *spp.*
　Armillaria mellea
　Botrytis cinerea
　Fomes annosus

(Monilia cinerea)
Sclerotinia laxa
SORGHUM
　Puccinia purpurea
Soya bean, see Glycine max
　　Merr.
Spinach, see Spinacia oleracea
　　L.
Spinach beet, see Beta vulgaris L.
SPINACIA OLERACEA L.
　(*spinach*)
　Colletotrichum spinaciae
　Heterosporium variabile
　Peronospora effusa
SPIRAEA
　Stereum purpureum
SPIRANTHES
　Uredo lynckii
Spruce, see Picea
STACHYS
　Oidium *sp.*
STATICE, *see* **LIMONIUM**
Stock, see Matthiola
STRANSVAESIA
　Armillaria mellea
Strawberry, see Fragaria
Sugar beet, see Beta vulgaris
　　L.
Sumach, see Rhus
Sunflower, see Helianthus
　　annuus L.
Swede, see Brassica napo-
　　brassica Mill.
Sweet pea, see Lathyrus odo-
　　ratus L.
Sweet sultan, see Centaurea
Sweet william, see Dianthus
　　barbatus L.
Sycamore, see Acer pseudo-
　　platanus L.
SYMPHORICARPUS
　Ascochytula symphoricarpi
SYMPHYTUM
　Erysiphe cichoracearum
　Melampsorella symphyti

SYRINGA VULGARIS L.
(*lilac*)
Armillaria mellea
Ascochyta *sp.*
Botrytis cinerea
Diaporthe perniciosa
Heterosporium syringae
Oidium *sp.*
Phoma syringae
(Phomopsis perniciosa)
Phyllosticta syringae
Phytophthora syringae
Stereum purpureum

TAGETES (*African marigold*)
Corticium solani
Phytophthora cryptogea
(Rhizoctonia solani)
Tall oat grass, see Arrhenatherum elatius Mert. & Koch
TAXUS BACCATA L. (*yew*)
Armillaria mellea
(Cytospora taxifolia)
Diplodia taxi
(Phyllostictina hysterella)
Physalospora gregaria *var.*
foliorum
Polyporus sulphureus
Sphaerulina taxi
Teazle, see Dipsacus fullonum L.
THUJA
Armillaria mellea
Didymascella thujina
Fomes annosus
Hypholoma fasciculare
(Keithia thujina)
THUNIA, *see also* ORCHIDS
Hypodermium orchidearum
TIGRIDIA
Penicillium gladioli
TILIA (*lime*)
Armillaria mellea
Ascochyta tiliae
Gloeosporium tiliae
Phyllosticta tiliae

Ustulina deusta
Timothy grass, see Phleum
pratense L.
Tobacco, see Nicotiana
Tomato, see Lycopersicon
esculentum Mill.
TOPINE, *a hybrid* Helianthus
tuberosus × H. macrophyllus
or merely a luxuriant strain of
H. tuberosus
Botrytis cinerea
Sclerotinia sclerotiorum
TRAGOPOGON PORRIFOLIUS L. (*salsify*)
Cystopus cubicus
Erysiphe cichoracearum
Helicobasidium purpureum
Puccinia hysterium
(Rhizoctonia crocorum)
Ustilago tragopogi-pratensis
Trefoil, see Medicago lupulina L.
TRICHOSANTHES
Colletotrichum concentricum
TRICUSPIDARIA (Crinodendron)
Armillaria mellea
TRIFOLIUM (*clover*)
Ascochyta imperfecta
Ascochyta trifolii
Botrytis anthophila
Botrytis cinerea
Botrytis *sp.*
Cercospora *sp.*
Corticium solani
Cymadothea trifolii
Erysiphe polygoni
Helicobasidium purpureum
Kabatiella caulivora
Mycosphaerella carinthiaca
Peronospora trifoliorum
Pleospora herbarum
(Polythrincium trifolii)
Pseudopeziza trifolii
Pythium *sp.*
(Rhizoctonia crocorum)

(Rhizoctonia solani)
Sclerotinia spermophila
Sclerotinia trifoliorum
Sphaerulina trifolii
Stagonospora meliloti
(Stemphylium botryosum)
Stemphylium sarcinaeforme
Thielaviopsis basicola
Typhula trifolii
Uromyces flectens
Uromyces jaapianus
Uromyces trifolii
Uromyces trifolii-repentis
TRITICUM (*wheat*)
  Alternaria *sp.*
  Botrytis cinerea
  (Calonectria nivalis)
  Cercosporella herpotrichoides
  Cladosporium herbarum
  Claviceps purpurea
  Corticium solani
  Dilophospora alopecuri
  Erysiphe graminis
  Fusarium avenaceum
  Fusarium culmorum
  Fusarium equiseti
  (Fusarium graminearum)
  (Fusarium nivale)
  (Fusarium sambucinum)
  Gibberella pulicaris
  Gibberella zeae
  Gibellina cerealis
  Griphosphaeria nivalis
  (Helminthosporium sativum)
  Leptosphaeria culmicola
  Leptosphaeria culmorum
  Leptosphaeria herpotrichoides
  Leptosphaeria nigrans
  Leptosphaeria nodorum
  Leptosphaeria tritici
  Ophiobolus graminis
  Ophiobolus graminis *var.* avenae
  Ophiobolus herpotrichus
  Ophiobolus sativus
  Passalora graminis

Puccinia glumarum
Puccinia graminis
Puccinia triticina
Pythium arrhenomanes
Pythium graminicolum
Pythium tardicrescens
Pythium torulosum
Pythium volutum
(Rhizoctonia solani)
Rhizoctonia *sp.*
(Septoria nodorum)
Septoria tritici
Tilletia caries
Tilletia foetida
Typhula incarnata
Ustilago nuda
(Ustilago tritici)
Verticillium *sp.*
Wojnowicia graminis
TRITONIA (*Montbretia*)
  Botrytis *sp.*
  Mystrosporium adustum
  Penicillium gladioli
  Rosellinia necatrix
  Sclerotinia gladioli
TROLLIUS
  Urocystis anemones
TROPAEOLUM
  Coleosporium tropaeoli
  Cronartium flaccidum
  Pythium *sp.*
TSUGA
  Armillaria mellea
  Didymascella tsugae
  Fomes annosus
  (Keithia tsugae)
  Phomopsis pseudotsugae
TULIPA
  Botrytis cinerea
  Botrytis tulipae
  Cercospora *sp.*
  Cercosporella *sp.*
  Corticium solani
  Fusarium avenaceum
  Helicobasidium purpureum

Penicillium *sp.*
Phytophthora cryptogea
Phytophthora erythroseptica
Puccinia prostii
Pythium ultimum
Pythium *sp.*
(Rhizoctonia crocorum)
(Rhizoctonia solani)
Rosellinia necatrix
Sclerotium tuliparum
TURF, *see also other references
under* GRASSES
Badhamia foliicola
Cladochytrium caespitus
Colletotrichum graminicola
Corticium fuciforme
Corticium solani
Fuligo septica
Fusarium culmorum
(Fusarium nivale)
Griphosphaeria nivalis
Helicobasidium purpureum
(Helminthosporium sativum)
Leptosphaeria nigrans
Marasmius oreades
Mortierella *sp.*
Ophiobolus graminis
Ophiobolus graminis *var.* avenae
Ophiobolus sativus
Passalora graminis
(Rhizoctonia crocorum)
(Rhizoctonia monteithianum)
(Rhizoctonia solani)
Sclerotinia homoeocarpa
Sclerotinia trifoliorum
Spumaria alba
Ustilago hypodytes
Turnip, see Brassica rapa L.

ULMUS (*elm*)
Armillaria mellea
Auricularia auricula-judae
Ceratostomella ulmi
Fomes ulmarius
(Graphium ulmi)

Nectria cinnabarina
Polyporus squamosus
Rosellinia necatrix
Septogloeum ulmi
Stereum purpureum
Systremma ulmi
Taphrina ulmi
Ustulina deusta
URSINIA
Botrytis cinerea

VACCINIUM
Calyptospora goeppertiana
VALERIANA
Erysiphe cichoracearum
Uromyces valerianae
VALERIANELLA
Oidium valerianellae
VANDA, *see also* ORCHIDS
Colletotrichum orchidearum
Vegetable marrow, see Cucur-
bita pepo L.
VERBASCUM
Armillaria mellea
Erysiphe cichoracearum
Sclerotinia sclerotiorum
VERBENA
Phytophthora parasitica
Thielaviopsis basicola
VERONICA, *including* HEBE
Armillaria mellea
Peronospora grisea
Septoria exotica
Sphaerotheca fuliginea
Vetch, see Vicia *spp.*
VIBURNUM
Armillaria mellea
Botrytis cinerea
Phyllosticta tinea
Septoria *sp.*
VICIA FABA L. (*broad and
field bean*)
Aphanomyces euteiches
Ascochyta fabae
Ascochyta *sp.*

45

Botrytis cinerea
Botrytis fabae
Cercospora zonata
Corticium solani
Fusarium *sp.*
Helicobasidium purpureum
Peronospora viciae
Phytophthora *sp.*
Pleospora herbarum
Pythium *sp.*
(Rhizoctonia crocorum)
(Rhizoctonia solani)
Sclerotinia sclerotiorum
Sclerotinia trifoliorum
Sclerotinia trifoliorum *var.* fabae
(Stemphylium botryosum)
Thielaviopsis basicola
Uromyces fabae
VICIA *spp.* (*vetches*)
Botrytis cinerea
Botrytis fabae
Peronospora viciae
Sclerotinia trifoliorum
Uromyces fabae
VINCA (*periwinkle*)
Puccinia vincae
Vine, see Vitis vinifera L.
VIOLA ODORATA L.
   (*violet*)
Alternaria violae
Myrothecium roridum
Phyllosticta violae
Puccinia violae
Ramularia lactea
Rosellinia necatrix
Sclerotium delphinii
Septoria violae
Thielaviopsis basicola
Urocystis violae
VIOLA *spp.*, *including pansy*
Aphanomyces euteiches
Cercospora *sp.*
Corticium solani
(Fusarium sambucinum)
Fusarium *sp.*

Gibberella pulicaris
Helicobasidium purpureum
Myrothecium roridum
Oidium *sp.*
Peronospora violae
Phoma *sp.*
Phyllosticta violae
Phytophthora cactorum
Phytophthora parasitica
Puccinia aegra
Pythium mammillatum
Pythium oligandrum
Pythium violae
Ramularia agrestis
Ramularia deflectens
Ramularia lactea
(Rhizoctonia crocorum)
(Rhizoctonia solani)
Thielaviopsis basicola
Urocystis violae
Violet, see Viola odorata L.
VITIS VINIFERA L. (*grape vine*)
Armillaria mellea
Botrytis cinerea
Cercospora roesleri
Coniothyrium diplodiella
Elsinoe ampelina
(Gloeosporium fructigenum)
Glomerella cingulata
Guignardia bidwelli
Plasmopara viticola
Roesleria pallida
Septoria badhami
(Sphaceloma ampelinum)
Uncinula necator
Verticillium dahliae

Wallflower, see Cheiranthus
Walnut, see Juglans regia L.
Watercress, see Nasturtium officinale R.Br.
Water lily, see Nymphaea
WEIGELIA
Gloeosporium diervillae

46

Wellingtonia, see Sequoiaden-
  dron
Western hemlock, see Tsuga
Western red cedar, see Thuja
Weymouth pine, see Pinus
  strobus L.
Wheat, see Triticum
Whitebeam, see Sorbus *spp.*
Willow, see Salix
Wineberry, see Rubus *spp.*
Winter cherry, see Solanum
  capsicastrum Link
WISTARIA
  Septoria wistariae

Yew, see Taxus baccata L.
YUCCA
  Cercospora concentrica
  Coniothyrium concentricum

ZANTEDESCHIA
  Corticium rolfsii
  Corticium solani
  Gloeosporium *sp.*

Helicobasidium purpureum
Phyllosticta richardiae
Phytophthora richardiae
(Rhizoctonia crocorum)
(Rhizoctonia solani)
Rosellinia necatrix
(Sclerotium rolfsii)
Thielaviopsis basicola
ZEA MAYS L. (*maize*)
  Botrytis cinerea
  Fusarium culmorum
  (Fusarium moniliforme)
  Gibberella fujikuroi
  Puccinia sorghi
  Sphacelotheca reiliana
  Ustilago maydis
ZINNIA
  Alternaria zinniae
  Ascochyta *sp.*
  Botrytis cinerea
  Corticium solani
  Phytophthora cryptogea
  (Rhizoctonia solani)
  Sclerotinia sclerotiorum

PART II

PARASITES

PARASITES

Acanthostigma parasiticum (Hartig) Sacc., *Syll.* ix, 855; Bisby & Mason, *TBMS*, **24**, 1940, 176; syn. *Trichosphaeria parasitica* Hartig, *Allg. Forst. JagdZtg*, 1884.

*On silver fir (*Abies*), Argyll (Watson, *Scot. For. J.* 47, 1933, 71).

Acremonium sp.

*On mushroom (*Agaricus*) associated with chocolate brown patches, Kent, 1948; Dorset (Bullock, *Bull. Mushroom Growers' Ass.* **18**, 1950, 24).

Actinomyces scabies (Thaxt.) Güssow = *Streptomyces scabies*.

Actinomyces tumuli Millard & Beeley = *Streptomyces tumuli*.

Actinonema aquilegiae (Roum. & Pat.) Grove, *J. Bot., Lond.*, 1918, 343; Grove, *Coelomycetes*, ii, 269; syn. *Gloeosporium aquilegiae* Thüm., *Beiträge zur Pilzflora Siberiens*, no. 144; Sacc., *Syll.* iii, 700; *Ascochyta aquilegiae* Sacc., *Syll.* iii, 396.

*Leaf spot of *Aquilegia*. Fairly common in southern England. Listed in Clyde.

Actinonema padi Fr. = stat. conid. of *Gnomonia padicola*.

Actinonema rosae (Lib.) Fr. = stat. conid. of *Diplocarpon rosae*.

Aecidium convallariae Schum. = *Puccinia sessilis*.

Aecidium dracontii Schw. = *Uromyces ari-triphylli*.

Aecidium importatum P. Henn., *Verh. bot. Ver. Brandenb.* **37**, 1895, 12; Sacc., *Syll.* xiv, 388; Wilson & Bisby, *TBMS*, **37**, 1954, 63.

*Rust of *Peltandra*. On *P. virginica*, Royal Botanic Garden, Edinburgh, 1924; had been present for several years. The mycelium persists in the rootstocks.

Aecidium phillyreae DC. = *Zaghouania phillyreae*.

Aecidium pseudo-columnare Kühn = *Milesina kriegeriana*.

Aleuria vesiculosa (Bull. ex Fr.) Boud., *Histoire et classification des Discomycètes d'Europe,* 1907; Ramsbottom & Browne, *TBMS,* **34,** 1951, 46; syn. *Peziza vesiculosa* Bull., *Histoire des Champignons de la France,* 1791–8, 270; Sacc., *Syll.* viii, 83.

*In mushroom beds. Fruiting bodies of *Aleuria* sp., probably *A. vesiculosa,* covering casing soil in shelved mushroom beds, Chelmsford, Essex, 1952 (as *Peziza*); North Devon, 1938.

Alternaria brassicae (Berk.) Sacc., *Mich.* ii, 172; *Syll.* iv, 546; Wakefield & Bisby, *TBMS,* **25,** 1941, 97; Wiltshire, *Mycol. Pap.* **20,** 1947, 1–8.

*Dark leaf spot of brassicas. This disease has been given little attention but is widely distributed in England and Wales and doubtless common in some years. Occurs also in Scotland. Found mainly on cauliflower and broccoli; also on Brussels sprouts, cabbage and kale; occasional on kohl-rabi, marrow-stem kale and rape. Another species, *A. brassicicola* (q.v.), is more often involved, but the two fungi are rarely differentiated. The spores of *A. brassicae* are large, with a long beak; those of *A. brassicicola* smaller and unbeaked. What was probably this species was found on blackened stem bases of cauliflower, Ayrshire, 1952.

*Leaf spot of turnip (*Brassica rapa*). Occasional since 1925.

*On leaves and pods of radish (*Raphanus*). Occasional.

*Causing seedling rot of cress (*Lepidium*). Dorset, 1947. A species of *Alternaria,* possibly this, was found on a sample of cress seed examined at Harpenden in 1944.

Alternaria brassicae var. **nigrescens** Pegl. = *A. cucumerina.*

Alternaria brassicicola (Schw.) Wiltshire, *Mycol. Pap.* **20,** 1947, 10; syn. *A. oleracea* Milbrath, *Bot. Gaz.* **74,** 1922, 320; Wakefield & Bisby, *TBMS,* **25,** 1941, 97; *A. circinans* (Berk. & Curt.) Bolle, *Meded. Phytopath. Lab. Scholten,* Baarn, **7,** 1924, 26.

*Dark leaf spot of brassicas. Along with and more frequent than *A. brassicae* (q.v.); common and widely distributed in Britain on cauliflower, broccoli, cabbage, Brussels sprouts and kale. Also commonly rots cabbage seed pods in Scotland, and occurs on curds of broccoli (Kent, 1948); occasional on radish.

Schimmer (*Plant Path.* **2**, 1953, 16–17) referred to the common occurrence of this fungus on summer cauliflower seed and confirmed that it can be controlled by hot water seed treatment for 18 min. at 122° F.

Alternaria cheiranthi (Fr.) Bolle, *Meded. Phytopath. Lab. Scholten*, Baarn, **7**, 1924, 43; Wiltshire, *TBMS*, **18**, 1933, 158; Wakefield & Bisby, ibid. **25**, 1941, 97.

*On wallflower (*Cheiranthus*). Oakley, Hants, 1930; London, 1935; Essex, 1947; Suffolk, 1953. Parasitism uncertain.

The species was discussed by Wiltshire (*TBMS*, **18**, 1933, 135–60) who provided a revised description of it.

Alternaria circinans (Berk. & Curt.) Bolle = *A. brassicicola*.

Alternaria cucumerina (Ell. & Everh.) Elliott, *Amer. J. Bot.* **4**, 1917, 439; Wakefield & Bisby, *TBMS*, **25**, 1941, 97; syn. *A. brassicae* var. *nigrescens* Pegl., *Riv. Pat. veg.* **1**, 1893, 296; Sacc., *Syll.* xxii, 1410.

*Leaf spot of cucumber (*Cucumis*). Occurring under glass (Bewley, *Diseases of Glasshouse Plants*, 1923, 99). A species of *Alternaria* was associated with fruit lesions, Salop, 1949.

Alternaria dianthi Stev. & Hall, *Bot. Gaz.* **47**, 1909, 409; Sacc., *Syll.* xxii, 1410; Wakefield & Bisby, *TBMS*, **25**, 1941, 97.

*Stem rot of carnation (*Dianthus*). Occasional, but significance uncertain. Glos, Essex, Salop, Ches, Midlothian.

Corbett (*Gdnrs' Chron.* **81**, 1927, 150) isolated a pathogenic *Alternaria* from carnation which he regarded as distinct from *A. dianthi*. *Alternaria* sp. has been recorded on this host from Staffs, Middx, Kent, Sussex. Closer study may show that in Britain, as elsewhere, the species usually present is *A. dianthicola* (Neergaard, *Danish Species of* Alternaria *and* Stemphylium, 1945, 190) and not *A. dianthi*.

*Leaf spot of *Saponaria*. Longstanton, Cambs, 1953.

Alternaria dianthicola Neerg., see *A. dianthi*.

Alternaria godetiae Neerg., *Aarsberetn. Ohlsens Enkes plantepat. Lab.* **10**, 1944–5, 14.

*Stem blight of *Godetia*. Kelvedon, Essex, Sept. 1947 (W. C. Moore & F. Joan Moore, *TBMS*, **32**, 1950, 275–7). The disease is seed-borne.

Alternaria linicola Groves & Skolko, *Canad. J. Res.* C, **22**, 1944, 222; syn. *A. linicola* Neergaard, *Danish species of* Alternaria *and* Stemphylium, 1945, 300.

*Seedling disease of flax (*Linum*). Occasional and insignificant. Hants, Wilts, Yorks, Tay. See W. C. Moore (*TBMS*, **29**, 1946, 256) and Loughnane (*Nature, Lond.*, **157**, 1946, 266).

Alternaria matthiolae Neerg. = *A. raphani*.

Alternaria oleracea Milbrath = *A. brassicicola*.

Alternaria porri (Ell.) Neerg., *Aarsberetn. Ohlsens Enkes plantepat. Lab.* **3**, 1937–8, 4; syn. *Macrosporium porri* Ell., *Grevillea*, **8**, 1879, 12; Sacc., *Syll.* iv, 537.

*On onion (*Allium*). This or a closely related species was associated with black-stem lesions and collapse of onion-seed heads in Wilts (1941) and Som (1942).

Alternaria radicina Meier, Drechsl. & Eddy = *Stemphylium radicinum*.

Alternaria raphani Groves & Skolko, *Canad. J. Res.* **22**, 1944, 227; Wiltshire, *Mycol. Pap.* **20**, 1947, 14; syn. *A. matthiolae* Neergaard, *Danish Species of* Alternaria *and* Stemphylium, 1945, 177.

*Leaf spot of stock (*Matthiola*). Rare. Ashford, Kent, 1935. Ware (*Gdnrs' Chron.* **100**, 1936, 236–7) described this disease and attributed it to a species of *Alternaria* which Groves and Skolko (loc. cit.) considered to be *A. raphani*. London, 1954, on seed, leading to seedling rot.

*Leaf spot and seedling rot of radish (*Raphanus*). First seen in England on seed sample, 1948; on seed pods, Glos, 1949. The fungus causes low germination and seedling mortality.

Alternaria senecionis Neergaard, *Danish Species of* Alternaria *and* Stemphylium, 1945, 198.

*Leaf spot of cineraria (*Senecio*). Occasional. Kent, Sussex, Dorset, Som, Wilts, Worcs. Was first seen in Kent (Green & Hewlett, *Gdnrs' Chron.* **126**, 1949, 216–17: *J. R. hort. Soc.* **75**, 1950, 199–202). Cooper (*J. hort. Sci.* **31**, 1956, 229–33) studied the influence of cultural conditions on the disease.

Alternaria solani (Ell. & Mart.) Sor., *Z. PflKrankh.* **6**, 1896, 1–9; Sacc., *Syll.* iv, 530; em. Jones & Grout, *Bull. Torrey bot. Cl.* **24**, 1897, 254–8; Wakefield & Bisby, *TBMS*, **25**, 1941, 98.

*Alternaria blight (target spot) of potato (*Solanum*). Widely distributed in Britain. Sporadic and causes little damage.

The early history of the disease in Britain is uncertain (see *Report*, viii, 17), but the fungus was recorded in 1909 as *Sporidesmium solani varians* Vanha (*Mitt. Land-Versuch. Pflanzenkr. Brünn*, **2**, 1904, extra 8; Sacc., *Syll.* xviii, 616). In 1934 (Salaman & O'Connor, *Nature, Lond.*, **134**, 1935, 932) it was particularly abundant around Cambridge and over a wide area in north Scotland and the outer Hebrides. Since then it has been sporadic, developing late and causing little damage.

The tuber rot caused by this species (see, for example, *Tijdschr. PlZiekt.* **39**, 1933, 165) has not yet been recognized in Britain.

Dillon Weston (*TBMS*, **20**, 1936, 112–15) and Charlton (ibid. **36**, 1953, 349–55) studied sporulation in culture. Brian *et al.* (*J. gen. Microbiol.* **5**, 1951, 619–32) discussed the production of alternaric acid by this fungus, as well as (*AAB*, **39**, 1952, 308–21) its phytotoxic properties in relation to the etiology of disease caused by the species.

*Alternaria blight of tomato (*Lycopersicon*). Widely distributed in Jersey, the Isle of Wight, and in south-east and east England; often serious in hot summers. Rare elsewhere. Devon, 1949; Oxon, 1950; Glos, 1957.

It is very doubtful if this disease occurred in England before 1944 (Glasscock & Ware, *Nature, Lond.*, **154**, 1944, 642; *Agriculture, Lond.*, **51**, 1944, 417–20), especially in view of its sub-

sequent rapid spread. It was probably introduced with imported seed. In 1945 the disease was widespread in the south-east, with only an occasional report from Essex and Herts, but in 1947 it was common throughout East Anglia and often destructive there.
*On eggplant (*Solanum melongena*). Rare. Corn, 1945.

Alternaria violae Galloway & Dorsett, *Bull. U.S. Div. Veg. Physiol. Path.* no. 23, 1900, 11; Sacc., *Syll.* xvi, 1080.
*Spot disease of violet (*Viola*). Reported in England by Cooke and W. G. Smith (*J. R. hort. Soc.* **26**, 1901, 492–94) and by Cook (ibid. **58**, 1933, 122), but has apparently not been seen since. The identity of the diseases occurring in England and America is open to doubt.

Alternaria zinniae Pape, *Angew. Bot.* **24**, 1942, 69.
*Leaf spot and seedling blight of *Zinnia*. Not infrequent in England in southern districts; also War, Yorks.

Alternaria sp.
*Associated with black point of wheat grain. Unimportant. Hyde and Galleymore (*AAB*, **38**, 1951, 348–56) discussed sub-epidermal invasion of wheat grains by *Alternaria tenuis*.
*Associated with core rot of stored apples (Bristol) and occasionally found on apple fruits in the orchard (Worcs, eastern England).

Ansatospora acerina (Hartig) Hansen & Tompk. = *Centrospora acerina*.

Anthostomella pullulans (de Bary) Bennett, see *Dematium pullulans*.

Aphanomyces cochlioides Drechsl., *J. agric. Res.* **38**, 1929, 326; *Phytopathology*, **18**, 1928, 149.
*This or *Aphanomyces* sp. is a minor cause of blackleg in mangold and sugar beet. West Midlands, Sussex. The disease has also been attributed, probably incorrectly, to *A. laevis* de Bary, *Pringsh. Jahrb.* ii, 1860, 199; Sacc., *Syll.* vii, 276.

Aphanomyces euteiches Drechsl., *J. agric. Res.* **30**, 1925, 311.

*Root rot of pea (*Pisum*). Widely distributed throughout Britain and not uncommon in heavy or poorly drained soils. First recognized in Staffs and Cards (*Report*, v, 36).

*Root rot of bean. One record on dwarf and runner bean (*Phaseolus*) in Scotland (Forth); also on broad bean (*Vicia*), War, 1941; Salop, 1949.

*Root rot of lupin (*Lupinus*). Staffs, 1928 (*Report*, vii, 97); Notts, 1938.

*Root rot of sweet pea (*Lathyrus*). Not uncommon in England since 1930 (*Report*, vii, 37) and widely distributed in Scotland.

*Root rot of *Viola*. Widely distributed in Scotland. Harpenden, 1936 (as *Aphanomyces* sp.).

Aphanomyces laevis de Bary, see *A. cochlioides*.

Apiognomonia errabunda (Rob.) Höhnel, see *Gloeosporium nervisequum*.

Armillaria mellea (Fr.) Quél., *Les Champignons du Jura et des Vosges*, Paris, 1872; Sacc., *Syll.* v, 80.

*Armillaria root rot (honey fungus). This fungus is an omnivorous parasite, attacking and often killing trees, especially conifers, bushes and a wide range of herbaceous plants. It usually begins as a saprophyte on decaying tree stumps and from them passes to the roots of healthy plants. The honey fungus occurs throughout Britain and is particularly common on the sites of old woodlands and orchards.

Day (*Quart. J. For.* **21**, 1927, 9–21: *Forestry*, **3**, 1929, 94–103) discussed its parasitism and Campbell (*AAB*, **21**, 1934, 1–22) the significance of the black lines formed by it in woody tissues. Garrett (*Ann. Bot., Lond.*, n.s., **17**, 1953, 63–79; **20**, 1956, 193–209) and Townsend (*TBMS*, **37**, 1954, 222–33) studied the factors that control the formation of rhizomorphs and infection by them; and Garrett (*Canad. J. Microbiol.* **3**, 1957, 135–49) also tested the effect of a soil microflora selected by carbon disulphide fumigation on the survival of *A. mellea* in woody host tissues. Jones and

H. I. Moore (*Gdnrs' Chron.* **98**, 1935, 284) published a note on the disease and included a short list of hosts of the parasite.

Over 100 British hosts are listed below, with known localities and where possible an indication of prevalence, but the list is not claimed to be complete.

POTATO. Rare. Midlothian (Wilson, *Trans. R. Scot. arb. Soc.* **35**, 1921, 186–7); Yorks, 1941; Herts, 1945; Ches, 1953.

VEGETABLES

Carrot (*Daucus*). Som, 1938.

Mint (*Mentha*). Cumb, 1953.

Onion (*Allium*). War, 1932 (*Report*, vii, 63).

Rhubarb (*Rheum*). Not infrequent, usually on old orchard sites.

Tomato (*Lycopersicon*). Surrey, 1945 (*Report*, ix, 59).

FRUIT

Almond (*Prunus*). Occasional.

Apple (*Malus*). Common, especially in south and south-west England. Its spread in orchards was studied by Marsh (*Rep. Long Ashton Res. Sta. for 1951*, 116–21; *TBMS*, **35**, 1952, 201–7.

Cherry (*Prunus*). Occasional, on both ornamental and fruiting trees.

Currant (*Ribes*). Not infrequent on black currant (Marsh, *TBMS*, **35**, 1952, 201–7); occasional on red currant.

Fig (*Ficus*). Devon.

Gooseberry (*Ribes*). Frequent. See Blackman & Jones, *Rep. on Gooseberry Diseases in E. Sussex 1922–3*, Lewes, 1923, 12 pp.

Grape vine (*Vitis*). Occasional.

Loganberry (*Rubus*). Corn, 1944.

Medlar (*Mespilus*). Som, 1928.

Nectarine (*Prunus*). Devon.

Nut (*Corylus*). On cob nut, East Malling, Kent; on hazel nut, Glam.

Peach (*Prunus*). Devon.

Pear (*Pyrus*). Not uncommon.

Plum (*Prunus*). Frequent, notably in the south.

Quince (*Cydonia*). Harpenden, Herts, 1949.

Raspberry (*Rubus*). Not infrequent.

Strawberry (*Fragaria*). Kent, west of England, Forth.

Walnut (*Juglans*). Occasional; Devon.

Wineberry (*Rubus*). Bristol, Som.

HOP (*Humulus*). First seen in Kent and Worcs, 1935. Not uncommon in south-east England where fruit or other trees are grubbed to make room for hop gardens (Salmon & Ware, *J. S.-E. agric. Coll. Wye*, **40**, 1937, 18–26).

CONIFERS. Day (*Quart. J. For.* **21**, 1927, 9–21) discussed the parasitism of the fungus on conifers, and Ritchie (*Scot. For. J.* **46**, 1932, 132–42) its distribution on conifers in north-east Scotland.

Abies alba (silver fir). *A. nobilis*, Dee; and *A. pectinata*, Southampton, Hants.

Athrotaxis selaginoides. Exbury, Hants.

Cedrus sp. Fife, 1951.

Cupressus. Glam, 1932; Som, 1945. On *C. macrocarpa*, Devon, 1948.

Juniperus sp. (Chinese juniper). Westmorland.

Larix decidua (larch).

Metasequoia. Fleet, Kirkcudbright, 1956.

Picea abies (Norway spruce). *P. sitchensis* (Sitka spruce).

Pinus nigra (Austrian pine). *P. nigra* var. *poiretiana* (Corsican pine). Rayner (*Forestry*, **4**, 1930, 65–77) infected this and Douglas fir with pure cultures of the fungus. *P. strobus* (Weymouth pine). *P. sylvestris* (Scots pine).

Pseudotsuga taxifolia (Douglas fir).

Sequoiadendron (*Wellingtonia*) sp. Salop.

Taxus baccata (yew).

Thuja plicata. In a hedge, Surrey, 1929; on *Thuja* sp., Forest of Dean, 1939; Som, 1945; Nairn, 1952.

Tsuga heterophylla (western hemlock).

OTHER TREES AND ORNAMENTALS

Acer sp. (Japanese maple). Westmorland.

Acer pseudoplatanus (sycamore).

Alnus glutinosa (alder).

Andromeda sp.

Berberis sp. Devon.

Betula pendula (birch). Berks, 1937.

Calluna vulgaris (heather). Clyde (Alcock & Wilson, *Scot. For. J.* **41**, 1927, 224–5).

Camellia sp. Devon and Corn, 1951.

Canna indica. Devon, 1928.

Castanea sativa (chestnut).

Ceanothus sp. Glam.

Choisya sp. Devon, 1951.

Cotoneaster sp. Hants, Som, Clyde, Tay.

Crataegus oxyacantha (hawthorn). Yorks, Som.

Crinodendron sp. Argyll, 1954.

Cytisus sp. Devon, 1937.

Daphne mezereum. Cambridge, 1946.

Davidia sp. Glam.

Euonymus japonica. Glam, Som, Devon.

Fagus sylvatica (beech). Relatively resistant.

Forsythia sp. Glam, 1949.

Fraxinus excelsior (ash). Devon, 1953; Midlothian, 1956.

Freesia sp. On plants in boxes, Worcs, 1949.

Garrya elliptica. Westmorland.

Genista sp. Corn.

Ilex aquifolium (holly). Wilts, Yorks, East Lothian.

Iris kaempferi. New Forest, Hants (Wilson, *Gdnrs' Chron.* **91**, 1932, 65).

Laburnum vulgare. Yorks, Devon, Peebles.

Laurus sp. Yorks, 1935; Devon, 1956.

Lavandula sp. Salop, 1938.

Ligustrum vulgare (privet). Common and destructive to privet hedges in many districts.

Limonium (Statice) sp. Devon, 1926; on *L. latifolium*, Clyde, 1956.

Lupinus sp. On a number of occasions in the north; Devon.

Magnolia wilsonii. Southampton, Hants, 1935; on *Magnolia* sp., Devon, 1953.

Narcissus sp. Rare. Middx, 1930; Stirling, 1955.

Olearia hastii. Som, 1944.

Paeonia sp. Not infrequent.

Paulownia sp. Glam, Devon.

Populus sp. Rare. Glam (on Chinese poplar).

Prunus sp. On flowering almond, Devon; flowering cherry, Bristol.

Prunus cerasifera atropurpurea (*P. pissardii*), Lincs; *P. laurocerasus*, Yorks; *P. lusitanica*, Som; *P. triloba*, Cumb.

Quercus sp. Usually on otherwise unhealthy trees such as those attacked by the green oak tortrix moth, *Tortrix viridana* (Robinson, *Quart. J. For.* **21**, 1927, 25–7).

Rhododendron spp. Not infrequent. Royal Botanic Garden, Edinburgh (Boughey, *Gdnrs' Chron.* **104**, 1938, 84).

Ribes sp. (flowering currant). Glam, Northumb. On *R. sanguineum*, Bristol, Som.

Robinia pseudacacia. Devon, 1937.

Rosa spp. Occasional.

Salix spp. Common in cricket bat willow plantations.

Sambucus niger (elder). Derby.

Scabiosa caucasica. Banff, 1949; Suffolk, 1952.

Sorbus aucuparia (mountain ash). *S. torminalis* (service tree), Nairn, 1953.

Stransvaesia davidiana. Clyde.

Syringa vulgaris (lilac). Occasional in south-west England, Lancs, Cambs, East Lothian.

Tilia sp. (lime). Bristol, 1937.

Ulmus sp. (elm).

Verbascum. Som, 1952.

Veronica sp. Devon, Forth.

Viburnum mariesii, Southampton, Hants, 1935. *V. tinus*, Corn.

Ascobolus leveillei Boud., *Mémoire sur les Ascobolées*, 1869, Sacc., *Syll.* viii, 519.

*In mushroom compost, Sussex, 1935.

Ascochyta aquilegiae Sacc. = *Actinonema aquilegiae.*

Ascochyta armoraciae Fuckel = *Septoria armoraciae.*

Ascochyta aspidistrae Massee, *Gdnrs' Chron.* 3rd Ser. **17**, 1895, 454 and 462; *Dis. Cult. Pl. Trees*, 1910, 431; Sacc., *Syll.* xxii, 1027; Grove, *Coelomycetes*, i, 322.

*Leaf blotch of *Aspidistra*. On *A. lurida*. Grove (loc. cit.) considered it a doubtful species.

Ascochyta asteris (Bres.) Gloyer, *Tech. Bull. N.Y. St. agric. Exp. Sta.* no. 177, 1931.

*Leaf spot of China aster (*Callistephus*). First British record at Mildenhall, Suffolk, Sept. 1953 (Storey & F. Joan Moore, *Plant Path.* **3**, 1954, 106).

Ascochyta betae Prill. & Delacr., *Bull. Soc. mycol. Fr.* 1891, 24; Sacc., *Syll.* x, 306; Grove, *Coelomycetes*, i, 296.

*In trivial amount on mangold (*Beta*), Devon, 1928 (*Report*, vii, 42).

Ascochyta bohemica Kab. & Bub., *Hedwigia*, **44**, 1905, 352; Sacc., *Syll.* xxii, 1024.

*Leaf spot of *Campanula*. On *C. betulaefolia* and *C. raineri*, Maidenhead, Berks, 1939 (W. C. Moore, *TBMS*, **24**, 1940, 60); on *C. medium*, Harpenden, Herts, 1941 (*TBMS*, **31**, 1947, 89). *Ascochyta* ?*carpathica* f. *caulicola* Grove, *J. Bot., Lond.*, 1922, 46; *Coelomycetes*, i, 297 (=*Phyllosticta carpathica* q.v.) was recorded on decayed stem bases of Canterbury bells at Kirkham, Lancs, in 1954.

Ascochyta boltshauseri Sacc., *Z. PflKrankh.* **1**, 1891, 136; Sacc., *Syll.* x, 303; as *Stagonosporopsis hortensis* (Sacc. & Malbr.) Petr., *Ann. Mycol., Berl.*, **19**, 1921, 21; syn. *Stagonospora hortensis* Sacc. & Malbr., *Mich.* ii, 629; Sacc., *Syll.* iii, 446.

*On bean (*Phaseolus*). This species has not been definitely recorded in Britain, though it was probably this that has been found on seeds of *Phaseolus* in Scotland, and in 1956 what appeared to be this species was reported on runner bean in Suffolk and on a number of other bean crops in East Anglia. In

1953 an *Ascochyta* (?*Stagonospora*) that may have been this was reported on pods of dwarf bean in Norfolk.

Ascochyta brassicae Thüm., *Contr. ad floram mycologiam lusitanicam*, no. 602; Sacc., *Syll.* iii, 397; Grove, *Coelomycetes*, i, 296.

*Leaf spot of brassicas. Not common, occasionally destructive. First British record, Devon, 1925 (*Report*, vi, 23) on cabbage; Lincs, 1931, on turnip or swede; Kent, 1930, on marrow-stem kale; Salop, Ches, Montgomery.

Ascochyta cinerariae Tassi, *Boll. Orto. bot. Siena*, 1899, 31; Sacc., *Syll.* xvi, 930.

*On cineraria (*Senecio*). Associated with stem base rot (foot rot), Lancs, 1934; Beds, 1940; on leaves, Ches, 1950. See W. C. Moore, *TBMS*, 31, 1947, 89. Parasitism uncertain.

Ascochyta clematidina Thüm., see under *Ascochyta* sp. (p. 70).

Ascochyta cucumeris Fautr. & Roum., *Rev. mycol.* 13, 1891, 79; Sacc., *Syll.* x, 304; Grove, *Coelomycetes*, i, 300.

*On leaves of vegetable marrow (*Cucumis*), Kew, Ayrshire. Chiu and Walker (*J. agric. Res.* 78, 1949, 82) regarded this species as identical with the imperfect stage of *Mycosphaerella citrullina* (q.v.) and considered the valid name to be *M. cucumis* (Fautr. & Roum.) Chiu & Walker.

Ascochyta dahliicola (Brun.) Petr., *Ann. mycol., Berl.*, 25, 1927, 201; syn. *Phyllosticta dahliicola* Brun., *Bull. Soc. bot. Fr.* 34, 1887, 429; Sacc., *Syll.* x, 129; Grove, *Coelomycetes*, i, 14; Moore, *TBMS*, 31, 1947, 90.

*Leaf spot of dahlia. Not common. Kent, 1926; Berks, Herts, Solway, Clyde, Dee.

Ascochyta deutziae Bresad., *Hedwigia*, 1900, 326; Sacc., *Syll.* xvi, 927; Grove, *Coelomycetes*, i, 301.

*Leaf spot of *Deutzia*. On *D. gracilis*, West Kilbride and Saltcoats, Ayrshire, Nov. 1910 (*TBMS*, 4, 1913, 175).

Ascochyta dianthi Berk., *Outl.* 1860, 320; Sacc., *Syll.* iii, 398; Grove, *Coelomycetes*, i, 298. Sacc., *Syll.* x, 301 also lists *A. dianthi* (A. & S.) Lib., *Plantae cryptogamicae*, ii, 1832, 158.

*Leaf spot of sweet william (*Dianthus*). Frequent in south-west England, Sussex, Yorks, Scotland. Also reported on carnation, Yorks, 1928. Grove (*Coelomycetes*, i, 14 and 380) regarded *Phyllosticta dianthi* Westend. (*Exs.* n. 293; Sacc., *Syll.* iii, 43) as a young state of this, and both as early states of *Septoria dianthi* (q.v.).

Ascochyta fabae Speg., *Fung. Arg.* 1899, 321; Sacc., *Syll.* xvi, 928; Grove, *Coelomycetes*, i, 302.

*Leaf spot of broad and field beans (*Vicia*). Widely distributed and not uncommon, notably in south-west England and north Scotland. Sometimes attacks the pods and seed, but rarely injurious.

Ascochyta pisi was reported on beans in Sussex and Northants many years ago (Carruthers, *J. R. agric. Soc.* **10**, 1899, 658; **11**, 1900, 731). First reported as *A. fabae* in 1927 and Beaumont (*TBMS*, **33**, 1950, 345–9) confirmed that the pathogen is distinct from *A. pisi*. (See also *Ascochyta phaseolorum*.)

Ascochyta gerberae Maffei, *Riv. Pat. veg.* **6**, 1913, 257; Sacc., *Syll.* xxv, 323.

*Leaf spot of *Gerbera*. Not infrequent in south-west England but not reported elsewhere. First seen on *G. jamesonii* at Penzance, Corn (Beaumont & Gregory, *Gdnrs' Chron.* **102**, 1937, 28).

Ascochyta graminicola Sacc., *Mich.* i, 127; *Syll.* iii, 407; Grove, *Coelomycetes*, i, 323.

*Found scattered throughout Britain in late summer on fading leaves of *Alopecurus*, *Dactylis*, *Festuca*, *Holcus*, *Lolium*, *Phleum*, and *Poa*.

Ascochyta humuli Kab. & Bub., *Hedwigia*, **43**, 1904, 419; Sacc., *Syll.* xviii, 346; Grove, *Coelomycetes*, i, 304.

*Leaf spot of hop (*Humulus*). East Malling, Kent, 1926; also on wild hops in the neighbourhood. Unimportant.

Ascochyta idaei Oud., *Hedwigia*, **37**, 1898, 178; Grove, *Coelomycetes*, i, 313.

*On branches of raspberry (*Rubus*), Glam, Notts.

Ascochyta impatientis Bres., *Hedwigia*, **39**, 1900, 326; Sacc., *Syll.* xvi, 927.

*Leaf spot of *Impatiens*. On *I. balsamina* L., Newton Abbot, Devon, 1942 (W. C. Moore, *TBMS*, **31**, 1947, 90). *Ascochyta weissiana* Allesch., in Rabenh., *Krypt. Fl.* i, 6, 1899, 647, is an older name for a closely allied if not identical fungus.

Ascochyta imperfecta Peck, *Bull. N.Y. St. Mus.* **157**, 1912, 21; Sacc., *Syll.* xxv, 331.

*Black stem of lucerne (*Medicago*). In England widely distributed in the south and midlands. Dee, East Lothian. Develops in spring and early summer. Found at Edinburgh on many seed samples from the 1949 harvest. The species was not recognized in England until 1934 (Toovey, Waterston & Brooks, *AAB*, **23**, 1936, 705–17) or in Scotland until 1945, but the disease was undoubtedly present in Britain before then.

*On trefoil (*Medicago lupulina*): not uncommon. First seen in the field in Bucks (1949) after it had been detected on many seed samples from the 1948 harvest both in England and Scotland; also Suffolk, Essex, Herts, Herefs, Carms.

*On *Trifolium*. Occurs on white clover (Wilts, Som) and possibly this on *Trifolium subterranea*, Kent, 1950.

Sprague (*Phytopathology*, **19**, 1929, 917–32) considered the fungus on trefoil to be a distinct species, *Ascochyta medicaginis* Bres., while Remsburg and Hungerford (ibid. **26**, 1936, 1015–20) claimed that *Pleospora rehmiana* Staritz (Sacc., *Syll.* xxiv, 1033) is the perfect stage of *Ascochyta imperfecta*.

PARASITES

Ascochyta kabatiana Trott. in Sacc., *Syll.* xxv, 330; Grove, *Coelomycetes*, i, 300.

*Leaf spot of laburnum. Occasional. Devon, Bristol area, Clyde. According to Grove (loc. cit.) *Phyllosticta cytisi* Desm. (*Ann. Sci. nat.* **14**, 1847, 34; Sacc., *Syll.* iii, 10), reported from Kew Gardens, Highgate, Shere, Hampton-in-Arden, etc., is identical.

Ascochyta linicola Naum. & Vassilievski, in Naumoff, *Mycology*, Leningrad, 1926.

*Foot rot of flax (*Linum*). Was not seen in Great Britain until the 1939–45 War, when the acreage devoted to flax increased greatly. It became prevalent from about 1945 in Scotland, Wales (especially Pembroke), Som and Dorset; and was occasionally seen elsewhere in England. With the recent decrease in flax acreage it has become less prominent.

There has been considerable confusion about the identity of the parasite. Pethybridge *et al.* (*J. Dep. Agric. Ire.* **20**, 1920, 325–42; **21**, 1921, 167–87) studied a Phoma foot rot of flax in Ireland caused by a species of *Phoma* which Marchal and Verplancke (*Bull. Soc. Bot. Belg.* **59**, 1926, 19) later named *Phoma linicola* n.sp., though their description was based on Belgian material. Muskett and Colhoun (*Nature, Lond.*, **155**, 1945, 367) found the same disease again much later in Northern Ireland and discussed its control (ibid. **156**, 1945, 538). Dennis (*TBMS*, **29**, 1946, 11) studied the fungus in culture and W. C. Moore (*Report*, ix, 76) found it identical with Russian material of *Ascochyta linicola*, but considered the fungus was not a good *Ascochyta* and should be referred to *Phoma* or *Diplodina*. Kerr (*TBMS*, **36**, 1953, 61–73), who made a very careful study of the fungus, also doubted whether it belonged to the genus *Ascochyta* but recommended retaining the name *A. linicola* to avoid still further confusion. Other species of *Phoma* or *Ascochyta* occur on flax but they do not appear to be pathogenic.

Ascochyta lycopersici Brun. = stat. conid. of *Didymella lycopersici*.

Ascochyta medicaginis Bres., see *A. imperfecta.*

ASCOCHYTA

Ascochyta orobi Sacc., *Syll.* iii, 398; Grove, *Coelomycetes*, i, 308.
*Leaf spot of sainfoin (*Onobrychis*). Common in Glam; local elsewhere. Also occurs on the stems.

Ascochyta papaveris Oud., *Contributions à la Flore mycologique de Nowaja Semlja*, 1885, 12; Sacc., *Syll.* x, 301; Grove, *Coelomycetes*, i, 301 (as *A. papaveris* var. *dicentrae* Grove).
*On leaves of *Dicentra*. On *D. spectabilis*, Saltcoats, Ayrshire (Smith & Ramsbottom, *TBMS*, 6, 1918, 49).

Ascochyta paulowniae Sacc. & Brun., *Fungi Gallici*, n. 2241; Sacc., *Syll.* iii, 390.
*Leaf spot of *Paulownia*. Dorset, 1954.

Ascochyta phaseolorum Sacc., *Mich.* i, 164; *Syll.* iii, 398; Grove, *Coelomycetes*, i, 309.
*Blotch of runner bean (*Phaseolus*). Rare. Richmond, 1916 (Smith & Ramsbottom, *TBMS*, 6, 1920, 368); Hants, 1922; Berks, 1930; Kent, 1939; Ayrshire. Also reported rotting French bean seedlings, Argyll, 1952.

At one time not infrequently reported from western counties, but it had been confused with a bacterial blight (*Report*, iv, 40). Doubtfully distinct from *Ascochyta fabae* Speg. (q.v.).

Ascochyta philadelphi Sacc. & Speg., *Mich.* i, 165; Sacc., *Syll.* iii, 386; Grove, *Coelomycetes*, i, 309.
*Leaf spot of *Philadelphus*. On *P. coronarius*, Seamill, Ayrshire (*TBMS*, 3, 1909, 118), Solway, Forth, Clyde, Tay, Kew.

Ascochyta piniperda Lindau, in Engler's *Pflanzenfamilien*, i, l. Abt., 1900, 368; Sacc., *Syll.* xxii, 1045.
*On spruce (*Picea*), causing defoliation of current year's shoots of *P. abies* and *P. sitchensis* (Laing, *Scot. For. J.* 43, 1929, 48–52).

Ascochyta pinodella L. K. Jones, *Bull. N.Y. St. agric. Exp. Sta.* no. 547, 1927, 4.
*Foot rot and leaf and pod spot of garden and field pea (*Pisum*).

Known in Yorks, Beds and Sussex and common in the south-west, but general distribution uncertain. First recognized in 1931 in Yorks (*Report*, vii, 47). Lowings (*Gdnrs' Chron*. **133**, 1953, 40) reported it in Northern Ireland.

*Foot rot of sweet pea (*Lathyrus*). Clyde. Possibly this species, reported as *Ascochyta* sp., causing similar disease in England (*Report*, vii, 97); Corn, 1951; Essex, 1953; Cards, 1955. (See also *Ascochyta pisi*.)

Ascochyta pisi Lib., *Plantae cryptogamicae*, i, 1830, no. 12; Sacc., *Syll*. iii, 397; Grove, *Coelomycetes*, i, 309.

*Leaf and pod spot of field and garden pea (*Pisum*). Very common throughout Britain.

This species has been known in Britain since 1841, and for many years it was regarded as the sole cause of leaf and pod spotting in peas, but Sattar (*TBMS*, **18**, 1934, 276–301) showed that, in Britain as in America, the disease may also be caused by *A. pinodella* (q.v.) and *Mycosphaerella pinodes* (q.v.) which produce spots rather different in character, and are often more destructive because they also cause foot rot. The three pathogens are still not often distinguished and their relative frequency is therefore uncertain. *Ascochyta pisi* is, however, the commonest, and in England it is this species which is believed to be predominant in commercial seed, of which about 25 % of the samples are infected. In Scotland *Mycosphaerella pinodes* seems to be more common on the seed than *Ascochyta pisi*, with *A. pinodella* infrequent. Hickman (*Rep. Long Ashton Res. Sta. for 1940*, 50–4) discussed the prevalence and significance of seed infection, and Ogilvie and Mulligan (ibid. *for 1932*, 115–19; *for 1933*, 116–19) obtained partial control by seed treatment.

*Leaf spot of sweet pea (*Lathyrus*). Forth, Clyde.

Ascochyta primulae Trail, *Scot. Nat*. 3, 1887, 88; Sacc., *Syll*. x, 300; Grove, *Coelomycetes*, i, 311.

*Leaf spot of *Primula*. Tay, Dee.

Ascochyta ribesia Sacc. & Fautr., *Bull. Soc. mycol. Fr.* 1900, 22; Sacc., *Syll.* xvi, 926; Grove, *Coelomycetes*, i, 312.
*Leaf spot of gooseberry. Clyde, 1921; Ross, 1952.

Ascochyta scabiosae Rabenh., in Klotzsch, *Herbarium vivum mycologicum*, no. 1253; Sacc., *Syll.* iii, 400.
*Leaf spot of scabious. Hyde End and Hurst, Berks, Dec. 1950. First British record. Ches, 1957.

Ascochyta senecionis Fuckel, *Symb. Myc.* 1869, 386; Sacc., *Syll.* iii, 400; Grove, *Coelomycetes*, i, 313.
*On living leaves of *Senecio greyii*, Polperro, Corn.

Ascochyta symphoricarpi Passer. = *Ascochytula symphoricarpi*.

Ascochyta tiliae Kab. & Bub., *Hedwigia*, **46**, 1907, 293; Sacc., *Syll.* xxii, 1029; Grove, *Coelomycetes*, i, 317.
*On *Tilia*. On leaves of *T. platyphyllos*, Ayrshire (Smith & Ramsbottom, *TBMS*, **6**, 1920, 367).

Ascochyta trifolii Bond. & Truss., *J. Bol. Rasteni*, **7**, 1913, 215; Sacc., *Syll.* x, 399; Grove, *Coelomycetes*, i, 317. Jones and Weimer, *J. agric. Res.* **57**, 1938, 807, named it *Stagonospora recedens* (Massal.) n.c.
*Leaf spot of clover (*Trifolium*). On *T. pratense*, Aberystwyth. Grove (loc. cit.) regarded it as an immature state of *Stagonospora meliloti* (q.v.).

Ascochyta violae Sacc. & Speg., see *Phyllosticta violae*.

Ascochyta vulgaris Kab. & Bub., *Öst. bot. Z.* **54**, 1904, 22; Sacc., *Syll.* xviii, 343, var. **lonicerae** Grove, *Coelomycetes*, i, 305.
*On leaves of *Lonicera*. Common.

Ascochyta weissiana Allesch., see *A. impatientis*.

Ascochyta sp.

*Leaf spot of potato (*Solanum*). In trial plot of var. Pentland Ace, Corn, 1955.
*On broad bean (*Vicia faba*), Som, 1947; apparently distinct from *Ascochyta fabae* and *A. viciae*.
*On anemone, south-west England, 1938.
*Die-back of *Clematis*. Occasional. Devon, Surrey, Westmorland (as *Phoma* sp.). *Ascochyta clematidina* Thüm. has been regarded as the chief cause of Clematis die-back in America (Gloyer, *Techn. Bull. N.Y. St. agric. Exp. Sta.* **44**, 1915, 14 pp.), but other factors may be involved in Britain (*Report*, viii, 82).
*On lower leaves of *Delphinium*, Edinburgh, 1951.
*Shoot die-back of lilac (*Syringa*). Sussex, 1937; Inverness and west Scotland, 1950; Devon, 1954.
*Leaf spot of *Rhododendron*. Dumfries, 1952.
*Leaf and stem blight of *Zinnia*. Inverness.

Ascochytula symphoricarpi Died. 411; Grove, *Coelomycetes*, i, 331; syn. *Ascochyta symphoricarpi* Passer., *Diagnosi di Funghi nuovi*, iv, 465; Sacc., *Syll.* x, 296.

*Stem blight of snowberry (*Symphoricarpus*). On *S. racemosus*, Polperro, Corn; on *S. albus*, East Lothian, 1948.

Asterocystis radicis Wildeman = *Olpidium brassicae*.

Auricularia auricula-judae (L. ex Fr.) Schröt., *Kr. Fl. Schles.* i, 386; Sacc., *Syll.* vi, 766; Rea, *Brit. Basid.* 727.

*Jews' ear fungus. A saprophyte which sometimes acts as a weak, slow parasite on old trees of elder (*Sambucus*); occasional also on beech (*Fagus*) and elm (*Ulmus*).

Badhamia foliicola Lister, *J. Bot., Lond.*, **35**, 1897, 209; *Mycetozoa*, 3rd ed., 1925, 13.

*On lawn, causing brown patches by smothering effect, Eas Midlands, 1933.

Badhamia utricularis (Bull.) Berk., *Linn. Trans.* **21**, 1852, 153; Cooke, *Handb.* no. 1147.

*On mushroom (*Agaricus*): Sussex, spreading over casing soil and attacking mushroom stalks (Wood, *Mushroom News*, **3**, 1951, 128); Worcs, 1952. Overrunning mushrooms and reducing them to a slimy mass.

Botryodiplodia sp.

*Twig die-back of *Choisya*. On *C. ternata*, Cambridge; Wisley, Surrey. Possibly *Botryodiplodia theobromae* Pat. (Woodward, *TBMS*, **11**, 1926, 281–3).

Botryosphaeria dothidea (Fr.) Ces. & de Not., *Schema Sfer*, 1863, 212; Sacc., *Syll.* i, 460; Bisby & Mason, *TBMS*, **24**, 1940, 143.

*Briar scab. Massee (*Diseases Cult. Pl. Trees*, 174) recorded it as sometimes epidemic on garden roses, as well as frequent on wild rose. Cooke (*Fung. Pests Cult. Pl.*, 46) called it *B. diplodia* (Moug.), presumably in error.

Botryotinia convoluta (Drayt.) Whetz. = *Sclerotinia convoluta.*

Botryotinia fuckeliana (de Bary) Whetz. = *Sclerotinia fuckeliana.*

Botryotinia narcissicola (Greg.) Buchw. = *Sclerotinia narcissicola.*

Botryotinia polyblastis (Greg.) Buchw. = *Sclerotinia polyblastis.*

Botryotinia porri (van Beyma) Whetz., see *Botrytis byssoidea.*

Botryotinia squamosa Viennot-Bourg., see *Botrytis squamosa.*

Botrytis allii Munn, *Bull. N.Y. St. agric. Exp. Sta.* no. 437, 1917, 396; Sacc., *Syll.* xxv, 693; Wakefield & Bisby, *TBMS*, **25**, 1941, 69.

*Neck rot of onion (*Allium*). Sometimes develops in the field before lifting, but is mainly a storage disease occurring throughout Britain from October onwards. Severe when wet, sunless conditions in late summer and early autumn prevent proper ripening of the bulbs. Under these conditions it may destroy

10 % of the crop in England and Wales (*Report*, ix, 50). Shallots are also very susceptible.

Massee (*Gdnrs' Chron.* **16**, 1894, 160) briefly described the disease, but it was not studied carefully for many years. Wallace and Hickman (*AAB*, **32**, 1945, 200–5) investigated the influence of date of lifting and method of storing on the amount of neck rot in storage. Croxall (*Rep. Long Ashton Res. Sta. for 1945*, 143–7) failed to find a suitable bulb treatment which controlled it. The development of the sclerotia was described by Townsend and Willetts (*TBMS*, **37**, 1954, 213–21). (See also *Botrytis byssoidea*.)

Botrytis anthophila Bondartzev, *J. Bolestni Rasteni*, **7**, 1913, 3; **8**, 1914, 22; Sacc., *Syll.* xxv, 694; Wakefield & Bisby, *TBMS*, **25**, 1941, 69.

*Anther mould of red clover (*Trifolium*). Widely distributed in England and Wales, but not commonly reported; occasional in Scotland. The disease cannot be detected until the flowers emerge, when the pollen is found to be wholly or partially replaced by *Botrytis* spores; it was described by Silow (*TBMS*, **18**, 1933, 239–48). The fungus, which was found in a high proportion of seed samples of the 1955 English harvest (Noble, *Plant Path.* **6**, 1957, 38), is perhaps identical with the imperfect stage of *Sclerotinia spermophila* (q.v.).

Botrytis byssoidea J. C. Walker, *Phytopathology*, **15**, 1925, 709.

*Neck and bulb rot of onion (*Allium*). Storage rot of bulbs, Som, 1944; causing wilt in the field, War, 1945; on seed samples, Edinburgh. Is much less common than *Botrytis allii* (q.v.).

Cronshey (*Nature, Lond.*, **160**, 1947, 798) found what he thought to be this species on *Allium vineale* and obtained the perfect stage, which he identified with *Sclerotinia porri* van Beyma (*Meded. phytopath. Lab. Scholten*, **10**, 1927, 43–6), syn. *Botryotinia porri* (van Beyma) Whetz. (*Mycologia*, **37**, 1945, 680).

Botrytis cinerea Fr., *Syst. Myc.* iii, 396; Sacc., *Syll.* iv, 129; Wakefield & Bisby, *TBMS*, **25**, 1941, 69; stat. conid. of *Sclerotinia fuckeliana* (de Bary) Fuckel (q.v.).

This is an aggregate species with a very wide host range, and it may be doubted whether in temperate regions there is any plant not liable to infection by it, given suitably humid conditions. Though its genetic connexion with *Sclerotinia fuckeliana* has been shown, the species is so variable that some forms of it may later prove to be referable to other species of *Sclerotinia*. British records are therefore given here, and, though they cover over 100 different hosts, the list could no doubt soon be greatly lengthened.

Brierley (*Kew Bull.* 1918, 129–46) studied the microconidia of the fungus and (*Phil. Trans.* B, **210**, 1920, 83–114) described a form with colourless sclerotia. Paul (*TBMS*, **14**, 1929, 118–35) made a morphological and physiological study of a number of strains of *Botrytis cinerea*, and Townsend and Willetts (ibid. **37**, 1954, 213–21) followed the development of the sclerotia.

This species formed the subject of an investigation into the physiology of parasitism by Brown (*Ann. Bot., Lond.*, **29**, 1915, 313–48; **30**, 1916, 399–406; **31**, 1917, 489–98) and by Blackman and Welsford (ibid. **30**, 1916, 389–98). More recently Cole (*Ann. Bot., Lond.*, n.s., **20**, 1956, 15–38) discussed the part played by pectolytic enzymes in the pathogenicity of this fungus, and Hislop (*Rep. Long Ashton Res. Sta. for 1956*, 116–20) studied the effect of fungicides on its enzyme system.

The disease it causes is usually called grey mould.

CEREALS. The sclerotia have been seen rarely on the grain of wheat (Ches, 1946) and rye (Ches, 1950); and in 1953 the fungus was widely distributed in the east midlands and eastern England on the ears of wheat and barley, at times killing the young grain; on the ears in the east midlands, 1956.

Sometimes attacks maize (*Zea*) cobs before and after harvesting. Essex, Lincs, Glos, Glam.

POTATO. Leaf blotch. Common most years during spells of warm, wet weather in summer. See Pethybridge (*J. Dep. Agric. Ire.* **16**, 1916, 579–84). Also causing tuber rot, Scotland.

ROOTS AND FODDER CROPS. In association with *Sclerotium* sp. (or possibly *Sclerotinia*), the cause of wastage (clamp rot) in

sugar-beet clamps in eastern districts and elsewhere. In un-
covered clamps *Botrytis* usually predominates; in covered ones
Sclerotium (*Report*, viii, 28; ix, 30). Occasional also on mangold
(see, for example, Salmon, *J. S.-E. agric. Coll. Wye*, **18**, 1909,
328–33) and beetroot.

A minor cause of root rot in turnip and swede, Forth (see also
Potter, *J. Bd Agric.* 3, 1896, 120–31).

PULSE AND FORAGE CROPS

*Field and broad bean (*Vicia*). Plays a relatively minor role in
causing chocolate spot, the major cause of which is *Botrytis fabae*
(q.v.).

*Dwarf and runner bean (*Phaseolus*). On stems, leaves and pods.
Also on tic beans (Kent, 1938); causing foot rot of runner bean,
Jersey, 1954.

*Vetches (*Vicia*). Causing chocolate spot, Devon, 1935; Suffolk,
1948.

*Pea (*Pisum*). Occasionally associated with stem lesions and
chocolate spotting of the leaves.

*Clover (*Trifolium*). Sometimes attacks the flowering head of
red clover; sclerotia are occasionally found mixed with the seed.

*Sainfoin (*Onobrychis*). Rotting the seed (W. C. Moore, *TBMS*,
28, 1945, 130) and causing chocolate spotting of the foliage.

VEGETABLES

*Artichoke (*Helianthus*). On stored tubers of Jerusalem artichoke, Worcs, Bucks, Notts. Also on topine, Som, 1956.

Asparagus. Causing death of stem tips, Worcs, 1931. Severe in
seedling beds, Worcs, 1956.

Brassicas. Found most winters in Wales and southern England,
following downy mildew (*Peronospora parasitica*), frost or insect
injury. In some years causes serious damage to the hearts of
spring cabbage in April and May, following frost injury; stem
rot of seedling cauliflower, north Scotland, 1947; occasional on
marrow-stem kale.

*Carrot (*Daucus*). Killing stored roots, Yorks, Staffs.

*Cucumber (*Cucumis*). Common on young stems and fruits, but
it is doubtful if it is always the primary cause of stigma end rot.

*Eggplant (*Solanum melongena*). Causing fruit rot, London, 1953.

*Globe artichoke (*Cynara*). Rotting flower heads, Jersey, 1953.

*Leek (*Allium*). Causes grey spots on foliage and a rot of the lower leaves; infects the seed heads.

*Lettuce (*Lactuca*). Causes heavy losses in glasshouses and frames almost every year in all districts, especially from January to April. In the open is scarce in dry seasons, often serious in wet ones. The fungus often produces a form of marginal leaf scorch which can easily be confused with tip burn (non-parasitic) and marginal spot (*Pseudomonas marginalis*).

Grey mould of lettuce was investigated by Salam (*J. Pomol.* **12**, 1934, 15–35). Later, Brown (ibid. **13**, 1935, 247–59), Smieton and Brown (*AAB*, **27**, 1940, 489–501), and Last (ibid. **39**, 1952, 557–68) dealt with its control in frame lettuces using chloronitrobenzenes. Ogilvie and Croxall (*Rep. Long Ashton Res. Sta. for 1941*, 76–8) made observations on the disease in glasshouse lettuces, and Hopp (*Fruit Gr.* **86**, 1938, 903–4) studied the effect of ridging. Brown and Montgomery (*AAB*, **35**, 1948, 161–80) considered grey mould along with other problems in the cultivation of winter lettuce. The possibility of microbiological control of *Botrytis cinerea* has been discussed by Newhook (ibid. **38**, 1951, 169–202) and by Wood (ibid. **38**, 1951, 203–16). Rhodes *et al.* (ibid. **45**, 1957, 216) used the antibiotic griseofulvin in small plot trials.

*Onion (*Allium*). Associated with shrivelling and death of leaf tips of winter onions, especially in western districts.

*Parsnip (*Pastinaca*). Occasionally rots the roots in clamps. Devon (*Report*, iv, 57).

*Rhubarb (*Rheum*). Sometimes troublesome in forcing sheds in Yorks and elsewhere, especially when plants over-forced in poor ventilation.

*Seakale (*Crambe*). See Carruthers (*J. R. agric. Soc.* **69**, 1908, 319–20) and *Gdnrs' Chron.* **26**, 1886, 758.

*Tomato (*Lycopersicon*). Grey mould and fruit spot. The fungus attacks the stems through pruning wounds and is common and often destructive under glass in all districts. Water spot or stigmonose of the fruit, formerly attributed to insect injury or non-parasitic causes (Walton, *Gdnrs' Chron.* **101**, 1937, 12–14), was shown to be caused by *Botrytis* by Ainsworth, Oyler and Read (ibid. **102**, 1937, 380–1; *AAB*, **25**, 1938, 308–21).

*Vegetable marrow (*Cucurbita*). Attacks the young fruits at the blossom end and is sometimes destructive.

FRUIT

*Apple (*Malus*). Not uncommon as the cause of dry eye rot of the fruit in some seasons (Wilkinson, *Rep. Long Ashton Res. Sta. for 1941*, 72–5; *J. Pomol.* **20**, 1943, 84–8; Wormald, *Gdnrs' Chron.* **109**, 1941, 44). May also cause a stalk end rot of the fruit (Wormald, ibid. **111**, 1942, 220) and a fruit rot in storage. Occasionally found on the roots of laid-in stocks (Wormald, *TBMS*, **16**, 1932, 309–10) or causing cankers on 1–2-year-old trees (M. H. Moore, *Rep. E. Malling Res. Sta. for 1948*, 101).

*Cherry (*Prunus*). Killing flowering spurs of Morellos (Kent, 1926) and girdling young shoots (Kent, 1933).

Corylus. A wound parasite of cob nuts and filbert nuts in south-east England (M. H. Moore, *Rep. E. Malling Res. Sta. for 1946*, 120–1; *J. hort. Sci.* **25**, 1950, 213–24).

*Currant (*Ribes*). Not infrequent as a cause of die-back of red and black currants. It occasionally causes a leaf blotch, and in 1949 was apparently responsible for killing buds in Kent and a root rot in Yorks.

*Gooseberry (*Ribes*). Very common everywhere as the cause of die-back, and sometimes attacks the ripening fruit.

Die-back in gooseberry was described by A. L. Smith (*J. Bot., Lond.*, **41**, 1903, 19–23), by Salmon (*J. S.-E. agric. Coll. Wye*, **18**, 1909, 319–27), by Brooks and Bartlett (*Ann. mycol., Berl.*, **8**, 1910, 167–85), and by Blackman and Jones (*Rep. on Gooseberry Diseases in E. Sussex 1922–3*, 12 pp., 1923).

*Fig (*Ficus*). Causing fruit rot and twig blight. Only weakly parasitic (Brierley, *Kew Bull.* 1916, 225–9).

*Grape vine (*Vitis*). A frequent cause of fruit rot of grapes, and often serious under glass and in storage. Occasionally attacks the twigs.

BOTRYTIS

*Melon (*Cucumis*). Causing damage to fruits under glass, Hants, 1921.
*Peach and nectarine (*Prunus*). Rotting outdoor fruit, Middx, 1908; Surrey, 1931; occasionally kills young peach shoots under glass; spotting apricot fruits, Perth, 1952.
*Pear (*Pyrus*). Causing a fruit rot, Dorset, 1934; Som, 1945.
*Plum (*Prunus*). Killing young shoots of Victoria, Kent, 1926.
*Quince (*Cydonia*). Causing a brown rot of the fruit, Kent, 1924.
*Raspberry (*Rubus*). Rots the fruits in wet seasons; may attack the pedicels and buds, and damage the canes. Similar symptoms occasionally seen on loganberry and cultivated blackberry.
*Strawberry (*Fragaria*). Very common on the fruits in wet seasons (see, for example, Wilkinson, *Plant Path.* 3, 1954, 12). Control with captan, thiram and other fungicides has been demonstrated by M. H. Moore and others (*J. hort. Sci.* 30, 1955, 213–24). Marsh, Martin and Crang (ibid. 30, 1955, 225–33) studied the effects of spray residues on the processed fruit.
*Walnut (*Juglans*). Causing fruit rot, Moray.

HOP (*Humulus*). *Sometimes discolours the cones and causes leaf spots (Wormald & Cheal, *J. Minist. Agric.* 33, 1926, 456–8).

FLAX (*Linum*). Not uncommon in wet seasons on seedlings, and on stems and bolls of mature plants; is commonly seed-borne. For occurrence in Northern Ireland see Colhoun (*Nature, Lond.*, 153, 1944, 25).

ORNAMENTALS
Anemone. Causes corm rot as well as leaf and flower damage in south-west England. Described by Brenchley and Johnstone (*Plant Path.* 4, 1955, 54–7). Also crown rot, Dunbartonshire, 1956.
Antirrhinum. Commonly attacks flower spikes in wet seasons; also causes damping-off of seedlings.
Armeria. Attacking flower stalks, Bristol, 1952.
*Aster (*Callistephus*). Damping-off of seedlings.
Aucuba. Die-back of *A. japonica*, Clyde; parasitism proved by Trapp (*TBMS*, 20, 1936, 299–303).

77

Begonia. Leaf blotching of indoor and outdoor plants; frequent.

Campanula. Causing soft brown rot of crowns, Harpenden, Herts, 1939.

*Carnation (*Dianthus*). Rotting blooms, Kent, Som, Hunts, Forth, Dee.

Chaenomeles. Causing shoot canker and death of *C. lagenaria* (*Cydonia japonica*), Sussex, 1945; Kent, 1948.

Chrysanthemum. Common as cause of rotting of cuttings, flower buds and flowers under glass, especially under moist conditions or when frost is allowed in.

*Cineraria (*Senecio*). Stem rot; Forth, Orkney, Cumb.

Coleus. Attacking leaf stalks and stems, Glam.

Cyclamen. A very common cause of flower spot and leaf wilt.

Dahlia. Commonly attacks stems, flower buds and cuttings under moist conditions.

Daphne. Not infrequent as the cause of sudden die-back of shoots or whole bushes of *D. mezereum.*

Delphinium. Sometimes troublesome as the cause of stem rot in annual larkspur.

Erythronium. Associated with root rot of *E. dens-canis*, East Lothian, 1956.

Freesia. Grey mould, Spalding, Lincs, 1951; corm rot (*Botrytis* sp.), west Scotland, 1956.

Gardenia. Affecting stems and flower buds, Middx, 1937; Sussex, 1940.

Gladiolus. One cause, and probably a minor one, of Botrytis rot (see *Botrytis gladiolorum*).

Godetia. Stem rot, Lincs, 1948.

Gypsophila. On stems, Middx, 1953.

Hydrangea. Leaf spotting, probably common.

Iris. On basal parts of Spanish and Dutch iris (Dowson, *J. R. hort. Soc.* 53, 1928, 45) and of var. Imperator, Middx, 1936; causing leaf rot of var. Wedgwood, Devon, 1950; and flower spot, Hants, 1954, Sussex. (See also *Sclerotinia convoluta* and *Botrytis* sp.)

Ixia. Sometimes aggressive on this host in the Isles of Scilly; Corn, 1956.

78

Jasminum. Killing long branches of winter jessamine by infection near base of branch, Wye, Kent, 1940.

Laburnum. Associated with cankers on small branches, East Lothian, 1953.

Lavender (Lavandula). Frequent as a cause of blossom wilt in wet seasons. Described by Ware (*J. S.-E. agric. Coll. Wye*, **28**, 1931, 206–10).

Lilac (Syringa). Blossom wilt and die-back, Cambs, Glam, Yorks, Clyde. (See *J. Bd Agric.* **10**, 1903, 520.)

Lily (Lilium). Occasionally responsible for lily disease in *Lilium candidum*.

Magnolia. Killing buds on large tree, Croydon, 1945; on *M. grandiflora*, Devon, 1949.

Marigold (Calendula). On leaves and flower heads, Jersey, 1953.

Michaelmas daisy (Aster). Causing flower rot, Flints, 1956.

Myosotis. Causing flower-bud decay under glass, Corn, Devon.

Myrtillocactus. On *M. (Cereus) geometrizans*, Westmorland, 1951.

Narcissus. Sometimes causes leaf and petal spotting, notably in the Isles of Scilly and west Corn; also Ches, Wilts.

Oleander. Causing die-back under glass, Anglesey, 1956.

Orchids. Causing leaf spot and flower rot (Brierley, *Gdnrs' Chron.* **65**, 1919, 61–2).

Peony. Occasional. (See also *Botrytis paeoniae*.)

Pelargonium. Common cause of leaf spot and stem rot. Described by Buddin and Wakefield (*Gdnrs' Chron.* **75**, 1924, 25).

Poinsettia. Leaf spot of *Euphorbia pulcherrima*, Staffs, 1946.

Poppy (Papaver). Causing flower rot, Denbigh.

Primula. Common outdoors and under glass.

Prunus spp. Causing a shoot wilt of *P. triloba*. Reported from Harpenden, Herts (W. C. Moore, *TBMS*, **23**, 1939, 313–15), and probably common, especially if wet weather occurs during or shortly after the flowering period. Also on *P. besseyi*, Kent, 1948 and *P. pumila*, Kent, 1924.

Rhododendron. Associated with stem cankers, Argyll; with defoliation and bud drop, Glam, 1955.

Ribes sp. On *R. aureum*, Glam, 1931.

*Rose. In wet seasons does much harm to the flower buds, flower stalks and young shoots. Williams (*Rep. exp. Res. Sta. Cheshunt for 1927*, 41–2) described the stem-rot phase and Oyler (ibid. *for 1938*, 47–8) petal spotting. Deacon (*TBMS*, **17**, 1933, 331–3; *Rose Annu.* 1934, 62–6) also discussed the disease.

Solanum. Frequent as cause of wilting and death of branches of *S. capsicastrum*. Leics, Herts, Middx, Som, Northumb.

Sorbus. On *S. aria*, East Malling, Kent, 1924.

*Statice (*Limonium*).* Common on flowers and shoots. Corn, Worcs, Berks, Bucks, Herts, Cumb, Lancs.

*Stock (*Matthiola*).* Common.

*Sunflower (*Helianthus*).* In damp weather frequently attacks the underside of the flower heads, and sometimes destroys the inflorescence.

*Sweet pea (*Lathyrus*).* On bases of shoots under glass, Som, 1932.

*Tulip. Has been isolated on rare occasions from spotted tulips; occasional as cause of bulb rot, Lincs, 1955.

Ursinia. Middx, 1950.

Viburnum. Causing flower scorch of *V. fragrans*, Som, 1956.

*Wallflower (*Cheiranthus*).* A common cause of loss and winter killing (Ashworth, *J. R. hort. Soc.* **63**, 1938, 522–32).

Zinnia. Frequent in south and west England as the cause of foot rot, bud and flower rot, and blossom wilt; occasional elsewhere.

TREES

Aesculus. Killing a tree of *A. pavia*, Kew Gardens (Brierley, *Kew Bull.* 1917, 315–31).

*Conifers. Very troublesome as the cause of damping-off in coniferous nursery stock seedlings. Muskett (*Nature, Lond.*, **124**, 1929, 481; *J. Minist. Agric. N. Ire.* 3, 1931, 102–16) obtained good control with soil disinfectants, notably sulphuric acid.

On shoots and seedlings of Douglas fir (*Pseudotsuga*), as *Botrytis douglasii* Tub., Perth, Forfar, Clyde; and on seedlings of *Larix leptolepis* (Wilson, *TBMS*, **7**, 1921, 85); on shoots of seedling *Sequoiadendron* (*Wellingtonia*) and larch (*J. Bd Agric.* **10**, 1903, 17–21). See also Cooke in *Fung. Pests Cult. Pl.* 1906, 226.

Botrytis convoluta Whetz. & Drayt.=stat. conid. of *Sclerotinia convoluta.*

Botrytis croci Cooke, *Grevillea*, **16**, 1887, 10; Sacc., *Syll.* x, 536; Wakefield & Bisby, *TBMS*, **25**, 1941, 70.
*White mould of crocus. Rare. Kew, Surrey, 1887; Harpenden, Herts, 1939; and occasionally on imported corms (W. C. Moore, *Diseases of Bulbs*, 1939, 149).

Botrytis douglasii Tub., see *Botrytis cinerea* (Conifers).

Botrytis elliptica (Berk.) Cooke, *J. R. hort. Soc.* **26**, 1901–2, p. cxxviii; *Gdnrs' Chron.* **30**, 1901, 58; Sacc., *Syll.* iv, 145; Wakefield & Bisby, *TBMS*, **25**, 1941, 70.
*Lily disease. Widespread throughout Britain, especially in the south. It began to attract attention about 1880 and was studied by Marshall Ward (*Ann. Bot., Lond.*, **2**, 1888, 319–82). Among the species *Lilium candidum* suffers most in Britain. Occasional on *L. speciosum* under glass. The early history of the disease was given by Grove (*Gdnrs' Chron.* **81**, 1927, 178–9 and 197–8), and it was fully described by Cotton (*Lily Yearb. 1933*, 194–214) and W. C. Moore (*Diseases of Bulbs*, 1939, 42). Wright (*Trans. bot. Soc. Edinb.* **30**, 1928, 59–65) discussed the nomenclature of the parasite, and Taylor (*Lily Yearb.* **3**, 1934, 82–9) the origin of outbreaks from infected basal leaves in *L. candidum.*

Botrytis fabae Sard., *Bol. Pat. Veg. Entom. Agric. Madrid*, **4**, 1929, 93.
*Chocolate spot of field and broad bean (*Vicia*). Occurs in all parts of England and Wales and varies very considerably in intensity according to the season. The worst attacks may be expected when late spring frosts are followed by wet and not too cold weather in June or July. Serious general epidemics occur about one year in six, but local epidemics in southern or western districts are more frequent.

Also reported on vetches (Som, 1950) and seed tares, *Vicia sativa* (Edinburgh, 1952); isolated from bean seed at Bristol, 1947.

Chocolate spot is an old disease (*Gdnrs' Chron.* **9**, 1849, 345 and 547; **11**, 1851, 531) which was regarded as bacterial in origin and caused by *Bacillus lathyri* Manns & Taubenh. (Paine & Lacey, *J. Minist. Agric.* **29**, 1922, 175–7; *AAB*, **10**, 1923, 194–203) until Wilson (ibid. **24**, 1937, 258–88; *J. Minist. Agric.* **43**, 1937, 1047–9) proved that it was caused by a number of strains of *Botrytis* which he regarded as belonging to the group species *B. cinerea* Fr. A decade later, however, it became apparent that although a small-spored ($9–15 \times 6\cdot5–10\mu$) *Botrytis* of the usual *cinerea* type, with large sclerotia, may often be concerned, the commonest form is a large-spored one ($15–24 \times 11–18\mu$) with small sclerotia, which deserved specific rank and did not differ from *B. fabae* Sard. (Ogilvie & Munro, *Nature, Lond.*, **160**, 1947, 96; *Rep. Long Ashton Res. Sta. for 1946*, 95–100).

Ware and Glasscock (*Agriculture, Lond.*, **48**, 1941, 91–4) described a local epidemic of chocolate spot, and general accounts of the disease were given by W. C. Moore (*Agriculture, Lond.*, **51**, 1944, 266–9) and Dillon Weston (*Agric. Progr.* **20**, 1945, 5 pp.). Partial control by soil application of potash was claimed by Scott Watson (*J. Minist. Agric.* **43**, 1936, 178–81) and Cowie (*Fertil. Feed. St. J.* **21**, 1936, 182), but Glasscock, Ware and Pizer (*AAB*, **31**, 1944, 97–9) considered that the amount of available phosphorus in the soil was a more important factor. Leach (*TBMS*, **38**, 1955, 171) confirmed that *B. fabae* is the predominant species involved, and concluded that epidemics may follow any set of conditions that lead to premature senescence of the leaves, provided conditions are favourable for sporulation and infection. Last and Buxton (*Nature, Lond.*, **176**, 1955, 655; see also Buxton, Last & Nour, *J. gen. Microbiol.* **16**, 1957, 764–73) measured the effect of ultra-violet radiation on the infective capacity of spores of *B. fabae*, and Last and Hamley (*AAB*, **44**, 1956, 410–18) described a local lesion technique for measuring the infectivity of the conidia. Hogg (*N.A.A.S. quart. Rev.* no. 32, 1956, 87–92) discussed the effect of weather on the incidence of the disease.

*White spot of sweet pea (*Lathyrus*). Causing spotting of petals and leaves in outdoor plantation near badly affected field beans, Kent, 1955.

Botrytis galanthina (Berk. & Br.) Sacc., *Syll.* iv, 136; Wakefield & Bisby, *TBMS*, **25**, 1941, 70.

*Grey mould of snowdrop (*Galanthus*). Not uncommon in Britain. First recorded in England as *Polyactis galanthina* Berk. & Br. (*Ann. Mag. nat. Hist.* Ser. iv, **11**, 1873, 346). Later described by W. G. Smith (*Gdnrs' Chron.* 5, 1889, 275) and W. C. Moore (*Diseases of Bulbs*, 1939, 94). See also Massee (*J. R. hort. Soc.* **26**, 1901, 41–6) and Green (*Gdnrs' Chron.* **99**, 1936, 93–4).

Botrytis gemella (Bon.) Sacc., *Mich.* ii, 258; *Syll.* iv, 134.

*In mushroom (*Agaricus*) beds, forming a fine, mealy, snuff-coloured coating on the casing soil, Sussex, 1949, 1951. The fungus might well have been the conidial state of *Plicaria fulva* Schneider, *Zbl. Bakt.* II, **108**, 1954, 153.

Botrytis gladioli Kleb., see *B. gladiolorum*.

Botrytis gladiolorum Timmerm., *Meded. Lab. Bloembollenond.* Lisse no. 67, 1941, 15.

*Botrytis rot of *Gladiolus*. Widely distributed in Britain as a cause of leaf spot, which is often severe in wet summers. Causes core rot and soft rotting of the corms in storage. Other species of *Botrytis* may be involved in what is now the most destructive disease of gladiolus.

This *Botrytis* rot was first recognized in England in 1927, and for many years it steadily became worse (W. C. Moore, *Diseases of Bulbs*, 1939, 112). The disease has now been observed in many parts of the world and the symptoms are always the same, but opinions differ about the identity of the parasite. In Britain the chief pathogen was thought for a time (Hawker, *AAB*, **33**, 1946, 200–8) to be a strain of *B. cinerea*, but Pieris (*TBMS*, **32**, 1950, 291–304) concluded it was distinct from *B. cinerea* and identical with *B. gladiolorum*. *Sclerotinia draytoni* (q.v.) sometimes plays a part and may, indeed, be the perfect stage of *Botrytis gladiolorum* (McLellan, Baker & Gould, *Phytopathology*, **39**, 1949, 260–71). *B. gladioli* Kleb., *Z. Bot.* **23**, 1930, 251–72, may also be involved.

*Botrytis rot of *Acidanthera.* On imported corms of *A. murielae*, 1952. Also seen on species of *Acidanthera* in Devon, Corn, Som, Middx, Cambs, Yorks, Cumb and in Scotland at Selkirk. The presence of the fungus on the corms seems to encourage development of the mite *Pediculopsis* (Harrison, *Plant Path.* **1**, 1952, 119–20).

*On *Freesia.* One of the forms of *Botrytis* known to occur on gladiolus caused extensive flower spotting of freesias at Penzance, Corn, in 1940 (*Report*, viii, 84), and leaf or flower spotting caused by this species is now widely distributed. Devon, Glos, Som, Dorset, Worcs, War, Bucks, Ches, Norfolk, south-east England, Clyde, Guernsey.

Botrytis hyacinthi Westerd. & van Beyma, *Meded. phytopath. Lab. Scholten*, **12**, 1928, 1; Wakefield & Bisby, *TBMS*, **25**, 1941, 71.

*Fire of hyacinth. Not uncommon in south-west England; unimportant elsewhere (W. C. Moore, *Diseases of Bulbs*, 1939, 14); west Scotland, 1949.

Botrytis narcissicola Kleb. = stat. conid. of *Sclerotinia narcissicola*.

Botrytis paeoniae Oudem., *Meded. Kon. Akad. Wet. Amst.* 1897, 464; Sacc., *Syll.* xiv, 1052; Wakefield & Bisby, *TBMS*, **25**, 1941, 70.

*Peony blight. Occasional in England; Spalding, Lincs (Smith & Ramsbottom, *TBMS*, **6**, 1920, 371). Rather common in north Scotland. The symptoms are not unlike those caused by *B. cinerea*, but the two species are distinct. Massee (*Gdnrs' Chron.* **24**, 1898, 124–5; **25**, 1899, 351) described a *Botrytis* disease of peony.

Botrytis parasitica Cav. = *B. tulipae.*

Botrytis polyblastis Dowson = stat. conid. of *Sclerotinia poly-blastis*.

Botrytis squamosa Walker, *Phytopathology*, **15**, 1925, 710.

*Leaf rot of onion (*Allium*). Widely distributed in Britain on winter onions; sometimes damaging. First fully recognized in Worcs, 1941 (Hickman & Ashworth, *TBMS*, **26**, 1943, 153–7) and in Scotland in 1943. Cronshey (*Nature, Lond.*, **158**, 1946, 379) obtained a perfect stage in culture, as also did Viennot-Bourgin (*Ann. Inst. nat. Rech. agron., Paris*, Ser. C, **4**, 1953, 23–43), who named it *Botryotinia squamosa* n.sp. Dennis (*Mycol. Pap.* **62**, 1956, 157) transferred it to *Sclerotinia squamosa* (Viennot-Bourgin) Dennis n.c.

Botrytis tulipae Lind, *Danish Fungi*, 1913, 650; syn. *B. parasitica* Cav., *Rev. mycol.* **10**, 1886, 206; Sacc., *Syll.* x, 536; Wakefield & Bisby, *TBMS*, **25**, 1941, 70.

*Tulip fire. Widespread in Britain and often destructive in the open, especially during cold, wet spring weather. Not serious under glass unless the conditions are too humid. Fully described by Beaumont, Dillon Weston and Wallace (*AAB*, **23**, 1926, 57–88) and by W. C. Moore (*Diseases of Bulbs*, 1939, 20). Beaumont (*13th Rep. Dep. Pl. Path. Seale-Hayne Agric. Coll. for 1936*, 23–5) classified seventy varieties according to their susceptibility. Rhodes *et al.* (*AAB*, **45**, 1957, 217–21) tested the effect of the antibiotic griseofulvin in small plot trials.

Botrytis sp.

*Clover (*Trifolium*). Damaging flower heads of red and crimson clover, Dorset, Glam.

*Fire of *Erythronium*. Causing damage, Oban, 1953.

*Iris. On bulb of *I. lauri*, Bucks, 1926; not infrequent as cause of rot in rhizomes of *I. susiana* in storage (W. C. Moore, *Diseases of Bulbs*, 1939, 139); on bearded irises (*I. germanica*) in a number of districts (Green & Fox Wilson, *Gdnrs' Chron.* **104**, 1938, 114; Green, *Iris Yearb. 1941*, 73–6). (See also *Sclerotinia convoluta*.)

*Lily of the valley (*Convallaria*). Causing severe damage to *C. majalis*, Devon, 1930; Som, 1937. Reported from Devon, 1955–6, as *B. cinerea* f. *convallariae*.

Lychnis. Causing a bud wilt of perennial *Lychnis*. Not uncommon in northern England; Scotland, 1947.

Penstemon. Stem rot, Aberdeen, 1953.

Pyrethrum. Frequent as cause of flower stalk rot. Corn, Glam, Berks, Surrey, Lincs, Berwick, Perth.

*Tritonia (*Montbretia*). On corms, Bucks, 1934; on foliage, Cambs, 1953.

Abies. Damping-off of seedling trees of *A. nobilis*, Ayrshire, 1953.

Boydia insculpta (Oud.) Grove = *Vialaea insculpta*.

Bremia lactucae Regel, *Bot. Z.* **1**, 1843, 666; Sacc., *Syll.* vii, 244.

*Downy mildew of lettuce (*Lactuca*). Occurs in most districts, is often troublesome in the open as well as under glass, and is most common in the south and west of England. Uncommon in Scotland on outdoor lettuce, though frequent there on native Compositae. Reported on celtuce, Harpenden, Herts, 1945. The fungus occurs on common weeds such as groundsel (*Senecio*), sow and field thistles, but does not pass from them to lettuce.

W. G. Smith (*Gdnrs' Chron.* **20**, 1883, 600) described the disease. Macpherson (*J. Minist. Agric.* **38**, 1932, 998–1003) studied varietal resistance; both Jagger (*Rep. exp. Res. Sta. Cheshunt for 1926*, 35) and Ogilvie (*Rep. Long Ashton Res. Sta. for 1941*, 76–8; *for 1943*, 90–4; and *for 1945*, 147–50) discussed the occurrence of physiologic races; Wild (*TBMS*, **31**, 1947, 112–25) and Powlesland (ibid. **37**, 1954, 362–71) investigated the biology of the fungus, including its transmission; and Powlesland and Brown (*AAB*, **41**, 1954, 461–9) dealt with its control by fungicides.

*Downy mildew of globe artichoke (*Cynara*). Battle, Sussex, 1945; Corn, 1956.

*Downy mildew of cineraria (*Senecio*). Aberdeen, 1954; Northants, 1954.

*Downy mildew of cornflower (*Centaurea*). Rare. On *Centaurea cyanus* Hailsham, Sussex, 1949.
*Downy mildew of *Gaillardia*. Not infrequent. Wye, Kent, 1939, on *G. grandiflora*; Harpenden, Herts, 1944 on *G. aristata*; and subsequently in Devon, Som, Surrey, Perthshire.
*Downy mildew of *Helichrysum*. Corn, 1934 and 1936; Worcs, 1941; War, 1953.

Bremia sp.
*On cultivated *Senecio jacobaea*, Aberdeen, 1952.

Brunchorstia destruens Erikss. = stat. conid. of *Crumenula pinea*.

Brunchorstia pinea (Karst.) Höhnel = stat. conid. of *Crumenula pinea*.

Bulgaria inquinans (Pers.) Fr. = *Phaeobulgaria inquinans*.

Bulgaria polymorpha Wettst., see *Phaeobulgaria inquinans*.

Byssochlamys fulva Olliver & G. Smith, *J. Bot., Lond.*, **72**, 1933, 196; Brown & Smith, *TBMS*, **40**, 1957, 37.
*This fungus spoils and completely disintegrates processed fruits, especially strawberries and plums (Olliver & Rendle, *J. Soc. chem. Ind., Lond.*, **53**, 1934, 166–72; Gillespy, *Rep. Fruit Veg. Pres. Sta., Campden for 1938*, 68–78; *for 1940*, 54–61). Hull (*AAB*, **26**, 1939, 800–22) studied its temperature relations and pointed out that it commonly occurs on the surface of fruits in the field.

Calonectria nivalis Schaffnit = *Griphosphaeria nivalis*.

Calyptospora goeppertiana Kühn, *Hedwigia*, **8**, 1869, 81; Sacc., *Syll.* vii, 766; Grove, *Brit. Rust Fungi*, 372; Wilson & Bisby, *TBMS*, **37**, 1954, 63.
*Aecidia on *Abies alba* and *A. nordmanniana*: teleutospores on *Vaccinium vitis-idaei*. Grove (loc. cit.) reported it very rare (England, Wales, Scotland), while Wilson and Bisby (loc. cit.) listed it as doubtfully British. See Massee, *Kew Bull.* 1907, 1–3.

Camarographium abietis Grove = *Myxocyclus cenangioides.*

Camarosporium abietis Wilson & Anders. = *Myxocyclus cenangioides.*

Camarosporium karstenii Sacc. & Syd., in *Syll.* xiv, 966; Grove, *Coelomycetes*, ii, 100.

*On apple (*Malus*). Associated with die-back of small shoots and buds, Reading, Berks; on cider variety, Salop, 1948.

Centrospora acerina (Hartig) Newhall, *Phytopathology*, **36**, 1946, 894, syn. *Cercospora acerina* Hartig, *Unters. Forstb. Inst. Münch.* **1**, 1880, 58; Sacc., *Syll.* iv, 465; *Ansatospora acerina* (Hartig) Hansen & Tompk. *Phytopathology*, **35**, 1945, 220.

*On carrot (*Daucus*). Affecting the roots, Edinburgh (Srivastava, *Plant Path.* **6**, 1957, 113; *TBMS*, **41**, 1958, 223–6).
*On celery (*Apium*). Causing rot of roots and petioles in association with *Pythium* sp., Beds, 1952 (Storey & Wilcox, *Plant Path.* **2**, 1953, 72); rotting stored celery grown in Lincs and Cambs (Lowings, ibid. **4**, 1955, 106–7).

Cephalosporium asteris Dowson = *Verticillium vilmorinii.*

Cephalosporium costantinii Smith, *TBMS*, **10**, 1924, 90; Wakefield & Bisby, ibid. **25**, 1941, 53.

*On mushroom (*Agaricus*). Not reported except from one locality by Smith (loc. cit.).

Cephalosporium gramineum Nisikado & Itaka, see *Verticillium* sp., cereals.

Cephalosporium lamellicola Smith, *TBMS*, **10**, 1924, 93; Wakefield & Bisby, ibid. **25**, 1941, 54.

*Gill mildew ('Flock') of mushroom (*Agaricus*). Occasional, and does little harm.

Ceratophorum setosum Kirchn. = *Pleiochaeta setosa.*

Ceratostomella ulmi Buism., *Tijdschr. PlZiekt.* **38**, 1932, 3; as *Ophiostoma ulmi* (Buism.) Nannf., *Svenska SkogsvFören. Tidskr.* 1934, 397; Bisby & Mason, *TBMS*, **24**, 1940, 146; stat. conid. *Graphium ulmi* Schwarz, *Meded. phytopath. Lab. Scholten*, **5**, 1922, 50; Wakefield & Bisby, *TBMS*, **25**, 1941, 56.

*Dutch elm disease. For many years after about 1930 common and serious in the midlands and south, and sporadic elsewhere in England and Wales. Is apt to flare up in different areas from time to time but is now doing comparatively little damage. In Scotland known only in the extreme south-east, except for one record at Queenferry, Midlothian.

The first outbreak of this disease seen in England was at Totteridge, Herts, in July 1927 (Wilson & Wilson, *Gdnrs' Chron.* **83**, 1928, 31; Day, *Quart. J. For.* **21**, 1927, 123–9). Wilson (*Gdnrs' Chron.* **81**, 1927, 133–4) had previously given a short general account of it. The subsequent progress of the disease was given by Grove (ibid. **90**, 1931, 396); by Peace (*Forestry*, **6**, 1932, 125–42) and in the *Scot. For. J.* **46**, 1932, 194–6. Walker (*Phytopathology*, **29**, 1939, 551–3) published observations on the fructifications of the fungus in England. Peace (*AAB*, **41**, 1954, 155–64) confirmed the effectiveness of insecticidal dusts, especially DDT, in preventing feeding by bark beetles (*Scolytus*), which are mainly responsible for spreading the fungus. The treatment is, however, uneconomic except for small trees of special value. A resistant strain of elm (*Ulmus* 'Christine Buisman'), raised in Holland, proved resistant in this country, but is very susceptible to coral spot (*Nectria cinnabarina*) and had to be abandoned on account of this.

Cercospora acerina Hartig = *Centrospora acerina*.

Cercospora angraeci Feuilleaub. & Roum., *Rev. mycol.* **5**, 1883, 177; Sacc., *Syll.* iv, 478.

*Leaf spot of *Odontoglossum*. Chupp (*A Monograph of Cercospora*, 1953, 425) listed this species from England and added, 'a specimen apparently of this species was received from Enfield,

England on Odontoglossum. It did not resemble *Cercospora odontoglossi'* (q.v.).

Cercospora apii Fres., *Beitr. Myc.* iii, 1863, 91; Sacc., *Syll.* iv, 442; Wakefield & Bisby, *TBMS*, **25**, 1941, 88.

*Leaf blight of celery (*Apium*). Very rare. The only British record appears to be one from Norwich (*J. Bd Agric.* **14**, 1907, 417), and judged from the description by Massee in *Dis. Cult. Pl. Trees*, 486, it could have been an error for Septoria leaf spot. *Cercospora apii* was recorded at the B.M.S. Killarney Foray, 1936 (*TBMS*, **22**, 1938, 11) (cf. *Centrospora acerina*).

Cercospora beticola Sacc., *Nuovo G. bot. ital.* **8**, 1876, 189; Sacc., *Syll.* iv, 456; Wakefield & Bisby, *TBMS*, **25**, 1941, 89.

*Leaf spot of sugar beet (*Beta*). Found in trivial amount in most districts of England every year, and at times damaging in the south and south-west. Occasional in Scotland, Forth, Kincardine.

Occurs occasionally on mangold, garden beet and spinach beet.

Cercospora bloxami Berk. & Br., see *Cercosporella brassicae*.

Cercospora bolleana (Thüm.) Speg., *Mich.* i, 1879, 475; Sacc., *Syll.* iv, 475; Wakefield & Bisby, *TBMS*, **25**, 1941, 89.

*On fig (*Ficus*). Described by Massee (*Gdnrs' Chron.* **28**, 1900, 5). See also *J. R. hort. Soc.* **33**, 1908, clxv.

Cercospora calendulae Sacc., *Mich.* i, 1879, 267; *Syll.* iv, 446; Wakefield & Bisby, *TBMS*, **25**, 1941, 89.

*Leaf spot of *Calendula*. Yorks, 1909, 1911.

Cercospora cantuariensis Salm. & Worm., *J. Bot., Lond.*, **61**, 1923, 134; Wakefield & Bisby, *TBMS*, **25**, 1941, 89.

*Leaf spot of hop (*Humulus*). Rare. Canterbury, Kent, 1922 (Salmon & Wormald, *J. Minist. Agric.* **30**, 1923, 433–4); Reading district, Berks, 1927; Hants, 1930. Wormald (*TBMS*, **13**, 1928, 32–9) provided proof of parasitism. Chupp (*A Monograph of Cercospora*, 1953, 394) thought the species should be transferred to a new genus.

CERCOSPORA

Cercospora circumcissa Sacc., *Nuovo G. bot. ital.* **8**, 1876, 189; Sacc., *Syll.* iv, 460; Wakefield & Bisby, *TBMS*, **25**, 1941, 89.

*Shot-hole of plum and peach (*Prunus*). Most British records are old ones and the fungus has not been reported in recent years. On peach, causing defoliation and death of twigs under glass, Wales, 1903; on peach and plum, Wilts, 1909 (Carruthers, *J. R. agric. Soc.* **64**, 1903, 296; **70**, 1909, 353); on peach, Harrow and Redhill (*J. R. hort. Soc.* **29**, 1904/5, xlvi; **32**, 1907, lxxxii); on plum, Westbury and Southend; on peach, Ryde, I.W. (*J. Bd Agric.* **14**, 1907, 221; **15**, 1908, 440; **16**, 1909, 390). Also recorded from Yorks and Scotland (Dee). There is a description of the disease in *J. Bd Agric.* **17**, 1910, 211–14.

Cercospora concentrica Cooke & Ell., *Grevillea*, **5**, 1877, 90; Sacc., *Syll.* iv, 479; Massee, *Brit. Fung. Fl.* iii, 416; Wakefield & Bisby, *TBMS*, **25**, 1941, 89.

*Leaf spot of *Yucca*. On *Y. gloriosa* and *Y. filamentosa* in gardens (Massee, loc. cit.).

Cercospora depazeoides (Desm.) Sacc., *Nuovo G. bot. ital.* **8**, 1876, 187; Sacc., *Syll.* xv, 85; syn. *Exosporium depazeoides* Desm., *Ann. Sci. nat.* 3 Ser. **11**, 1849, 364.

*Leaf spot of elder (*Sambucus*). Recorded in Herts, Kent and Surrey in 1941 (W. C. Moore, *TBMS*, **29**, 1946, 251) and probably not uncommon.

Cercospora fabae Fautr. = *C. zonata*.

Cercospora melonis Cooke, *Gdnrs' Chron.* **20**, 1896, 271; Sacc., *Syll.* xviii, 598; Wakefield & Bisby, *TBMS*, **25**, 1941, 89; syn. *Corynespora melonis* (Cooke) Lind. in Rabenh., *Krypt. Fl.* 2, i, 9, 805; Sacc., *Syll.* xxii, 1435.

*Cucumber and melon blotch (*Cucumis*). Uncommon in England and Wales on cucumber, and usually causes only slight damage. Rare on melon, Sussex, 1929 (*Report*, vii, 88), Herts. Not recorded from Scotland.

This disease was first reported on melon leaves at Totteridge, Herts, in 1896 (Cooke, loc. cit.), and shortly afterwards (Cooke, *J. R. hort. Soc.* **26**, 1901–2, cxliv) was found on cucumber. It spread rapidly (*J. Bd Agric.* **9**, 1902, 196–8) and soon virtually wiped out the cucumber crop in north London. Massee (*J. R. hort. Soc.* **28**, 1903–4, 142–5) and Hall (*J. Bd Agric.* **12**, 1905–6, 19–21) endeavoured to find a direct cure, but it was following the introduction of the variety Butchers' Disease Resister, and the adoption of more hygienic measures, that the disease rapidly waned and by 1919 it was practically non-existent in the Lea Valley. It has only occasionally been seen since then (see, for example, Green, *Gdnrs' Chron.* **86**, 1929, 449; *J. R. hort. Soc.* **57**, 1932, 63–4). Güssow (*J. R. agric. Soc.* **65**, 1904, 270; *Z. PflKrankh.* **16**, 1906, 10–13) made the fungus the type of a new genus *Corynespora*, as *C. mazei*, and Wei (*Mycol. Pap.* **34**, 1950, 5) considered the correct name to be *Corynespora cassicola* (Berk. & Curt.) Wei.

Cercospora moricola Cooke, in Ravenel & Cooke, *Fungi Americani Exsiccati*, no. 591; *Grevillea*, **12**, 1883, 30; Sacc., *Syll.* iv, 475; Cooke, *Fung. Pests Cult. Pl.* 139; Wakefield & Bisby, *TBMS*, **25**, 1941, 89.

*On *Morus*. Apparently rare. Clevedon, Som.

Cercospora myrticola Speg., *An. Soc. cient. argent.* **16**, 1883, 167; Sacc., *Syll.* x, 643; syn. *C. myrti* Erikss., *Bidr. Känned. odl. växt-sjukdomar*, 1885, 79; Sacc., *Syll.* iv, 462; Wakefield & Bisby, *TBMS*, **25**, 1941, 89.

*Leaf spot of myrtle (*Myrtus*). Clyde, Sept. 1918 (Smith, *TBMS*, **6**, 1919, 157); two places in southern England, 1951 (Hewlett, *J. R. hort. Soc.* **77**, 1952, 416–18); Corn, 1952, and subsequently elsewhere.

Cercospora odontoglossi Prill. & Delacr., *Bull. Soc. mycol. Fr.* **8**, 1893, 271; Sacc., *Syll.* xi, 629; Wakefield & Bisby, *TBMS*, **25**, 1941, 89.

*Leaf spot of *Odontoglossum*. On *O. crispum* (Massee, *Dis. Cult. Pl. Trees*, 1910, 489); *Rep. exp. Res. Sta. Cheshunt for 1927*, 32; Dorset, 1956.

Cercospora resedae Fuckel, *Symb. Myc.* 1869, 353; *Hedwigia*, **5**, 1866, 30; Sacc., *Syll.* iv, 435; Wakefield & Bisby, *TBMS*, **25**, 1941, 89.

*On leaves of mignonette (*Reseda*). Formerly reported destructive at times to *R. odorata*, especially under glass (Cooke, *Fung. Pests Cult. Pl.* 1906, 22).

Cercospora roesleri (Catt.) Sacc., *Mich.* ii, 1880, 128; Sacc., *Syll.* iv, 458; Wakefield & Bisby, *TBMS*, **25**, 1941, 89.

*Leaf spot of grape vine (*Vitis*). Tay (Blairgowrie).

Cercospora zonata Wint., *Contr. ad floram mycologicam lusitanicam*, v, 1884, 22; *Hedwigia*, **23**, 1884, 191; Sacc., *Syll.* iv, 437; syn. *C. fabae* Fautr., *Rev. mycol.* **13**, 1891, 13; Sacc., *Syll.* x, 621; Wakefield & Bisby, *TBMS*, **25**, 1941, 89.

*Leaf spot of broad and field bean (*Vicia*). First seen in Kent and Oxon, 1927. Widely distributed, occasional, chiefly in the southern half of England. Described by Woodward (*TBMS*, **17**, 1932, 195–202).

Cercospora sp.

*White clover (*Trifolium*). Not uncommon in Wales. Corresponds with *Cercospora zebrina* Passer. in general features.
*Leaf spot of *Rhododendron*. Hants, 1951.
*On leaves of scabious, Ches, 1939.
*On tulip leaves, Dingwall, Ross-shire, 1952.
*On leaves of *Viola*, Berks, 1937.

Cercosporella antirrhini Wakef. = *Heteropatella antirrhini*.

Cercosporella brassicae (Fautr. & Roum.) Höhnel, *Ann. mycol., Berl.*, **20**, 1924, 193; Wakefield & Bisby, *TBMS*, **25**, 1941, 89; syn. *Cylindrosporium brassicae* Fautr. & Roum., *Rev. mycol.* 1891, 81; Sacc., *Syll.* x, 501.

*White spot of turnip and swede (*Brassica*). Widely distributed in Britain; more common on swede than turnip.

The fungus, formerly erroneously called *Cercospora bloxami* Berk. & Br. (see Mason, *TBMS*, **20**, 1936, 110–11), was described by Johnson (*J. Dep. Agric. Ire.* **5**, 1905, 438–42).

Cercosporella herpotrichoides Fron, *Ann. Sci. agron.*, Paris, **1**, 3–29; Wakefield & Bisby, *TBMS*, **25**, 1941, 89.

*Eyespot of cereals. Occurs throughout Britain, but most widespread in the east and midlands of England, and in the Lothians, Fife and Moray. Mainly on wheat and barley; formerly rare on oats but now more frequent; only occasional on rye. Sometimes seen on wild oats (*Avena ludoviciana*). Possibly this species on *Agrostis alba*, Yorks, 1944. The disease is a seasonal one and is rarely serious except where rotations are too short. Severe attacks were noted on oats in Ireland in 1955 (McKay, Loughnane & Kavanagh, *Nature, Lond.*, **177**, 1956, 193).

Eyespot was not recognized in Britain until 1935 (Glynne, *TBMS*, **20**, 1936, 120), but is probably of long standing. It has been studied almost exclusively by Glynne and her colleagues. They have published a number of papers on the relation of the disease to lodging in cereal crops (*Agriculture, Lond.*, **49**, 1942, 91–4; *AAB*, **29**, 1942, 254–64; **31**, 1944, 377–8); on the distribution of the disease in Scotland (ibid. **33**, 1946, 377–8); on the effect of nitrogen (ibid. **32**, 1945, 297–303); on previous cropping (ibid. **36**, 1949, 341–51); spore production (*TBMS*, **36**, 1953, 46–51); seeding rates and other cultural treatments (*AAB*, **38**, 1951, 665–8; *J. agric. Sci.* **46**, 1955, 407–16; **48**, 1957, 326–35); on the incidence of eyespot and the effect of the disease on yields (*AAB*, **40**, 1953, 221–4); and other factors (*Agriculture, Lond.*, **53**, 1946, 305–8).

Storey (*AAB*, **33**, 1947, 546–50) made observations on the disease in Yorks, and Batts and Fiddian (*Plant Path.* **4**, 1955, 25–8) also studied the effect of previous cropping on yields.

Cercosporella narcissi Boud., see *Ramularia vallisumbrosae*.

Cercosporella pastinacae Karst., *Hedwigia*, **23**, 1884, 63; Sacc., *Syll.* iv, 219; Wakefield & Bisby, *TBMS*, **25**, 1941, 89.

*Leaf spot of parsnip (*Pastinaca*). Rare. On leaves and petioles in Worcs, Glos, Cambs (Cotton, *Kew Bull.* 1918, 19). Possibly only a stage of *Ramularia pastinacae* (q.v.).

Cercosporella primulae Allesch., *Ber. bayer bot. Ges.* **2**, 1892, 18; Sacc., *Syll.* xi, 607; Moore, *TBMS*, **31**, 1947, 90.

*Leaf spot of *Primula*. Occasional. On *P. wanda* and hybrids of *P. juliae*, Staffs, 1928; Herefs; Southampton, Hants (W. C. Moore, *TBMS*, **25**, 1941, 208); Clyde, Tweed. Probably a stage of *Ramularia primulae* (q.v.).

Cercosporella sp.

*On leaves of *Cyclamen*, Surrey, 1938. (See *Ramularia* sp.)
*On statice (*Limonium*), Herts, 1935.
*On tulip, Scotland (Alcock & Foister, *Trans. bot. Soc. Edinb.* **30**, 1931, 338); Clyde, causing brown markings on outermost, fleshy, bulb scale.

Chaetomium cochlioides Palliser, in *N. Amer. Flora*, **3**, 1910, 61; Sacc., *Syll.* xxii, 118; Bisby & Mason, *TBMS*, **24**, 1940, 149.

*On tomato, associated with brown root rot, especially in early part of season (Ebben, *Rep. exp. Res. Sta. Cheshunt for 1949*, 28–32; Williams & Ebben, ibid. *for 1950*, 23–6), but doubtfully parasitic (Ebben & Williams, *AAB*, **44**, 1956, 425–36).

Chaetomium globosum Kunze, see *C. olivaceum*.

Chaetomium olivaceum Cooke & Ell., *Grevillea*, **6**, 1878, 96; Sacc., *Syll.* i, 225.

*Olive green mould of mushroom (*Agaricus*) beds. There have been some severe infestations in recent years, especially where the tray or two-zone system is used (Storey, *Grower*, **36**, 1951, 969–71). Yorks, Worcs, Lancs. *C. globosum* Kunze (*Mykologische Hefte*, i, 1817, 16; Sacc., *Syll.* i, 222) is sometimes involved, and the factors that influence fruiting and spread of this species have been studied by Basu (*J. gen. Microbiol.* **5**, 1951, 231–8).

Chalaropsis thielavioides Peyr., *Staz. sper. agr. ital.* **49**, 1916, 595;
   Sacc., *Syll.* xxv, 763; Mason, *Mycol. Pap.* **5**, 1941, 133;
   Wakefield & Bisby, *TBMS*, **25**, 1941, 62.

*Surface blackening of carrot (*Daucus*). Causing surface
blackening of stored roots, Liverpool, 1930 (*Report*, vii, 55);
Manchester, 1954, on carrots from eastern England.

*Graft disease of walnut (*Juglans*). Uncommon. East Malling,
Kent, 1930, damaging graft unions, layers and roots (Hamond,
J. Pomol. **13**, 1935, 81–107; *TBMS*, **19**, 1935, 158–9). The fungus
was recorded as *Cylindrosporium longipes* Preuss on walnut shells
in Scotland, 1878.

Chalaropsis sp.

*On bark of a felled beech tree, Sevenoaks, 1952. Parasitism
uncertain but also found, apparently as a wound parasite, in
standing trees. Associated perithecia of a species of *Ceratocystis*
(*Ophiostoma*) shown to be genetically connected with it (Vincent,
Nature, Lond., **172**, 1953, 963–4).

Chrysomyxa abietis Unger, *Beiträge zur vergleichenden Pathologie*,
   1840, 24; Sacc., *Syll.* vii, 762; Wilson & Bisby, *TBMS*, **37**,
   1954, 63.

*Needle rust of spruce (*Picea*). Uredo-stage mainly on Norway
spruce (*P. abies*), but also on Sitka spruce (*P. sitchensis*) and
P. rubens. No alternate host known. Unknown before about
1900, now fairly common in northern Britain (Murray, *Scot. For.*
7, 1953, 52–4) and occasionally severe.

Chrysomyxa ledi (Alb. & Schw.) de Bary var. **rhododendri** (DC.)
   Savile = *C. rhododendri*.

Chrysomyxa rhododendri de Bary, *Bot. Ztg*, **37**, 1879, 809; Sacc.,
   Syll. vii, 760; Grove, *Brit. Rust Fungi*, 384; Wilson & Bisby,
   TBMS, **37**, 1954, 63. Savile, *Canad. J. Res.* **28**, 1950, 325
   reduced it to varietal rank as *C. ledi* (Alb. & Schw.) de Bary
   var. *rhododendri* (DC.) Savile.

Rhododendron rust. Not uncommon. On *R. hirsutum*, Douglas Castle, Lanarks, 1913; on *Rhododendron* sp., Ches, 1930; on *R. roylei*, Penzance, Corn, 1937 (Ashworth, *J. R. hort. Soc.* **63**, 1938, 487–8); on *R. keleticum* and *R. hippophaeoides* × *racemosum*, Seal, Kent, 1948; on *R. cinnabarinum*, Corn; on *R. ferrugineum*, Aberdeen, 1951–2; on *R. ponticum*, Ayrshire, Kincardine, Aberdeensh, Inverness. On *Azaleodendron* Dr Masters, Windlesham, Surrey, 1953; widespread in a large collection of rhododendrons, particularly on *R. roylei* and its hybrids, Trewithen, Corn, 1954–5. The rust also occurs in Northern Ireland on *R. ponticum*.

The aecidial stage occurs sporadically on Norway spruce (*Picea abies*) and occasionally on Sitka spruce (*P. sitchensis*) in Scotland. The rust can survive on rhododendron in the absence of spruce, but not conversely. It is possible that the fungus in Scotland is a different strain or even species from that occurring in southern England.

Chrysophlyctis endobiotica Schilb. = *Synchytrium endobioticum*.

Cladochytrium caespitis Griff. & Maubl., *Bull. Soc. mycol. Fr.* 1910, 320; Sacc., *Syll.* xxi, 847.

*Root rot of seedling grasses. A frequent cause of failure, especially of fescue and *Agrostis* in newly sown lawns and bowling greens. Also on cocksfoot (*Dactylus*), meadow foxtail (*Alopecurus*) and *Agropyron repens*.

The species described by Massee (*Kew Bull.* 1913, 205–7; *J. Bd Agric.* **20**, 1913, 701–3) as *Cladochytrium graminis* Büsgen was thought by Ivimey Cook (*Ann. Bot., Lond.*, **48**, 1934, 177–85) to belong here. Cook also described the disease elsewhere (*J. Bd Greenkeep. Res.* **5**, 1937, 18–22).

Cladochytrium graminis Büsgen, see *C. caespitis*.

Cladosporium album Dowson = *Ramularia deusta*.

Cladosporium carpophilum Thüm. = *Fusicladium carpophilum*.

Cladosporium cucumerinum Ell. & Arth., *Bull. Agric. Stat. Indiana*, 1889, 9–10; Sacc., *Syll.* x, 601; Wakefield & Bisby, *TBMS*, **25**, 1941, 84.

*Gummosis of cucumber (*Cucumis*). Widely distributed in England and Wales under glass and on ridge cucumbers in the open. Has increased noticeably in southern counties in recent years; Jersey, 1951. Not listed from Scotland. Also occurs on gherkins, and has been seen rarely on vegetable marrow; on custard marrow, Bodmin, Corn, 1945; on bush marrow, Swindon, Wilts, 1951; and on melon fruit, Smarden, Kent, 1946; Sussex, 1952.

This disease was reported as new to Britain when found by Cooke (*Gdnrs' Chron.* **34**, 1903, 100 and 172) at Bristol and Hillingdon, but it had evidently been seen many years previously by M. J. B[erkeley] (ibid. **7**, 1847, 355). It became rather prominent late in the 1939–45 War when there was a big expansion of the acreage of outdoor cucumbers, and since then has tended to increase under glass.

Bond (*AAB*, **25**, 1938, 277–307) studied the infection phenomena of the fungus.

Cladosporium fulvum Cooke in Ravenal & Cooke, *Fungi Americani Exsiccati*, no. 599; *Grevillea*, **12**, 1883, 32; Sacc., *Syll.* iv, 363; Wakefield & Bisby, *TBMS*, **25**, 1941, 84.

*Leaf mould of tomato (*Lycopersicon*). Occurs under glass in all districts and is very destructive when ventilation is inadequate. May begin in April and is usually general by early June. Occasionally affects the fruit. Slight attacks are sometimes seen in the open. First recognized in England in 1887 and in Scotland, 1908.

The disease was first described in England by Plowright (*Gdnrs' Chron.* **2**, 1887, 532). There is a considerable amount of information about it scattered through the *Reports of the Experimental and Research Station, Cheshunt* (1926–42). Bond (*AAB*, **23**, 1936, 11–29; **25**, 1938, 277–307) studied infection phenomena and varietal resistance, and Small (ibid. **17**, 1930, 71–80) the relation of temperature and humidity to the disease. Day (*Plant Path.* **3**, 1954, 35–9) distinguished six physiologic races of the fungus in England and Wales; Dovaston (*Scottish Agric.* **34,** 1954–5, 172–3) estimated crop loss from moderate infection in fourteen different varieties; Clarke (*Sci. Hort.* **11,** 1955, 140–9) discussed some of the factors involved in his work on breeding resistant varieties in Eire, and Day (*Nature, Lond.,* **179,** 1957, 1141–2) induced virulence in a non-pathogenic race by treatment with a mutagen.

CHEMICAL CONTROL. Massee (*J. R. hort. Soc.* **28,** 1903, 142–5) tried to prevent the disease by watering the plants with copper sulphate. Various fungicides and fumigants were tried by Small (*AAB,* **18,** 1931, 305–12): shirlan by Bewley and Orchard (ibid. **19,** 1932, 185–9) and by Holmes Smith (*Gdnrs' Chron.* **93,** 1933, 46); colloidal copper in petroleum oil emulsions by Read (*AAB,* **23,** 1936, 183–9); and organic fungicides, including the successful zineb, by Beaumont (*Plant Path.* **3,** 1954, 21–5).

Cladosporium herbarum Fr., *Syst. Myc.* iii, 370; Sacc., *Syll.* iv, 350; Wakefield & Bisby, *TBMS,* **25,** 1941, 84.

The taxonomic aspects of this species have been dealt with by Brett (*TBMS,* **30,** 1948, 141–51), and Hyde and Williams (ibid. **36,** 1953, 260–6) studied the incidence of spores of the fungus in the air at Cardiff. See also Richards (ibid. **39,** 1956, 431–41).

*On wheat (*Triticum*). Commonly associated with blackening of the ears in August–September, especially after summer rain. Doubtfully parasitic, and not responsible for 'thinning out', 'deaf ears' or 'whiteheads' (Bennett, *AAB,* **15,** 1928, 192–212).

*On pea (*Pisum*). Very common on the haulm in wet seasons and probably mainly secondary. Sometimes develops on the pods and seeds. Until 1920 the fungus was sometimes named *C. pisi* Cug. & Mac.

A species of *Cladosporium* not different from *C. pisicola* (q.v.) has been seen in Kent (*Report,* viii, 35) and near Edinburgh (1947) associated with definite brown lesions on the leaves, but it was not regarded as specifically distinct from *C. herbarum.*

*On apple (*Malus*). Occasional as the cause of a slow rot in stored fruit (Kidd & Beaumont, *TBMS,* **10,** 1924, 112).

*On blackberry (*Rubus*). The apparent cause of a style and stigma rot of cultivated blackberries in Ayrshire, 1945, resulting in a total loss of crop (fide H. F. Dovaston).

Cladosporium orchidearum Cooke & Massee, *Grevillea*, 16, 1888, 80; Sacc., *Syll.* x, 605; Wakefield & Bisby, *TBMS*, 25, 1941, 85.

*Spot of orchids. Occurs on species of *Oncidium*, *Dendrobium*, *Coelogyne*, *Cypripedium* and other orchids. Described by W. G. Smith (*Gdnrs' Chron.* 8, 1890, 410). Said to be parasitic but very doubtfully distinct from *Cladosporium herbarum* (Brierley, ibid. 65, 1919, 61; W. C. Moore, *TBMS*, 25, 1941, 207).

Cladosporium paeoniae Passer., in Just's *Jber.* 1876, 235; Myc. univ. no. 670; Sacc., *Syll.* iv, 362; Cooke, *Fung. Pests Cult. Pl.* 1906, 19; Wakefield & Bisby, *TBMS*, 25, 1941, 85.

*Red spot of peony. Aberdeen, 1952.

Cladosporium pisicola Snyder, *Phytopathology*, 24, 1934, 890.

*On pea (*Pisum*). Was described as the cause of a leaf disease but is doubtfully distinct from *C. herbarum* (q.v.).

Cladosporium sp.

*On hop (*Humulus*). Causing browning of the cones (Salmon & Ware, *J. S.-E. agric. Coll. Wye*, 38, 1936, 53). Kent, occasional; Worcs, 1947.

*On *Freesia*. Associated with rot of young cormels, Lanarks.

Clasterosporium carpophilum (Lév.) Aderh., *Landw. Jb.* 30, 1901, 815; Wakefield & Bisby, *TBMS*, 25, 1941, 90; syn. *Coryneum beijerinckii* Oudem., *Hedwigia*, 22, 1883, 115; Sacc., *Syll.* iii, 774; Grove, *Coelomycetes*, ii, 336.

*Shot-hole of peach, cherry, nectarine and almond (*Prunus*). Not infrequent, especially on peach. Much of the shot-hole attributed to this fungus in the past was probably non-parasitic in origin, but this fungus does cause shot-hole in a number of districts, including Devon, Clyde and Tay, and occasionally a bud rot and die-back of young shoots. See M. J. B[erkeley] in *Gdnrs' Chron.* 24, 1864, 938 and Cooke in *J. R. hort. Soc.* 33, 1908, 527–8.

Clasterosporium **putrefaciens** (Fuckel) Sacc. = *Sporodesmium*
putrefaciens.

Claviceps purpurea (Fr.) Tul., *Ann. Sci. nat.* 3 Ser. **20**, 1853, 45;
Sacc., *Syll.* ii, 564; Bisby & Mason, *TBMS*, **24**, 1940, 202.

The conidial stage ('Honeydew') was formerly called *Sphacelia
segetum* Lév., *Mem. Soc. Linn.* **5**, 1827, 578; Sacc., *Syll.* iv, 666.
*Ergot of cereals. In England and Wales slight most years on
rye (*Secale*), chiefly in Yorks and the west; formerly occasional
only on wheat (notably Rivet), but has become more common in
recent years; infrequent on barley (*Hordeum*) and very rare on
oats (*Avena*). Also common on wild and cultivated grasses,
including rye grasses (*Lolium*), fescue (*Festuca*), cocksfoot
(*Dactylus*), tall oat grass (*Arrhenatherum*), meadow foxtail
(*Alopecurus*), couch grass (*Agropyron*), brome grasses (*Bromus*)
and occasionally *Melica nutans* and *Alopecuris myrosurioides*.
Found on Canary grass (*Phalaris canariensis*), Essex, 1954 (Ives,
Plant Path. **2**, 1953, 34–5). Occurs on wheat, barley, rye and many
grasses in Scotland. There is evidence that different physiologic
races exist. The disease is noticeable on cereals only after the
flowering period has been cool and wet.

Observations on ergot were made many years ago by A. S.
Wilson (*Gdnrs' Chron.* **4**, 1875, 774–5 and 807–8) and by Carruthers (*J. R. agric. Soc.* **10**, 1874, 443–8). Details of the occur-
rences of ergot in England were given in *Report*, viii, 6, 9 and 10,
and by Dillon Weston and Taylor (*J. agric. Sci.* **32**, 1942, 457–64)
who also devised a method for dealing with ergot in barley.
Willis (*Plant Path.* **2**, 1953, 34–5) discussed its distribution in
wheat variety trials in Herts. Robertson and Ashby (*Brit. med. J.*
3503, 1928, 302–3) and Morgan (*J. Hyg., Camb.*, **29**, 1929, 51)
gave some details of ergot poisoning among rye bread consumers
at Manchester. Batts (*Plant Path.* **5**, 1956, 76) infected wheat
with ergots from *Alopecurus.*

Clinterium obturatum Fr. = *Sporonema obturatum.*

Clitocybe dealbata (Fr.) Gillet, *Les Hyménomycètes*, Alençon, 1874, 152; Sacc., *Syll.* v, 157; Rea, *Brit. Basid.* 276.

*Occasionally invades mushroom beds. Hants (Buddin & Ware, *Gdnrs' Chron.* 93, 1933, 246), Herts, Staffs. See also Cooke, ibid. 14, 1893, 299.

Clitopilus cretatus (Berk. & Br.) Sacc., *Syll.* v, 702; Rea, *Brit. Basid.* 311.

*Cat's ear fungus. Not infrequent as a mushroom bed invader. Kent (Ware, *Gdnrs' Chron.* 97, 1935, 325), Hants, Dorset, Herts, Lancs.

Coleosporium cacaliae Fuckel, *Symb. Myc.* 43; Sacc., *Syll.* vii, 753; Grove, *Brit. Rust Fungi*, 325; Wilson & Bisby, *TBMS*, 37, 1954, 63.

*On *Emilia* (*Cacalia*) *hastata* and *E. suaveolens*. Uredo- and teleuto-stages. Rare. Batheaston, Oxford, Aberdeen (St Fergus). An introduced species. The aecidia on *Pinus montana* have not been seen in Britain.

Coleosporium campanulae Lév., *Ann. Sci. nat.* 3 Ser. 8, 373; Sacc., *Syll.* vii, 753; Grove, *Brit. Rust Fungi*, 328; Wilson & Bisby, *TBMS*, 37, 1954, 63.

*Campanula rust. Uredo- and teleuto-stages fairly common on species of *Campanula* in mid- and south England, including *C. latifolia*, *C. trachelium*, *C. persicifolia*, *C. planifera alba* and *C. carpatica* var. *turbinata*. Widely distributed in Scotland on *C. glomerata*, *C. persicifolia* and *C. rotundifolia*; also on *C. isophylla*, Perthshire, 1956.

The aecidial stage occurs on the leaves of *Pinus sylvestris*.

Coleosporium euphrasiae Wint., *Pilze Deutschl.* 246; Sacc., *Syll.* vii, 754; Grove, *Brit. Rust Fungi*, 326; Wilson & Bisby, *TBMS*, 37, 1954, 63.

*Aecidia on Scots pine (*Pinus sylvestris*). The uredo- and teleutospores occur on weed hosts (*Euphrasia, Bartsia, Rhinanthus*). Common.

Coleosporium melampyri Karst., *Myc. fenn.* iv, 1878, 62; Sacc., *Syll.* vii, 754; Grove, *Brit. Rust Fungi*, 327; Wilson & Bisby, *★TBMS*, **37**, 1954, 63.

*Aecidia on Scots pine (*Pinus sylvestris*). Uredo- and teleutospores on *Melampyrum* spp. Not uncommon.

Coleosporium narcissi Grove, *J. Bot., Lond.*, **60**, 1922, 121; Wilson & Bisby, *TBMS*, **37**, 1954, 63.

*Rust of *Narcissus*. Very rare. On leaves of *N. poeticus ornatus*, Lincs, 1922.

Coleosporium petasites Lév., *Ann. Sci. nat.* 1847, 373; Sacc., *Syll.* vii, 752; Grove, *Brit. Rust Fungi*, 323; Wilson & Bisby, *TBMS*, **37**, 1954, 63.

*Rust of *Petasites*. Uredo- and teleutospores on *P. japonicus* and *P. palmatus*, Tweed (*TBMS*, **9**, 1924, 142). Also commonly on *P. officinalis*. The aecidia occur on leaves of Scots pine (*Pinus sylvestris*) and *P. nigra* subsp. *laricio*.

Coleosporium senecionis Fr., *Summ. Veg. Scand.* 512; Sacc., *Syll.* vii, 751; Grove, *Brit. Rust Fungi*, 320; Wilson & Bisby, *TBMS*, **37**, 1954, 64.

The aecidia occur on *Pinus sylvestris* and *P. austriaca* and the uredo- and teleutospores on various members of the Compositae.

*On *Calendula*. Uncommon. First seen in 1937 in Berks and Devon, in gardens where groundsel attacked. Recurred in Devon the next two seasons (*Report*, viii, 81). Also in Scotland, Dee.

*On *Cineraria* (*Senecio cruentis*). Fairly common in many districts. Described by Chittenden (*J. R. hort. Soc.* **33**, 1908, 511–13). Also on *S. smithii* Argyll, Shetland. It occurs on wild species, and Cooke (Cooke, *Fung. Pests Cult. Pl.* 1906, 52) listed it on *S. pulcher* and *S. sarracenicus* in gardens.

*On *Chrysanthemum carinatum* in a garden at Norwich, 1945 (Ellis, *Trans. Norfolk Norw. Nat. Soc.* **16**, 1946, 172).

*On *Notonia grandiflora*, Kew Gardens, 1894.

*Needle rust of Scots pine (*Pinus sylvestris*). Common but seldom damaging.

Coleosporium sonchi Lév., *Ann. Sci. nat.* 1847, 373; Sacc., *Syll.* vii, 752; Grove, *Brit. Rust Fungi*, 324; Wilson & Bisby, *TBMS*, **37**, 1954, 64.

*Aecidia on the needles of Scots pine (*Pinus sylvestris*). The uredo- and teleutospores occur on *Sonchus* spp.

Coleosporium tropaeoli (Desm.) Palm, in *Vesterg. Micromycetes selecti scandinavici*, no. 1456; Sacc., *Syll.* vii, 862 (as *Uredo tropaeoli* Desm.); Wilson & Bisby, *TBMS*, **37**, 1954, 64.

Tropaeolum rust. Very rare. On *T. peregrinum*, Shere, Surrey, Oct. 1865 (as *Uredo tropaeoli*), Yorks, Guernsey.

Coleosporium tussilaginis (Pers.) Lév., *Mém. Uréd.* 1854, 136; Sacc., *Syll.* vii, 752; Grove, *Brit. Rust Fungi*, 322; Wilson & Bisby, *TBMS*, **37**, 1954, 63.

*Aecidia on leaves of Scots pine (*Pinus sylvestris*) and Corsican pine (*P. nigra* var. *poiretiana*). Uredo- and teleutospores on *Tussilago farfara*. Common.

As north European species of *Coleosporium* cannot be distinguished on morphological grounds, Wilson and Bisby (loc. cit.) include as forms of this species those with their uredo- and teleuto-stages on other hosts, viz. *cacaliae, campanulae, euphrasiae, melampyri, narcissi, petasites, senecionis, sonchi* and *tropaeoli*. (See also under these names.)

Colletotrichum atramentarium (Berk. & Br.) Taubenh., *Mem. N.Y. bot. Gdn*, **6**, 1916, 549; as *Vermicularia atramentaria* Berk. & Br. in Grove, *Coelomycetes*, ii, 244; Sacc., *Syll.* iii, 227.

*Black dot of potato (*Solanum*). Very common throughout Britain on decaying roots and haulm at or near digging time, but usually harmless. Sometimes causes premature death of haulm in August after a hot dry spell. First recognized in England in 1924 (*Report*, v, 29), but known some years before (Pethybridge, *TBMS*, **6**, 1919, 107). See also Cheal (*Gdnrs' Chron.* **84**, 1928, 508; **86**, 1929, 493).

*Black dot of tomato (*Lycopersicon*). Common under glass in all parts of England and destructive where soil conditions are unsuitable or drainage is faulty. Serious in Yorks and elsewhere in the north; occurs in several districts in Scotland. Was first encountered in the Lea Valley in 1919 and described there by Bewley and Shearn (*AAB*, **11**, 1924, 244–51). Is more active on old than on seedling roots (Ebben, *Rep. exp. Res. Sta. Cheshunt for 1949*, 28–32).

*On roots of winter cherry (*Solanum capsicastrum*). Ches, War, 1950.

Colletotrichum cinctum Stonem. = stat. conid. of *Glomerella cincta*.

Colletotrichum circinans (Berk.) Vogl., *Ann. Accad. Agric. Torino*, **49**, 1907, 175; syn. *Vermicularia circinans* Berk., *Gdnrs' Chron.* **11**, 1851, 595; Sacc., *Syll.* iii, 233; Grove, *Coelomycetes*, ii, 239.

*Onion smudge. Formerly apparently common but now rarely seen. First recorded 1851 (M.J.B., *Gdnrs' Chron.* **11**, 1851, 595 and 646). War, 1945; Corn, 1950; Glos, 1953. See Cooke, *Fung. Pests Cult. Pl.* 1906, 104.

Colletotrichum concentricum Massee, *Kew Bull.* 1913, 198; Sacc., *Syll.* xxv, 568; Grove, *Coelomycetes*, ii, 236.

*On fruit of snake gourd (*Trichosanthes anguina*), Kew Gardens.

Colletotrichum fuscum Laubert, *Gartenwelt*, **31**, 1937, 674.

*Anthracnose of *Digitalis*. On *D. lanata*, Dartford, Kent, July 1952 (Spilsbury, *TBMS*, **36**, 1953, 335–42).

Colletotrichum gloeosporioides Penz., in *Mich.* ii, 1882, 450; Sacc., *Syll.* iii, 735; Grove, *Coelomycetes*, ii, 240 (as *Vermicularia gloeosporioides*).

*Fruit rot of apple (*Malus*). Very rare (Kidd & Beaumont, *TBMS*, **10**, 1924, 108).

Colletotrichum graminicola (Ces.) Wilson, *Phytopathology*, **4**, 1914, 110; Sacc., *Syll.* xxv, 570.

*On *Poa annua*, causing a root and shoot rot. Widely distributed in Great Britain and Northern Ireland (Drew Smith, *J. Sports Turf Res. Inst.* **8**, 1954, 344–53; **9**, 1955, 74–5). Probably first recognized in lawn at Huddersfield, Yorks, Nov. 1936.

Colletotrichum holci (Syd.) Grove, *J. Bot., Lond.*, **56**, 1918, 341; syn. *Vermicularia holci* Syd., *Hedwigia*, **38**, 1899, 137; Sacc., *Syll.* xvi, 894; Grove, *Coelomycetes*, ii, 242.

*Leaf spot of *Holcus*. Clyde (Smith & Ramsbottom, *TBMS*, **6**, 1920, 369), Hereford, Epping (Duke, ibid. **13**, 1928, 178).

Colletotrichum lagenarium (Passer.) Ell. & Halsted, *Bull. Torrey bot. Cl.* **20**, 1893, 250; Sacc., *Syll.* iii, 719; Grove, *Coelomycetes*, ii, 231. For many years this fungus was reported under the name *C. oligochaetum* Cav.

*Anthracnose of cucumber (*Cucumis*). Not now common but found under glass in Scotland as well as farther south. From about 1917–25 was a serious disease in the Lea Valley. Occasional on vegetable marrow, Clyde, Dee, Bristol; and melon, Yorks, War.

Described as new to England in 1911 (*J. Bd Agric.* **18**, 670–1). A full account of it was given by Bewley (*J. Minist. Agric.* **29**, 1922, 469–72, 558–62). Stevens (*Mycologia*, **23**, 1931, 134–9) obtained the perfect stage by exposing cultures to ultra-violet radiation, and called it *Glomerella lagenarium* (Passer.) Stevens.

Colletotrichum lindemuthianum (Sacc. & Magn.) Bri. & Cav., *Fungh. Parass.* no. 50, 1899; Sacc., *Syll.* iii, 717; Grove, *Coelomycetes*, ii, 234.

*Anthracnose of dwarf bean (*Phaseolus*). Widely distributed in England and Wales, fairly common, and apt to be severe in cool, wet summers, especially in the south. Occasional on runner bean. The fungus has not been reported in Scotland for many years.

The disease was first recognized about 1880 (Cooke, *Grevillea*, **10**, 1881, 48; Berkeley, *Gdnrs' Chron.* **14**, 1880, 272) and has attracted attention from time to time (Massee, ibid. **23**, 1898,

293; Alcock, ibid. **84**, 1928, 293). Dey (*Ann. Bot., Lond.*, **33**, 1919, 305–12) studied its infection phenomena.

The perfect stage is perhaps *Glomerella lindemuthianum* Shear (Shear & Wood, *Bull. U.S. Bur. Pl. Ind.* no. 252, 1913), but perithecia have been seen only in culture.

Colletotrichum linicola Pethybr. & Lafferty, *Sci. Proc. R. Dublin Soc.* **15**, 1918, 359; Grove, *Coelomycetes*, ii, 232.

*Seedling blight of flax (*Linum*). First reported 1919 (*Report*, iii, 60), now slight in many districts of Britain.

In Eire the disease was investigated by Pethybridge *et al.* (*J. Dep. Agric. Ire.* **20**, 1920, 325–42; **21**, 1921, 167–87; **22**, 1922, 103–20). Muskett and Malone (*AAB*, **28**, 1941, 8–13) devised the Ulster method for examining flax seed for the presence of seed-borne parasites, and Muskett and Colhoun (*Ann. Bot., Lond.*, **6**, 1942, 219–27) a technique for evaluating seed disinfectants for the control of seed-borne diseases. Using these methods or modifications of them, Muskett and Colhoun (*AAB*, **30**, 1943, 7–18; **31**, 1944, 295–300; **32**, 1945, 34–7) compared different methods of flax seed disinfection. For a general summary of this work see Muskett and Colhoun (ibid. **33**, 1946, 331–3).

Colhoun (*AAB*, **33**, 1946, 260–3) studied the relation between the degree of contamination of flax seed with *Colletotrichum linicola* and *Polyspora lini*, and the incidence of disease in the field. He has also tested the field reaction of flax varieties to these two fungi and to *Phoma* sp. and *Melampsora lini* (ibid. **35**, 1948, 582–97). Colhoun and Muskett (ibid. **35**, 1948, 429–34) made observations on the longevity of the seed-borne parasites of flax in stored seed.

For a general summary of the seed-borne pathogens of flax and their control see Muskett in *Proc. IInd Int. Congr. Crop Prot. Lond. 1949*, 297–303, 1951.

Colletotrichum malvarum Southw., *J. Mycol.* **6**, 1890, 45; Sacc., *Syll.* x, 468; Grove, *Coelomycetes*, ii, 233.

*On leaves of *Lavatera*. Occasional, widely distributed. On malvaceous plant, Perthsh, 1908 (Smith & Rea, *TBMS*, **3**, 1908,

38); on *L. trimestris*, Alton, Hants; Langley, Bucks and else-where in the south (Chittenden, *J. R. hort. Soc.* **35**, 1909, 213–15); Devon, 1936–7; Windsor, 1941; on *L. rosea*, Norfolk, 1920; Ayrshire, 1954; Essex, 1948.

Colletotrichum oligochaetum Cav., see *C. lagenarium*.

Colletotrichum orchidearum Allesch., Rabenh., *Krypt. Fl.* i, 7, 1903, 563; Sacc., *Syll.* xviii, 467; Grove, *Coelomycetes*, ii, 233.

*Orchid spot. On *Oberonia* and other orchids (Smith & Rams-bottom, *TBMS*, **5**, 1917, 430); on pseudobulbs of *Coelogyne*, Kew; on *Vanda* sp., St Albans, Herts, 1926; probably this species on leaves of *Phalaenopsis sanderiana*, Crawley, Sussex, 1950; on *Odontoglossum*, Lancs, 1950. The species was intercepted on *Coelogyne ochracea* and *Dendrobium* imported from Bengal in 1946. This, or a closely allied species, is common on moribund orchid tissues.

Colletotrichum phomoides (Sacc.) Chester, *Rep. Del. agric. Exp. Sta.* **6**, 1894, 111; Sacc., *Syll.* iii, 718 (as *Gloeosporium pho-moides* Sacc.); Grove, *Coelomycetes*, ii, 235.

*Ripe rot of tomato (*Lycopersicon*). Occurs on the fruit. Rare. Cambridge, 1919 (Brooks & Searle, *TBMS*, **7**, 1921, 173); on a few fruits, Reading, Berks, 1941 (*Report*, viii, 54); on one fruit, Warfield, Salop, 1952; on one fruit Auchincruive, Ayrshire, 1954.

Swank (*Phytopathology*, **43**, 1953, 285–7) obtained a peri-thecial stage in culture and named it *Glomerella phomoides*.

Colletotrichum rhodocyclum (Mont.) Petrak = stat. conid. of *Glomerella phacidiomorpha*.

Colletotrichum spinaciae Ell. & Halst., *J. Mycol.* **6**, 1890, 34; Sacc., *Syll.* x, 469.

*On spinach (*Spinacia*). On seed of prickly spinach harvested in Lincs, 1953 (Baker, *Plant Path.* **3**, 1954, 139). What was thought to be this species was associated with leaf spotting in Sussex, 1937.

Colletotrichum trichellum (Fr.) Duke, *TBMS*, **13**, 1928, 173; Sacc., *Syll*. iii, 224 (as *Vermicularia trichella* Grev., *Scot. Crypt. Fl*. 1828, 345); Grove, *Coelomycetes*, ii, 242.

*Leaf spot of ivy (*Hedera*). Not uncommon throughout Britain.

Colletotrichum trifolii Bain & Essary, *J. Mycol*. **12**, 1906, 193; Sacc., *Syll*. xxii, 1201.

*Anthracnose of lucerne (*Medicago*). Not uncommon. Kent, Sussex, Norfolk, Dorset, Merioneth, Carms, Cards.

First seen in Kent in 1950 (Glasscock, *Plant Path*. **1**, 1952, 102). Occurs notably on the variety Du Puits.

Colletotrichum sp.

*On *Pelargonium*. Yorks, 1936.

Collybia velutipes (Curt. ex Fr.) Quél., *Les Champignons du Jura et des Vosges*, Paris, 1872, 59; Rea, *Brit. Basid*. 332.

*Stem toadstool of red currant (*Ribes*). Said to be parasitic on red currant in Cambs, Beds, Berks, Northumb (*Report*, ii, 87) and not infrequent on bushes not thriving well (ibid. v, 79). The nutritional requirements for fruiting of this species were studied by Plunkett (*Ann. Bot., Lond.*, **17**, 1953, 193–217).

Colpoma quercina (Fr.) Wallr., *Fl. crypt*. no. 2339; Sacc., *Syll*. ii, 803.

*Die-back of oak (*Quercus*). Mainly on branches of young trees, coppiced oak and terminal branches of older trees. Twyman (*TBMS*, **29**, 1946, 234–41) described the fungus and the histology of infection.

Coniosporium corticale Ell. & Everh. = *Cryptostroma corticale*.

Coniothecium chomatosporum Corda, *Ic. Fung*. i, 2; Sacc., *Syll*. iv, 510; Wakefield & Bisby, *TBMS*, **25**, 1941, 98.

*On apple (*Malus*). A very common fungus on apple bark and in cracks on russeted apple fruits. It was formerly regarded as the cause of rough scab (branch blister) of apple (Massee, *Kew Bull*.

PARASITES

1915, 104–7), but is now considered to be saprophytic. Rough scab is thought to be non-parasitic in origin (M. H. Moore, *Rep. E. Malling Res. Sta. for 1938*, 255; *Sci. Hort.* 7, 1939, 88); its incidence is affected by the rootstock and is greatly reduced by potash manuring (M. H. Moore & Bennett, *AAB*, **39**, 1952, 588–98).

Coniothyrium concentricum (Desm.) Sacc., *Mich.* i, 204; Sacc., *Syll.* iii, 317; Grove, *Coelomycetes*, ii, 14.

*Leaf spot of *Yucca*. Common in Britain where the plant grows.

Coniothyrium concentricum (Desm.) Sacc., *Mich.* i, 204, var. **agaves** Sacc., *Syll.* iii, 317.

*Leaf spot of *Agave*. Clyde.

Coniothyrium diplodiella Sacc., *Syll.* iii, 310; Grove, *Coelomycetes*, ii, 11.

*On fruit and fruit stalks of grape vine (*Vitis*) under glass. Was formerly recorded occasionally in this country, but has not been seen for many years.

Coniothyrium fuckelii Sacc. = stat. conid. of *Leptosphaeria coniothyrium*.

Coniothyrium hellebori Cooke & Massee, *Grevillea*, **15**, 1887, 108; Sacc., *Syll.* x, 261; Grove, *Coelomycetes*, ii, 6.

*Leaf spot of Christmas rose (*Helleborus*). Not uncommon in southern and especially south-west England; common in north Scotland.

Dennis and Foister (*TBMS*, **25**, 1942, 292) pointed out that in fresh material the spores are frequently colourless, in which state the fungus agrees with *Phyllosticta atrozonata* Voss, which Grove (loc. cit.) lists as a synonym. The species described as *P. helleborella* Sacc. var. *nigra* Cooke (*Grevillea*, **14**, 1885, 73; *Gdnrs' Chron.* **6**, 1889, 476 and 479) is probably the same.

Coniothyrium ilicis Smith & Ramsb., *TBMS*, **5**, 1917, 426; Sacc., *Syll.* xxv, 232; Grove, *Coelomycetes*, ii, 6.

*On leaves of holly (*Ilex*), St Annes, Lancs, 1916; Sutton Cold-field, War, 1922, causing an epidemic.

Coniothyrium minitans Campb., see *Sclerotinia trifoliorum.*

Coniothyrium phormii Cooke, *Grevillea*, 7, 1879, 96; Sacc., *Syll.* iii, 166 (as *Phoma phormii* (Cooke) Sacc.); Grove, *Coelomycetes*, ii, 13.

*On *Phormium tenax*, Polperro and St Ives, Corn; Mumbles, Glam.

Coniothyrium pirinum (Sacc.) Sheldon, *Torreya*, 7, 1907, 142; Sacc., *Syll.* iii, 7 (as *Phyllosticta pirina* Sacc.).

*Leaf spot of apple (*Malus*). Rare. Ross-shire, 1944 (Dennis & Wakefield, *TBMS*, 29, 1946, 155–8).

Coniothyrium rosarum Cooke & Harkn., *Grevillea*, 12, 1884, 92; Sacc., *Syll.* iii, 307.

*Graft disease of rose. Caused much trouble in a nursery in Herts, 1927, and believed to have been introduced on scions imported from U.S.A., where the disease was first described (Vogel, *Phytopathology*, 9, 1919, 403). The species is doubtfully distinct from *C. fuckelii* (q.v.), and Grove (*Coelomycetes*, ii, 2) lists the two as identical. Bewley (*Sci. Hort.* 6, 1938, 97) published a note on the disease.

Coniothyrium tumefaciens Güssow, *J. R. hort. Soc.* 34, 1908, 229; Sacc., *Syll.* xxii, 968; Grove, *Coelomycetes*, ii, 9.

*Canker of blackberry (*Rubus*). Kent, 1908. Grove (loc. cit.) considered the species a doubtful one, and Massee (*Dis. Cultiv. Pl. Trees*, 1915, 14) was convinced the disease was bacterial and caused by *Bacterium tumefaciens*.

Coniothyrium vagabundum Sacc., *Syll.* iii, 310; Grove, *Coelomycetes*, ii, 4.

*On gooseberry (*Ribes*). First recorded in Worcs, 1906, as *Coniothyrium ribicolum* Brun. (Smith & Rea, *TBMS*, 2, 1907, 168).

Was described by A. L. Smith (*Gdnrs' Chron.* **42**, 1907, 341) as the cause of disease, but its parasitism is uncertain.

Coniothyrium wernsdorffiae Laub., *Arb. biol. Abt. Landw. Forstw.* **4**, 1905, 458; Sacc., *Syll.* xviii, 303.

*Brand canker of rose. First seen in January 1954 on briars (*Rosa canina*) newly imported from Germany into Scotland (Gray, *Plant Path.* 3, 1954, 105) and subsequently found on many varieties in northern Scotland.

Corticium arachnoideum Berk., see *C. centrifugum*.

Corticium centrifugum (Lév.) Bres., *Ann. mycol., Berl.*, **1**, 1903, 96; Sacc., *Syll.* xvii, 174.

*Fish eye rot of apple (*Malus*). Not seen occurring naturally in Britain, but found causing a decay of slightly scabbed apples imported from Canada in 1930 (*Report*, vii, 70). Has been seen on stored apples in Northern Ireland (Colhoun & Muskett, *Gdnrs' Chron.* **97**, 1935, 418–19; Colhoun, *AAB*, **25**, 1938, 92). The fungus has been studied abroad by Butler (*J. agric. Res.* **41**, 1930, 269–94). Rea (*Brit. Basid.* 676) listed it as a synonym of *Corticium arachnoideum* Berk., which is not uncommon on stumps and fallen branches.

Corticium chrysanthemi Plowr. = *C. sambuci*.

Corticium fuciforme (Berk.) Wakefield, *TBMS*, **5**, 1916, 481; syn. *Isaria fuciformis* Berk., *J. Linn. Soc.* **13**, 1872, 175; Sacc., *Syll.* iv, 595.

*Red thread of turf grasses. Becomes conspicuous most years in September–October on bowling greens, golf fairways, lawns and pastures, in many parts of Britain. *Festuca rubra* is particularly susceptible. Also found on *Agrostis tenuis, Poa annua, Lolium perenne*, etc. Descriptions of the disease were given by W. G. Smith (*Gdnrs' Chron.* **17**, 1882, 377–8); by Bennett (*J. Bd Greenkeep. Res.* **4**, 1935, 32–9), including control; by Libbey (ibid. **5**, 1938, 269–70); and by Drew Smith (*J. Sports Turf Res.*

Inst. 8, 1954, 253–8, 365–77), who used some modern organic fungicides for controlling it.

There may be a genetic connexion between this fungus and a pink hyphomycete commonly associated with it, which has been called (probably erroneously) *Geotrichum roseum* Grove, Sacc., *Syll.* iv, 40 (see *Report*, viii, 40).

Corticium praticola Kotila, see *C. solani* (p. 114).

Corticium rolfsii (Sacc.) Curzi, *Boll. Staz. Pat. veg. Roma*, **11**, 1931, 365; stat. mycel. *Sclerotium rolfsii* Sacc., *Ann. mycol.*, *Berl.*, **9**, 1911, 257; Sacc., *Syll.* xxii, 1500.

This fungus has not yet been found occurring naturally in Britain.

*With *Sclerotium delphinii* (q.v.) isolated from Iris bulbs, var. Wedgwood, imported from France in 1935 (W. C. Moore, *Diseases of Bulbs*, 1939, 137); on imported Arum (*Zantedeschia*) corms, 1926; on bulb of *Amaryllis* imported from Jamaica, 1936. Abeygunawardena and Wood (*TBMS*, 40, 1957, 221–31) have studied the fungus in culture in this country.

Corticium sambuci (Pers.) Fr., *Epicr.* 1, 565; Sacc., *Syll.* vi, 656 (as *Hypochnus sambuci*); Rea, *Brit. Basid.* 677; syn. *C. chrysanthemi* Plowr., *TBMS*, **2**, 1905, 91; Sacc., *Syll.* xxi, 409.

*On *Chrysanthemum*, attacking the stem bases, King's Lynn (Plowright, loc. cit.).

Corticium solani (Prill. & Delacr.) Bourd. & Galz., *Bull. Soc. mycol. Fr.* **27**, 1911, 248; Sacc., *Syll.* xi, 130; xxi, 414 (as *Hypochnus solani* Prill. & Delacr.); stat. mycel. *Rhizoctonia solani* Kühn, *Krankheiten der Kulturgewächse*, 1858, 224; Sacc., *Syll.* xiv, 1175. The valid name appears to be *Pellicularia filamentosa* (Pat.) Rogers, *Farlowia*, **1**, 1943, 113.

Different strains of this species attack a wide range of cultivated plants, producing mainly damping-off, root rot and stem canker diseases, but also a variety of other symptoms. The perfect stage has not been seen on many of the hosts, but is prominent on a few in certain seasons. Reference is made below to the occurrence of the fungus on cereals, root plants, vegetables, ornamentals and some other crop plants, but the list is not claimed to be complete.

The problem of strains has been discussed by Briton-Jones (*TBMS*, **9**, 1924, 200–10) and by Storey (*AAB*, **28**, 1941, 219–28). Flentje (*TBMS*, **39**, 1956, 343–56) induced twenty-eight isolates of *Rhizoctonia solani* to form the perfect stage. All but four yielded *Pellicularia filamentosa*; the other four, including one from lettuce in England, yielded *P. praticola* (Kotila) Flentje, syn. *Corticium praticola* Kotila, *Phytopathology*, **19**, 1929, 1059. Flentje (*TBMS*, **40**, 1957, 322–36) has also studied the host-parasite relations of various strains, and, with Saksena (ibid. **40**, 1957, 95–108), their pathogenicity.

Blair (*AAB*, **30**, 1943, 118–27) studied the behaviour of the fungus in the soil, and Townsend and Willetts (*TBMS*, **37**, 1954, 213–21) the development of its sclerotia.

CEREALS

*Root rot ('purple patch') of cereals. Occasional on wheat, barley and oats, but of little importance; seen in Norfolk and elsewhere in England (Dillon-Weston & Garrett, *AAB*, **30**, 1943, 79). Occasional also on turf grasses in England and recorded on Italian ryegrass (*Lolium*) in Scotland (Inverness, 1953).

*Sharp eyespot of wheat (*Triticum*). Widely distributed in Britain, but rarely affects more than 1 % of the straws in a crop (Glynne & Ritchie, *Nature, Lond.*, **152**, 1943, 161). Occasional on oats, Beds (Glynne, ibid. **166**, 1950, 232), barley (Fife, 1945) and rye (Yorks, 1948).

POTATO

*Black scurf. A very common blemish of the tubers everywhere. The fungus may blacken and kill the young shoots before emergence, and this becomes of economic importance when cold weather retards growth during the first few weeks after planting. Later it causes stem canker, and some years in June or July the perfect stage develops profusely as a white 'collar' fungus on the stem bases. It is possible that tuber pitting may sometimes be caused by this species (see, for example, *J. agric. Res.* **9**, 1917, 421;

Z. PflKrankh. **50**, 1940, 225), though care should be taken to ensure that the pitting is not a result of damage by ants (Dillon-Weston, *Plant Path.* **2**, 1953, 55–6); and it is unlikely that *C. solani* is a primary cause of jelly end rot, as is sometimes suspected.

An early account of black scurf in this country is that by Güssow (*J. R. agric. Soc.* **66**, 1905, 173–7). Small (*AAB*, **30**, 1943, 221–6; **32**, 1945, 206–9) followed up the use of clean and contaminated tubers in the field, Johnson (ibid. **23**, 1936, 152–63) tried the effect of treating contaminated soil with mercuric chloride, and Graham, Srivastava and Foister (*Plant Path.* **6**, 1957, 149–52) mercurial and other tuber dips.

ROOTS AND FODDER CROPS

*Causing damping-off of swede (*Brassica*) seedlings and occasionally rotting of the roots. Scotland (Dennis, *Nature, Lond.*, **147**, 1941, 87); War, 1954.

*This or *Rhizoctonia* sp. is sometimes associated with a seedling disease (black leg) of sugar beet (*Beta*), notably in Lincs, Notts and Yorks, leading to uneven braird ('chicken and hen').

*Has been recorded causing damping-off of rape (*Brassica*) (Kent, Sussex, Berks), mustard (Som, Kent, Lancs) and marrow-stem kale (Ayrshire).

PULSE

*On roots of broad bean (*Vicia*) in Yorks, Cambs, Berks, Mon; on pea roots (*Pisum*) (Yorks, Glam, Som); and causing foot rot of dwarf and runner beans (*Phaseolus*) in Clyde.

PASTURE AND FORAGE CROPS

*Occasional on roots of red and white clover (*Trifolium*) (Glam) and lucerne (*Medicago*) (Glam, Notts).

VEGETABLES

*Damping-off and wire stem of *Brassica* seedlings. Common, especially on cauliflower.

*On carrot roots (*Daucus*) (Cumb, Norfolk) and seedlings (Dorset).

*On celery (*Apium*). Not uncommon. The perfect stage of the fungus is occasionally prominent on the stems and petioles, and the mycelial stage may girdle the roots.

*Damping-off of cress (*Lepidium*). Very common.

*Foot rot of cucumber (*Cucumis*). Occasionally reported since 1938.

*Associated with root damage to horse radish (*Armoracia*). Ches, 1954.

*Root rot of leek (*Allium*). Ches, Som, Mon (or as *Rhizoctonia* sp.).

*Damping-off and root rot of lettuce (*Lactuca*). Common and widely distributed. Studied by Abdel-Salam (*J. Pomol.* **11**, 1933, 259–75) and El-Helaly (*Proc. Linn. Soc. Lond. 1939–40 Session*, **152**, pt. I, 1940). Brown and Montgomery (*AAB*, **35**, 1948, 161–80) and Last (ibid. **39**, 1952, 557–68) showed that it can be controlled with chloronitrobenzene dusts. Wood (ibid. **38**, 1951, 217–30) investigated the possibilities of microbiological control.

*Damping-off in onion (*Allium*). Kent, Hants, War, Som, east midlands.

*Radish (*Raphanus*) canker. Not infrequent. Bucks, Berks, Middx, Surrey, Worcs, Mon, Lancs. See W. C. Moore, *TBMS*, **29**, 1946, 255.

*On seakale (*Crambe*), causing a black rot of the leaf stalks in forcing pits, and a natural disbudding of plants in the field (Brown, *J. Pomol.* **15**, 1937, 81).

*Damping-off of tomato (*Lycopersicon*) seedlings. Frequent. Not uncommon as the cause of foot rot in older plants (Small, *AAB*, **14**, 1927, 290–5).

FRUIT

*On strawberry (*Fragaria*), causing fruit rot (Clyde) or root rot (Som, 1946).

ORNAMENTAL PLANTS AND TREES

Antirrhinum. Damping-off and collar rot. Worcs, Som, Dorset, Kent.

Azalea. Damping-off and leaf blight of *A. indicum* and *A. japonicum*. Norwich, 1953–4 (Storey, *Plant Path.* **4**, 1955, 71).

Similar symptoms have been attributed in America (*Plant Dis. Reptr.* **39**, 1955, 860) to *Cylindrocladium scoparium* Morgan.

*Begonia. Root rot. Solway, Clyde.

*Cactus. Damping-off and root-base rot. Devon, Herts, 1955.

*Calceolaria. Stem rot. Glam, War.

*Carnation (*Dianthus*). Stem rot. Bath, Som, 1950. Also *Rhizoctonia* sp. rotting cuttings, Berks, 1945.

*China aster (*Callistephus*). Damping-off and foot rot.

*Chrysanthemum. Damping-off of cuttings. Frequent throughout Britain.

*Cineraria (*Senecio*). Wilt.

*Conifers. Muskett (*Nature, Lond.,* **124**, 1929, 481–2; *J. Minist. Agric. N. Ire.* **3**, 1931, 102–16) studied the control of seedling disease in Sitka spruce (*Picea*) and Douglas fir (*Pseudotsuga*) in Northern Ireland by using soil disinfectants.

*Cyclamen. Root rot of seedlings. Som, 1951; Herts, 1955.

*Dahlia. On underground shoots, Berks, 1934.

*Delphinium. Foot rot. Som, 1950.

*Flax (*Linum*). On seedlings, Yorks, Suffolk, Kent.

*Freesia. On seedlings, Worcs, 1949; Devon.

*Gerbera. Root rot. Pensilva, Corn, Aug. 1956; Devon, 1950.

*Gladiolus. On roots, Cambs, 1939.

*Godetia. Foot rot. Jersey, 1953.

*Heliotropium. Causing rot of cuttings, Herts (Sheard, *Rep. exp. Res. Sta. Cheshunt for 1944*, 26).

*Hydrangea. On cuttings, Middx, 1940; and as *Rhizoctonia* sp. causing damping-off, Yorks, 1956.

*Iris. Root rot. War, 1935; Yorks, 1935, 1937; Ayr, 1952 (*Rhizoctonia* sp.). On bulbs, Corn, 1950; Guernsey, 1956–7, sometimes also associated with bulb decay.

*Lilium. Crown rot of *L. auratum* (Clyde) and root rot of *L. longiflorum* (Som, 1950). *Corticium* sp. has been reported as a cause of stem rot of *L. tigrinum* in Lancs, 1945.

*Lobelia. Damping-off. Common.

*Myosotis. Root rot. Cardiff, 1949.

*Orchid. Occurs as an endophyte of *Orchis purpurella* (Downie *Nature, Lond.,* **179**, 1957, 160).

*Petunia. Root rot. East Sussex, 1955 (*Rhizoctonia* sp.).

*Primula. Foot rot. Staffs, Argyll, Forth.

*Pyrethrum. Root rot. Suffolk, Yorks, Northumb, Forth. *Rhizoctonia* sp. has been found rotting the shoots in Ches (1937) and Worcs (1941).

*Ranunculus. Damping-off in garden varieties, Berks, 1951, and as *Rhizoctonia* sp., Sussex, 1956.

*Reseda. Root rot of *R. odorata*. Forth.

*Salvia. Staffs and Kent, 1956.

*Saponaria. Glos, 1951.

*Schizanthus. Wilt. Staffs, 1937.

*Stock (*Matthiola*). Damping-off and stem rot. Common.

*Sweet Pea (*Lathyrus*). Root rot, Scotland.

*Tagetes (African marigold). Folkestone, Kent, 1949.

*Tulip. Associated with russeting of bulbs, Wilts, 1947.

*Viola. Root rot. Common (see also *Pythium violae*).

*Wallflower (*Cheiranthus*). Root and stem rot. Not uncommon.

*Zantedeschia (Arum lily). Root rot. Frequent.

*Zinnia. Damping-off. Not infrequent.

Coryneopsis microsticta (Berk. & Br.) Grove = stat. conid. of *Griphosphaeria corticola*.

Coryneopsis rubi (Sacc.) Grove, see *Hendersonia rubi*.

Coryneopsis rubi f. **rubi-idaei** Brun., see *Hendersonia rubi*.

Corynespora cassicola (Berk. & Curt.) Wei, see *Cercospora melonis*.

Corynespora melonis (Cooke) Sacc. = *Cercospora melonis*.

Coryneum beijerinckii Oudem. = *Clasterosporium carpophilum*.

Coryneum laurocerasi Prill. & Delacr., *Bull. Soc. mycol. Fr.* **6**, 1890, 180; Sacc., *Syll.* x, 481; Grove, *Coelomycetes*, ii, 335.

*On leaves of cherry laurel (*Prunus laurocerasus*), Potterne, Wilts, causing serious loss of foliage.

Coryneum microstictum Berk. & Br., see *Griphosphaeria corticola*.

Coryneum microstictum var. mali Kidd & Beaum., see *Griphosphaeria corticola*.

Cristulariella depraedens (Cooke) Höhnel, *S.B. Akad. Wiss. Wien*, 125, 1916, 27; Cooke, *J. Quekett micr. Cl.* 2, 1885, 138–43 (as *Polyactis depraedens*); Wakefield & Bisby, *TBMS*, 25, 1941, 71.
*On leaves of sycamore (*Acer*), sometimes causing defoliation, Norfolk, *c*. 1879; Dunkeld, *c*. 1935; Clyde, 1945.

Cronartium asclepiadeum Fr. = *C. flaccidum*.

Cronartium flaccidum (Alb. & Schw.) Wint., *Pilze Deutschl.* i, 1884, 236; Sacc., *Syll.* vii, 598; Wilson & Bisby, *TBMS*, 37, 1954, 64; syn. *C. asclepiadeum* Fr., *Obs. Myc.* 1, 1815, 220; Sacc., *Syll.* vii, 597; Grove, *Brit. Rust Fungi*, 313.
*Aecidia on branches of Scots pine (*Pinus sylvestris*), Devon, 1927. Uredo- and teleutospores on *Paeonia officinalis*, Norfolk (before 1889), Sussex, 1924; and on *Tropaeolum majus*, Norfolk, 1934. Very uncommon.

Cronartium quercuum Miyabe, See *Uredo quercus*.

Cronartium ribicola J. C. Fischer, *Hedwigia*, 11, 1872, 182; Sacc., *Syll.* vii, 598; Grove, *Brit. Rust Fungi*, 316; Wilson & Bisby, *TBMS*, 37, 1954, 64. For the authority of this species see Sydow, *Ann. mycol., Berl.*, 32, 1934, 115–17.
*Black currant rust. Uredo- and teleuto-stages widely distributed in Britain on black currant (*Ribes*) from August onwards, and in some years (e.g. 1917, 1941) severe in many districts. Occasional on gooseberry (Oxford, 1917, *Report*, i, 28), *Ribes sanguineum*, white currant, and red currant (ibid. vii, 82). Tweed, Clyde, Tay.
The aecidia occur on all species of five-needle pines, including Weymouth pine (*Pinus strobus*) (Boyce, *J. For.* 24, 1926, 893–6), *P. monticola, P. lambertiana, P. cembra* and *P. parviflora*, but *P. peuce* shows a high degree of resistance.

The fungus was first seen in Britain at King's Lynn, Norfolk, in 1892 and is the most serious rust on British forest trees. Adam, Dickinson and Marsh (*Rep. Fruit Veg. Pres. Res. Sta., Campden for 1945*, 40–50) studied the effect of dithiocarbamate spray residues on canned black currants. Hahn (*Trans. bot. Soc. Edinb.* 30, 1929, 137–46) demonstrated the resistance of the Norwegian Red Dutch currant variety.

Crumenula pinea (Karst.) Ferdinands. & Jørgens., *Skovtraeernes Sygdomme* 1939, 196; stat. conid. *Brunchorstia pinea* (Karst.) Höhnel, *S.B. Akad. Wiss. Wien*, Abt. I, **124** (*Fragm. Mykol.* 17), 1915, 95; syn. *B. destruens* Erikss., *Bot. Zbl.* 1891, 298; Sacc., *Syll.* x, 431.

The perfect stage has not been seen in Britain.

*Die-back of pine. Widely distributed in Great Britain on the Austrian pine (*Pinus nigra austriaca*) and other species (Waldie, *Trans. R. Scot. arb. Soc.* 40, 1926, 120–5). Other species attacked are Corsican pine (*P. nigra poiretiana*), *P. cembra*, *P. mugo*, *P. pinaster* and perhaps *P. strobus*. It appears to be in part responsible for the debility of Corsican pine in high-rainfall areas and at high elevations in Britain. First British record on *P. sylvestris* in the Spey Valley, as *Brunchorstia destruens* (van Vloten, *Scottish For. J.* **43**, 1929, 157–8). The disease was discussed by Leven (*Quart. J. For.* **26**, 1932, 225–31).

Cryptodiaporthe castanea (Tul.) Wehm., *TBMS*, **17**, 1933, 284; syn. *Diaporthe castanea* Sacc., *Syll.* i, 624; stat. conid. *Fusicoccum castaneum* Sacc., *Syll.* iii, 249; Grove, *Coelomycetes*, i, 247.

*'Javart' disease of chestnut (*Castanea*). On dead twigs and branches of *C. sativa*. Kent, Sussex, Surrey, Hants, Worcs and probably common in south-east England. Definitely acting as a parasite, Midhurst, Sussex (Day, *Quart. J. For.* **24**, 1930, 114–17).

Cryptodiaporthe salicella (Fr.) Wehm., *TBMS*, **17**, 1933, 281; Bisby & Mason, ibid. **24**, 1940, 159; syn. *Diaporthe spina* Fuckel, *Symb. myc.* 1869, 210; Sacc., *Syll.* i, 685.

*On stems of willow (*Salix*), York, 1927; Suffolk, 1931. Parasitism doubtful (*Report*, vii, 103). (See also *Discella carbonacea*.)

Cryptodiaporthe salicina (Curr.) Wehm., see *Discella carbonacea*.

Cryptomyces maximus (Fr.) Rehm, *Discomycetes*, in Rabenh., *Krypt. Fl.* iii, 107; Sacc., *Syll.* viii, 707; Ramsbottom & Browne, *TBMS*, **34**, 1951, 105.
*On willow (*Salix*). Reported causing twig girdling of *S. fragilis*, Scotland (Alcock & Maxwell, *Trans. R. Scot. arb. Soc.* **39**, 1925, 34–7; see also *TBMS*, **11**, 1926, 161–7); on cricket-bat willow (*S. alba* var. *coerulea*), Norfolk, 1950 (*Quart. J. For.* **46**, 1952, 1). (See also *Physalospora miyabeana*.)

Cryptosporella umbrina (Jenkins) Jenkins & Wehm., *Phytopathology*, **25**, 1935, 888; Bisby & Mason, *TBMS*, **24**, 1940, 144; syn. *Diaporthe umbrina* Jenkins, *J. agric. Res.* **15**, 1918, 596; Sacc., *Syll.* xxiv, 755.
*Brown canker of rose. Rare. Glos, Oct. 1931 (Ogilvie, *TBMS*, **17**, 1932, 153; see also *Mycologia*, **24**, 1932, 485–8); Ayrshire, 1953; Lewis, Aberdeen and Kincardineshire, 1956.

Cryptosporiopsis corticola (Edgert.) Nannf. = *Myxosporium corticola*.

Cryptosporium minimum Laub., *Zbl. Bakt.* II, **19**, 1907, 163; Sacc., *Syll.* xxii, 1234; Grove, *Coelomycetes*, ii, 302.
*Black blotch of raspberry (*Rubus*). Frequent, and weakly parasitic, but does not itself cause appreciable damage. Though attributed for many years to this fungus Eaton (*J. hort. Sci.* **25**, 1950, 128–31) has shown that the black blotches are caused primarily by the toxic action of iron compounds washed down to the canes by rain from rusty iron wire upon which the affected canes are trained. *C. minimum* often colonizes the damaged tissues.
*Black blotch of rose. Not common. Known since 1925.

Cryptostroma corticale (Ell. & Everh.) Gregory & Waller, *TBMS*, **34**, 1951, 594; syn. *Coniosporium corticale* Ell. & Everh., *J. Mycol.* **5**, 1889, 69; Sacc., *Syll.* x, 570.

*Sooty bark of sycamore (*Acer*). Was generally distributed in parks and cemeteries in the London area, with isolated attacks in neighbouring counties and further afield, including Guildford, Surrey and Mundford, Norfolk. Also on a felled sycamore pole, Som. Not known farther north or in Scotland. So far virulent only in Wanstead Park (north-east London). After a few years it began to abate and in most localities the fungus is evidently secondary. Seen also on two trees of field maple (*Acer campestre*).

Sooty bark was first noticed in 1945 on a dead sycamore tree in Wanstead Park (Gregory, Peace & Waller, *Nature, Lond.*, **164**, 1949, 275) where it caused extensive damage between 1948 and 1952. It was investigated by Gregory (*TBMS*, **34**, 1951, 579–97), but the pathogenicity of the fungus was not proved. The reason for the virulent attack in this area only is not known and by 1953 (Robertson, *Rep. For. Res., Lond., for 1953–4*, 1954, 57) the disease was tending to die out. Townrow (ibid. *for 1952–3*, 1954, 118–20) showed that the growth of the fungus is markedly affected by temperature, and Peace (*Quart. J. For.* **49**, 1955, 197–204), who gave details of special surveys made in 1949–50, considered the disease to be of no economic importance.

Cucurbitaria laburni (Pers. ex Fr.) de Notaris, *Erb. Critt. Ital.* no. 875; Sacc., *Syll.* ii, 308; Bisby & Mason, *TBMS*, **24**, 1940, 189.

*Branch canker of *Laburnum*. Common throughout Britain. Though formerly regarded as a wound parasite, Green (*TBMS*, **16**, 1932, 289–303) was unable artificially to induce disease with it, even through wounds.

Cucurbitaria piceae Borthwick, *Notes R. bot. Gdn Edinb.* **4**, 1909, 261; Sacc., *Syll.* xxii, 289.

*On buds of *Picea pungens* and on Sitka and Norway spruce, Scotland.

Cumminsiella mirabilissima (Peck) Nannf., *Fungi Exsicc. suecic. Praes. upsal.* xxxi–xxxii, 1947, 3; Wilson & Bisby, *TBMS*, **37**, 1954, 64; syn. *Puccinia mirabilissima* Peck, *Bot. Gaz.* **6**, 1881, 226; Sacc., *Syll.* vii, 620.

*Mahonia rust. Widely distributed and fairly common on *Mahonia aquifolium*, particularly in Scotland, Wales, mid- and south England. First seen in Scotland in 1922 (Wilson, *Trans. bot. Soc. Edinb.* **28**, 1922–3, 164; *TBMS*, **9**, 1924, 136) and later in Northumb (Miller, *Gdnrs' Chron.* **88**, 1930, 131) and Wales (Pethybridge, ibid. **88**, 1930, 312). Wilson (ibid. **87**, 1930, 132–3; *Ann. mycol., Berl.*, **28**, 1930, 225–9) discussed its distribution in Europe.

Curvularia trifolii (Kauff.) Boed., *Bull. Jard. bot. Buitenzorg*, **13**, 1933, 128.

*On *Freesia*. A species of *Curvularia* resembling *C. trifolii* was found in abundance forming a sooty covering to the corms and leaf bases of dying *Freesia* seedlings at Bristol in 1949. Parasitism uncertain.

Cylindrocarpon album (Sacc.) Wollenw., *Fusarium autographice delineata*, no. 652, 2nd ed. 1926; *Z. Parasitenk.* **1**, 1928, 153; Sacc., *Syll.* iv, 698 (as *Fusarium album* Sacc.); Wakefield & Bisby, *TBMS*, **25**, 1941, 63.

*Fruit rot of apple (*Malus*). Reported parasitic on stored apples by Kidd and Beaumont (*TBMS*, **10**, 1924, 117) under the name *Ramularia heteronema* (Berk. & Br.) Wollenw.

Cylindrocarpon mali (Allesch.) Wollenw. = stat. conid. of *Nectria galligena*.

Cylindrocarpon radicicola Wollenw., *Fusarium autographice delineata*, nos. 651–2, 1st ed. 1924; *Z. Parasitenk.* **1**, 1928, 166; Wakefield & Bisby, *TBMS*, **25**, 1941, 63.

Associated with root rot in a number of plants, but doubtfully parasitic.

*Strawberry (*Fragaria*). Common.

*Azalea. Associated with wilt, Kelsall, Ches, July 1956.

Chrysanthemum. Root and stem base rot. Staffs, Lancs, Ches, Northumb.

Gloxinia (Sinningia). Associated with corm rot (W. C. Moore & Tomlinson, *Plant Path.* **2**, 1953, 71).

Lilium. Ogilvie (*Lily Yearb.* **6**, 1937, 106–8) considered it non-parasitic on this host.

Narcissus. Root rot is widespread and troublesome in the Isles of Scilly and parts of Devon and Corn. Occasionally seen elsewhere. *Cylindrocarpon radicicola* is usually associated with it, but is not the primary cause, though it may play a part in association with root-lesion eelworms (*Pratylenchus* spp.).

Cylindrocladium scoparium Morgan, *Bot. Gaz.* **17**, 1892, 191; Sacc., *Syll.* xi, 600.

*Shoot wilt of *Prunus*. A species of *Cylindrocladium*, closely resembling *C. scoparium*, has caused wilting of the shoots in layer rows of plum, cherry, peach and apricot at East Malling, Kent (Wormald, *J. Pomol.* **20**, 1943, 80–3; *TBMS*, **27**, 1944, 71–80). Lupins were also affected.

*Associated with disease in rose (Bewley, *Rose Ann.* 1933, 110).

Cylindrosporium brassicae Fautr. & Roum. = *Cercosporella brassicae*.

Cylindrosporium chrysanthemi Ell. & Dearn., see *Septoria chrysanthemella*.

Cylindrosporium concentricum Grev., see *Gloeosporium concentricum*.

Cylindrosporium padi Karst., *Symbolae ad Mycologiam Fennicam*, xv, 159; Sacc., *Syll.* iii, 738; Grove, *Coelomycetes*, ii, 293.

*Leaf spot of plum (*Prunus*). Tay, Scotland.

Cylindrosporium pastinacae Lind = stat. conid. of *Phyllachora pastinacae*.

Cymadothea trifolii Wolf, *Mycologia*, 27, 1935, 71; syn. *Dothidella trifolii* Bayliss-Elliott & Stansf., *TBMS*, 9, 1924, 227; Bisby & Mason, *TBMS*, 24, 1940, 206; stat. conid. *Polythrincium trifolii* Fr., *Syst. Myc.* ii, 435; Sacc., *Syll.* iv, 350; Wakefield & Bisby, *TBMS*, 25, 1941, 86.

*Black blotch of clover (*Trifolium*). Widely distributed throughout Great Britain on crimson, red and white clover, but not very common. Does no appreciable damage. Known since 1892 (Whitehead, *Rep. Insects and Fungi injurious to crops 1892*, Bd Agric. 1893, 8). Life history studied by Bayliss-Elliott and Stansfield (*TBMS*, 9, 1924, 218–28).

Cystopus candidus (Pers. ex Chev.) Lév., *Ann. Sci. nat.* 3 Ser. 8, 1847, 371; Sacc., *Syll.* vii, 234.

*White blister of crucifers. Widely distributed in Britain, but showing marked seasonal occurrence. Common on cabbage, etc., especially in Wales and south-west England; frequent on *Lunaria*, turnip and swede; occasional on marrow-stem kale, cress, horse-radish, mustard, wallflower, *Alyssum* and *Arabis*. On *Aubretia* at several places in east Scotland, 1954 (Henderson, *Plant Path.* 4, 1955, 110) and in Jersey, 1954–5. Described many years ago by Berkeley (*J. hort. Soc.* 3, 1848, 265–71). Napper (*J. Pomol.* 11, 1933, 81–100) investigated specialization of parasitism by this species.

Cystopus cubicus (Strauss ex Unger) Lév., *Ann. Sci. nat.* 3 Ser. 8, 1847, 371; syn. *C. tragopogonis* Schröt., *Kr. Fl. Schles.* i, 234; Sacc., *Syll.* vii, 234.

*White blister of salsify (*Tragopogon*). Not infrequent. Som, Devon, Bucks, Cambs. The same species also occurs on *Scorzonera*.

Cystopus tragopogonis Schröt. = *C. cubicus*.

Cytospora ambiens Sacc. = stat. conid. of *Valsa ambiens*.

Cytospora chrysosperma Fr., *Syst. Myc.* ii, 542; Sacc., *Syll.* iii, 260; Grove, *Coelomycetes*, i, 272.

*On poplar (*Populus*). Very common on twigs, but usually on those that have died from some other cause.

*On willow (*Salix*). Common, but not usually markedly parasitic. Killing young trees, Glos, 1935.

Cytospora juglandina Sacc., *Syll.* iii, 267; Grove, *Coelomycetes*, i, 269.

*On walnut (*Juglans*), associated with die-back, Glos, 1930; Kew Gardens.

Cytospora leucostoma Sacc. = stat. conid. of *Valsa leucostoma*.

Cytospora rhodophila Sacc. = stat. conid. of *Valsa rhodophila*.

Cytospora taxifolia (Cooke & Massee) em. Pilat & Macal. = stat. conid. of *Sphaerulina taxi*.

Cytospora sp.

*Associated with die-back of apple trees (*Malus*), Ches, Glam, west and south-west England.

*Associated with die-back of apricot (*Prunus*), Yorks.

*Causing canker and die-back of peach shoots (*Prunus*) (see, for example, Cayley, *AAB*, **10**, 1923, 257).

Cytosporella fructorum Marchal, see *Phacidiella discolor*.

Cytosporina ribis Magn., *Ann. mycol., Berl.*, **1**, 1903, 508; Sacc., *Syll.* xviii, 406; Grove, *Coelomycetes*, i, 452.

*On black currant (*Ribes*), associated with die-back of shoots, Kent, 1941; Scotland. Doubtfully parasitic.

*Die-back of red currant. Plentiful at Long Ashton, Pershore and Wisbech, 1919 (Wiltshire, *Rep. Long Ashton Res. Sta. for 1919*, 30–3).

*Die-back of gooseberry. A weak parasite formerly given some attention (Brooks & Bartlett, *Ann. mycol., Berl.*, **8**, 1910, 167–85; Blackman & Jones, *Rep. on Gooseberry Diseases in E. Sussex 1922–23*, Lewes, 1923, 12 pp.; *J. Bd Agric.* **16**, 1909, 34–5).

Dactylium dendroides Fr., *Syst. Myc.* iii, 413; Sacc., *Syll.* iv, 189; Wakefield & Bisby, *TBMS*, **25**, 1941, 90.

*Cobweb disease of mushroom (*Agaricus*). The fungus spreads in cobweb fashion over mushrooms from the casing soil. More common than at one time thought, especially in southern England. Gathercole (*M.G.A. Bull.* no. 88, 1957, 138–41) and Wood (*Mushroom News*, **6**, 1957, 256–7) tested pentachloronitrobenzene for controlling it.

Daldinia concentrica (Fr.) Ces. & de Not., *Schema Sfer.* 1863, 198; Sacc., *Syll.* i, 393; Bisby & Mason, *TBMS*, **24**, 1940, 150.

*Branch rot of ash (*Fraxinus*). Frequent. The formation of conidia and the growth of stromata were described by Bayliss-Elliott (*TBMS*, **6**, 1920, 269–73) and Ingold (ibid. **29**, 1946, 43–51; **39**, 1956, 378–80) studied spore discharge.

Dasyscypha calyciformis (Willd.) Rehm, *Discomycetes* in Rabenh., *Krypt. Fl.* I, 3, 834; Sacc., *Syll.* viii, 467 and 1143; Ramsbottom & Browne, *TBMS*, **34**, 1951, 83.

*On spruce (*Picea*). On trunk of *P. abies*, Keir, Stirlingsh; Peebles (Wilson, *TBMS*, **7**, 1922, 79).

Dasyscypha calycina (Schum.) Fuckel = *Trichoscyphella wilkommii*.

Dasyscypha wilkommii (Hartig) Rehm, see *Trichoscyphella wilkommii*.

Dematium pullulans de Bary, *Beiträge zur Morphologie und Physiologie der Pilze*, ii, 1866, 182; Sacc., *Syll.* iv, 351 (as a synonym of *Cladosporium herbarum* (Pers.) Link).

*Sooty mould of *Erica*. A sooty mould, provisionally identified as this species, was troublesome on several acres of cultivated *E. lusitanica* in Corn in 1933 (*Report*, viii, 84); on leaves, stems and petals, Corn, 1954.

Hoggan (*TBMS*, **9**, 1923, 100–7) studied this fungus, and Bennett (*AAB*, **15**, 1928, 371–91) considered its perfect stage to be *Anthostomella pullulans* (de Bary) Bennett.

Dematophora necatrix Hartig = stat. conid. of *Rosellinia necatrix.*

Dermatella prunastri Dowson, see *Dermea prunastri.*

Dermea prunastri (Pers. ex Fr.) Fr., *Summ. Veg. Scand.* 1849, 362; Groves, *Mycologia*, **38**, 1946, 406; Sacc., *Syll.* viii, 556 (as *Cenangium prunastri* (Pers.) Fr.).

*Die-back of plum (*Prunus*). Cambs, Worcs, Hants, Devon. At most a weak parasite. Dowson (*New Phytol.* **12**, 1913, 207–16) described this fungus, under the name *Dermatella prunastri* Pers., as the cause of die-back in greengage plums. Groves (loc. cit.) pointed out that Dowson and not Persoon was the first to use this combination.

Dialonectria galligena (Bres.) Petch = *Nectria galligena.*

Dialonectria peziza (Tode ex Fr.) Cooke, *Grevillea*, **12**, 1884, 110; Petch, *TBMS*, **21**, 1938, 263; Bisby & Mason, ibid. **24**, 1940, 196; syn. *Nectria peziza* (Tode) Fr., *Summ. Veg. Scand.* 1849, 388; Sacc., *Syll.* ii, 501.

*In mushroom (*Agaricus*) beds. Frequent in 1956 and mostly confined to peat casing, especially where peat steamed before use (*Mushroom News*, **6**, 1956, 180).

Diaporthe castanea Sacc. = *Cryptodiaporthe castanea.*

Diaporthe eres Nits., *Pyrenomycetes germanici*, 1870, 246; Sacc., *Syll.* i, 631; Wehmeyer, *TBMS*, **17**, 1933, 252; syn. *D. scobina* Nits., *Pyrenomycetes germanici*, 293; Sacc., *Syll.* i, 676; Bisby & Mason, *TBMS*, **24**, 1940, 164; stat. conid. *Phomopsis scobina* Höhnel, *S.B. Akad. Wiss. Wien*, **115**, 1906, 681; Grove, *Kew Bull.* 1917, 64; *Coelomycetes*, i, 188; Sacc., *Syll.* x, 147 (as *Phoma scobina* Cooke).

*On twigs and petioles of ash (*Fraxinus*). Rather common in Britain, usually saprophytic but, according to Macdonald and Russell (*Trans. bot. Soc. Edinb.* **32**, 1937, 341–52), also parasitic, causing twig die-back and stem cankers. Grove (loc. cit.) considered *Phomopsis controversa* (Sacc.) Trav., also recorded on *Fraxinus excelsior*, to be identical, but Macdonald and Russell (loc. cit.) regarded the two as distinct species, differing among other things in pathogenicity, *Phomopsis scobina* being the more active.

Diaporthe leiphaemia (Fr.) Sacc., *Mycologiae Venetae Specimen*, 1873, 135; Sacc., *Syll.* i, 615; Wehmeyer, *TBMS*, **17**, 1933, 275; Bisby & Mason, ibid., **24**, 1940, 162.

*On oak (*Quercus*). Normally saprophytic but occasionally parasitic on trees weakened by other causes. On trees weakened by drought, Harling, Norfolk.

Diaporthe perniciosa Marchal, *Bull. Soc. Bot. Belg.* **54**, 1921, 109; Bisby & Mason, *TBMS*, **24**, 1940, 163; stat. conid. *Phomopsis perniciosa* Grove, *Coelomycetes*, i, 214. Wehmeyer (*The Genus Diaporthe Nitschke and its Segregates*, 1933, 89) lists this as a synonym of *D. eres* Nits.

This fungus was at one time regarded as the main cause of die-back in plum and other fruit trees (Cayley, *AAB*, **10**, 1923, 253–75; Marsh & Nattrass, *Rep. Long Ashton Res. Sta. for 1927*, 93–8; Amos, Hatton & Mackenzie, *Rep. E. Malling Res. Sta. for 1925*, II Suppl. 13, 1927, 33–7), but much of this is now known to have been bacterial canker (*Pseudomonas mors-prunorum* Wormald). Most of the records referred to below were made before the bacterial disease had been differentiated. Cayley (*J. Genet.* **13**, 1923, 353–70) described the behaviour of *D. perniciosa* in culture.

*On fruit trees. This species has been found in association with die-back of apple trees (Briton-Jones, *J. Pomol.* **4**, 1925, 162–83) and causing a late fruit rot in stored apples (Kidd & Beaumont, *TBMS*, **10**, 1924, 101–4). It has also been recorded as causing a fruit rot of peach (Cumb, Salop, Glos, Worcs) and quince (Som, Kent); in association with branch die-back of pear (occasional) and black currant (Worcs, Herefs) and with a crown rot of raspberry and blackberry (Bristol area, 1925), and on medlar (Wilts, 1923).

*On lilac (*Syringa*) at Cambridge, but Deighton (*TBMS*, **12**, 1927, 70–3) could not establish its parasitism on this host.

Diaporthe scobina Nits. = *D. eres.*

Diaporthe spina Fuckel = *Cryptodiaporthe salicella.*

Diaporthe taleola (Fr.) Sacc., *Fung. Ven.* 4, 1875, 12; Sacc., *Syll.* i, 626; Wehmeyer, *TBMS*, **17**, 1933, 278; Bisby & Mason, *TBMS*, **24**, 1940, 164.

*On oak (*Quercus*). Normally saprophytic but occasionally parasitic on trees weakened by other causes. Associated with bark lesions and premature death of young trees in Bardney Forest, Lincs.

Diaporthe umbrina Jenkins = *Cryptosporella umbrina.*

Didymascella tetraspora (Phill. & Keith) Maire, *Bull. Soc. Hist• nat. Afr. N.* **18**, 1927, 117; syn. *Keithia tetraspora* (Phill. & Keith) Sacc., *Syll.* x, 50; Ramsbottom & Browne, *TBMS*, **34**, 1951, 105; Phillips, *British Discomycetes*, 1887, 388 (as *Phacidium tetrasporum* Phill. & Keith).

*Juniper. On upper leaves of *Juniperus communis*, Forres (Phillips & Keith, *Gdnrs' Chron.* 14, 1880, 308).

Didymascella thujina (Durand) Maire, *Bull. Soc. Hist. nat. Afr. N.* **18**, 1927, 117; syn. *Keithia thujina* Durand, *Mycologia*, **5**, 1913, 9; Sacc., *Syll.* xxiv, 1263; Ramsbottom & Browne, *TBMS*, **34**, 1951, 105.

*Leaf blight of *Thuja*. Widely distributed in the British Isles. It causes a serious nursery disease of *T. plicata* and *T. occidentalis.* Common also on plantation *Thujae* but seldom damaging outside the nursery. First English record (on *T. plicata*), Horsham, Sussex, 1919 (*Report*, iii, 61).

Pethybridge (*Quart. J. For.* 13, 1919, 93–7) and Miles (*Gdnrs' Chron.* **72**, 1922, 353) dealt with its occurrence in Ireland, and Alcock (*Scot. For. J.* **42**, 1928, 77–9) in Britain. Peace (*Rep. For. Res., Lond., 1954*, 144–8) discussed the possibility of control by raising disease-free plants in isolated nurseries; and Pawsey (*TBMS*, **40**, 1957, 166) overwintering in the ascospore stage.

Didymascella tsugae (Farlow) Maire, *Bull. Soc. Hist. nat. Afr. N.* **18**, 1927, 117; syn. *Keithia tsugae* (Farlow) Durand, *Mycologia*, **5**, 1913, 10; Sacc., *Syll.* viii, 668 (as *Propolidium tsugae* (Farlow) Sacc.).

*Leaf spot of *Tsuga*. On one tree of *T. canadensis*, Tweed (Wilson, *Scot. For. J.* **51**, 1937, 46–7).

Didymella applanata (Niessl) Sacc., *Syll.* i, 546; Bisby & Mason, *TBMS*, **24**, 1940, 164.

*Spur blight of raspberry (*Rubus*). Very common and widespread, but rarely serious except in Scotland and northern England. Sometimes also causes leaf blotching. Not infrequent on loganberry.

First recorded in Scotland in 1931 and described there by Foister and Gregor (*Scot. J. Agric.* **21**, 1938, 163–6). The fungus may follow midge attack (see under *Leptosphaeria coniothyrium*).

Didymella lycopersici Kleb., *Z. PflKrankh.* **31**, 1921, 1–16; Bisby & Mason, *TBMS*, **24**, 1940, 165; stat. conid. *Diplodina lycopersici* Hollós, *Ann. hist.-nat. Mus. hung.* **5**, 1907, 461; Sacc., *Syll.* xxii, 1040; syn. *Ascochyta lycopersici* Brun., *Bull. Soc. bot. Fr.* **34**, 1887, 430; Sacc., *Syll.* x, 304; Grove, *Coelomycetes*, i, 314.

*Stem and fruit rot of tomato (*Lycopersicon*). A major disease of the crop in all parts of England and Wales, both under glass and in the open. A few outbreaks have been seen in Scotland. Serious also out-of-doors in Jersey, and common there and in Guernsey under glass.

Stem rot has been known in England since 1885 (*Grevillea*, **13**, 94). Between 1906 and 1909 it was very destructive in the Lea Valley (Herts), but subsequently decreased in virulence, and until 1938, though still about, was of no great significance. The following year it flared up again and, apart from a minor lessening in 1946–8, has been very destructive since 1942.

The disease was described by Massee (*Kew Bull.* 1909, 292–3) and by Brooks and Price (*New Phytol.* **12**, 1913, 13–21), and various strains of the fungus were carefully studied by Brooks and

Searle (*TBMS*, **7**, 1921, 173–97). After that it received no attention until the beginning of the 1939–45 War. Small (*Rapp. aux États de Jersey pour 1938*, 31–4; *pour 1939*, 22–32) discussed its occurrence in Jersey and gave a good general account of the disease on outdoor crops (*Agriculture, Lond.*, **50**, 1943, 64–7). Oyler and Read (*Gdnrs' Chron.* **112**, 1942, 120) began studying it at Cheshunt under glass, and there are many references to it in the *Reps. exp. Res. Sta. Cheshunt* from 1942 onwards. Williams, Sheard and Read (*J. hort. Sci.* **28**, 1953, 278–94) summarized the main results of all this work.

The perithecial stage is rarely seen. It was found by Hickman (*Nature, Lond.*, **154**, 1944, 708) during an investigation of the disease in Worcs (*J. Pomol.* **22**, 1946, 69–75), and Fisher (*Plant Path.* **4**, 1955, 71) found it again in War in 1954 and in 1956.

Ogilvie (*Gdnrs' Chron.* **118**, 1945, 71–2) discussed seed transmission of the disease, and Fisher (*Nature, Lond.*, **174**, 1954, 656; *Rep. nat. Veg. Res. Sta. Warwick for 1953*, 15–16; *for 1954*, 60) showed how infection of the seed takes place. In Jersey nearly one-third of the seed samples are infected, but the fungus mostly dies within a year and seed transmission is unimportant there (Phillips, *TBMS*, **39**, 1956, 319–29) compared with soil transmission when crop remains are ploughed in (Phillips, ibid. **39**, 1956, 330–40).

Croxall (*Rep. Long Ashton Res. Sta. for 1944*, 161–6) paid special attention to its possible control by copper sprays, and Dennis (*TBMS*, **29**, 1946, 11) studied the cultural characteristics of the fungus. Taylor (*J. Sci. Food Agric.* **7**, 1956, Suppl. S 82–S 88) referred to the extended use of ethyl mercury phosphate in controlling the disease in the Vale of Evesham, while Croxall, Norman and Gwynne (*Plant Path.* **6**, 1957, 27–31) obtained evidence that the susceptibility of tomato plants was increased when they were watered with 2:4:6-trichlorophenoxyacetic acid (2, 4, 6-T). Williams and Hack (*AAB*, **45**, 1957, 304–11) investigated the effect of soil steaming and other treatments on the incidence of stem rot. Day *et al.* (*Plant Path.* **5**, 1956, 150–1) could find no basis for resistance among forty lines of five species of *Lycopersicon*.

*Cucumber canker. Occasional. See Collinge, *Rep. econ. Biol.*
1, 1911, 45–6; Cumb, 1910; Herts, 1911; Sussex, 1920–1; Ches,
1923; Yorks and Northumb, 1950. It is probable that the fungus
concerned, at least in the early attacks, was *Mycosphaerella
citrullina* (q.v.).

*Leaf stalk rot of strawberry. Under glass, Cheshunt, Herts,
1925 (Small, *J. Pomol.* **7**, 1928, 212–15; *Rep. exp. Res. Sta.
Cheshunt for 1926*, 38–9; *for 1927*, 45–6). In the open, Ipplepen,
Devon, 1951 and 1953.

*Causing rot in *Gloxinia* (*Sinningia*) just above soil level,
Bowscar, Cumb, June, 1931.

*Stem canker of winter cherry (*Solanum capsicastrum*), Lancs,
1938. On *Solanum* sp., Yorks, 1950 (as *Diplodina* sp.).

Didymella sepincoliformis (de Not.) Sacc., *Syll.* i, 551; Bisby &
Mason, *TBMS*, **24**, 1940, 165.

*This species was regarded by Deacon (*Rose Annu. for 1941*,
113–15) as the cause in England of a wilt and die-back of young
lateral and terminal shoots of *Rosa* spp., including *R. willmottiae,
R. ecae, R. canina* and *R. arvensis.*

Didymella sp.
*On chrysanthemum stem cankers, Ches, May, 1935.

Didymellina dianthi C. C. Burt, *TBMS*, **20**, 1936, 214; Bisby &
Mason, ibid. **24**, 1940, 165; stat. conid. *Heterosporium echinu-
latum* (Berk.) Cooke, *Grevillea*, **5**, 1877, 122; Sacc., *Syll.* iv, 481;
Wakefield & Bisby, *TBMS*, **25**, 1941, 92.

*Ring spot or fairy ring of carnation (*Dianthus*). Fairly common
under glass and in the open. Described by M. J. B[erkeley]
(*Gdnrs' Chron.* **30**, 1870, 382) and W. G. Smith (ibid. **26**, 1886,
244).

*Leaf spot of sweet william (*Dianthus*). The imperfect stage of
the fungus occurs regularly in many districts. Disease described
by Burt (loc. cit. 207–15). The fungus from sweet william does
not infect carnation.

Didymellina iridis (Desm.) Höhnel, see *D. macrospora.*

Didymellina macrospora Kleb., *Ber. dtsch. bot. Ges.* **42**, 1924, 60; stat. conid. *Heterosporium gracile* Sacc., *Syll.* iv, 480; Wakefield & Bisby, *TBMS*, **25**, 1941, 93. The perfect stage has not yet been seen in Britain.

*Leaf spot of *Iris*. Very common everywhere in England and south Scotland. Sometimes severe on *I. germanica* and the cultivated rhizomatous forms. Also occurs in epidemic form most years on bulbous irises in south-west England and the Isles of Scilly, with severe attacks in some seasons in Flints and elsewhere in Wales.

First recorded in England by Cooke (*Gdnrs' Chron.* **15**, 1894, 718; *J. R. hort. Soc.* **26**, 1901, 450) and later studied by Ramsbottom (ibid. **40**, 1915, 481). W. C. Moore (*Diseases of Bulbs*, 1939, 128) gave a full description of the disease and included reasons for preferring the name *Didymellina macrospora* to *D. iridis* (Desm.) Höhnel (see Tisdale, *Phytopathology*, **10**, 1920, 148–63) for the perfect stage.

*Leaf spot of *Gladiolus*. Reported on this host from several districts in Scotland (Tweed, Forth, Clyde, Tay) and occasionally from England (*Gdng ill.* 17 Sept. 1927, 588; Corn, 1950; Devon, 1951).

*Leaf spot of *Schizostylis*. Falmouth, Devon, 1949. *Heterosporium* sp. has been reported on this host in Scotland, Wigtownsh, 1955.

Dilophospora alopecuri (Fr.) Fr., *Summ. Veg. Scand.* 1849, 419; Sacc., *Syll.* iii, 600 (as *D. graminis* Desm.); Grove, *Coelomycetes*, i, 449.

*Twist of wheat (*Triticum*). Uncommon and localized in England and Wales. Reported from time to time since 1862 (Berkeley, *Gdnrs' Chron.* **22**, 1862, 1009), usually on wheat, twice on oats (*Report*, iv, 17; Devon, 1929), occasionally on grasses, including *Alopecurus* and *Holcus*. May cause damage to wheat. In Scotland known on meadow grass (Trail, *Scot. Nat.*, n.s., **4**, 305), *Agrostis* (Moray), *Holcus* (widely distributed), *Dactylis* and *Cynosurus cristata* (Clyde, Dumfries).

The disease is sometimes associated on the same plants with

Anguina tritici (Steinb.) Filipjev. Sampson and Western (*TBMS*, **22**, 1938, 168–73) could find no evidence for regarding *Dilophospora alopecuri* and *Mastigosporium album* (q.v.) as genetically related.

Dilophospora graminis Desm. = *D. alopecuri*.

Diplocarpon earlianum (Ell. & Everh.) Wolf, *J. Elisha Mitchell sci. Soc.* **39**, 1924, 141; syn. *Fabraea fragariae* Kleb., *Ber. dtsch. bot. Ges.* **42**, 1924, 192; stat. conid. *Marssonina fragariae* (Sacc.) Kleb., loc. cit.; syn. *Marsonia fragariae* Sacc., *Malpighia*, 1896, 276; Sacc., *Syll.* xiv, 1021; Grove, *Coelomycetes*, ii, 279.

The perfect stage has not been found in Britain.

*Leaf scorch of strawberry (*Fragaria*). Less common than *Mycosphaerella fragariae* (q.v.), but widely distributed throughout Britain and sometimes damaging in the north.

Diplocarpon rosae Wolf, *Bot. Gaz.* **54**, 1912, 231; Sacc., *Syll.* xxiv, 911; Bisby & Mason, *TBMS*, **24**, 1940, 207; stat. conid. *Actinonema rosae* (Lib.) Fr., *Summ. Veg. Scand.* 424; Sacc., *Syll.* iii, 408; Grove, *Coelomycetes*, ii, 270.

The perfect stage has not been found in Britain.

*Black spot of rose. Widely distributed throughout Britain, common, and in some districts severe in wet summers.

The life history of the fungus was studied by Alcock (*Kew Bull.* 1918, 193–7). Green (*J. R. hort. Soc.* **56**, 1931, 18–30) investigated the incidence and control of the disease, as well as varietal resistance to it (ibid. **57**, 1932, 58–62). Downes (*Gdnrs' Chron.* **92**, 1932, 358–9) and Shelley (*Rose Annu.* 1936, 118–20) also discussed its control. Bewley (*Sci. Hort.* **6**, 1938, 97–101) published notes on the disease.

Diplodia aucubae West., *Bull. Acad. Belg.* II Ser. **12**, no. 7; Sacc., *Syll.* iii, 361.

*On *Aucuba japonica*, Whitchurch, Glam, 1945.

Diplodia griffoni Sacc. & Trav., *Syll.* xx, 1228; xxii, 994; Grove, *Coelomycetes*, ii, 54.

*Associated with die-back of apple (*Malus*). Sussex (Alcock, *TBMS*, **8**, 1923, 190).

Diplodia opuntiae Sacc., *Mich.* ii, 267; *Syll.* iii, 344.

*Cactus scab. On *Phyllocactus*, Isleworth, Middx (Massee, *Gdnrs' Chron.* **38**, 1905, 125).

Diplodia pinea (Desm.) Kickx, *Fl. Crypt. Flandr.* **1**, 1867, 397; Sacc., *Syll.* iii, 359; Grove, *Coelomycetes*, ii, 50.

*On leaves and branches of *Pinus*. Once considered rather important in southern England (Bancroft, *Kew Bull.* 1911, 60–2; see also *J. Bd Agric.* **14**, 1907, 164–6), but has attracted no attention in recent years.

Diplodia taxi (Sow.) de Not., *Micromycetes Italici*, Decade IV; Sacc., *Syll.* iii, 359; Grove, *Coelomycetes*, ii, 60.

*On leaves of yew (*Taxus baccata*). Widely distributed.

Diplodina castaneae Prill. & Delacr., *Bull. Soc. mycol. Fr.* **9**, 1893, 276; Sacc., *Syll.* xi, 527; Grove, *Coelomycetes*, i, 333.

*On chestnut (*Castanea*). This species occurs in the chestnut areas of south-east England and appears to cause symptoms closely similar to those of the Javart disease (see *Cryptodiaporthe castanea*), but it is no longer regarded as synonymous with *Fusicoccum castaneum*.

Diplodina delphinii Laskaris, *Phytopathology*, **40**, 1950, 620.

*Stem rot of *Delphinium*. A species of *Diplodina* closely related to, and probably identical with this has been widely distributed in southern England since 1941 as the cause of a stem rot of annual larkspur. Also seen on perennial *Delphinium* in Worcs and Hants.

Diplodina eurhododendri Voss, *Mat. Pilzfl.Krains.* **5**, 229; Sacc., *Syll.* x, 312; Grove, *Coelomycetes*, ii, 359.

*Leaf spot of *Rhododendron*. Ches, Corn. Howarth and Chippendale (*Gdnrs' Chron.* 86, 1929, 471) found this species, but their inoculation experiments showed it to be innocuous.

Diplodina lycopersici Hollós = stat. conid. of *Didymella lycopersici*.

Diplodina passerinii Allesch., *Rab. Krypt. Fl.* I, 6, 1900, 678; Sacc., *Syll.* x, 300 (as *D. decipiens* Passer.); Grove, *Coelomycetes*, i, 332.
*Wilt of *Antirrhinum*. Cambs, Bucks, Surrey, Herefs, War, Staffs, Clyde, Forth.
*Wilt of cornflower (*Centaurea*). On *C. cyanus*, Newcastle, Northumb, 1950.
*On *Delphinium* and larkspur. Frequent.
*Wilt of *Godetia*. Cambridge (Taylor, *AAB*, 28, 1941, 91–101); Worcs, 1953.
*Foot rot of sweet sultan (*Centaurea*). On *C. moschata*, Cambridge; Suffolk, 1950.

Diplodina salicis West., see *Discella carbonacea*.

Discella carbonacea (Fr.) Berk. & Br., *Ann. mag. nat. Hist.* 5, 1850, 377; Sacc., *Syll.* iii, 687; Grove, *Coelomycetes*, ii, 148.
*Branch canker of willow (*Salix*). On *S. caerulea*, Kent, 1931 (*Report*, vii, 103); Melrose, Scotland, 1951.
The fungus is common throughout Britain on dead twigs of *Salix*. It is said to be the pycnidial stage of *Diaporthe salicella* Sacc. = *Cryptodiaporthe salicina* (Curr.) Wehm., *TBMS*, 17, 1933, 282. Grove (*Coelomycetes*, i, 337) considered *Diplodina salicis* West. (*Cinq. Not.* 19; Sacc., *Syll.* iii, 411) to be identical.

Dothichiza populea Sacc. & Briard, Sacc., *Syll.* iii, 672.
*Canker and die-back of poplar (*Populus*). The fungus occurs widely in Britain but its status as a parasite is uncertain. May severely damage over-lush, newly planted trees. Much of the cankering formerly attributed to it was probably bacterial canker

(*Pseudomonas syringae* forma *populae*). Hiley (*Bull. For. Comm., Lond.*, no. 5, 1923, 47–50) reported the disease in Lincs (on *P. nigra italica*) and Berks. See also ibid. no. 19, 1952, 43.

Dothidella ribesia (Pers. ex Fr.) Thiess. & Syd. = *Plowrightia ribesia.*

Dothidella trifolii Bayliss-Elliott & Stansf. = *Cymadothea trifolii.*

Dothidella ulmi (Fr.) Wint. = *Systremma ulmi.*

Echidnodes aulographoides (Bomm., Rouss. & Sacc.) Robertson, *TBMS*, **33**, 1950, 107; syn. *Lembosia aulographoides* Bomm., Rouss. & Sacc., *Syll.* ix, 1107.
*On living twigs of *Rhododendron*, Dawick, Peeblesh, 1944–5. At most weakly pathogenic.

Elsinoe ampelina Shear, *Phytopathology*, **19**, 1929, 677; Bisby & Mason, *TBMS*, **24**, 1940, 207; stat. conid. *Sphaceloma ampelinum* de Bary, *Annal. Oenolog.* **4**, 1874, 165; syn. *Gloeosporium ampelophagum* (Passer.) Sacc., *Mich.* i, 217; *Syll.* iii, 719; Grove, *Coelomycetes*, ii, 228.
The perfect stage has not been found in Britain.
*Vine anthracnose. Comparatively rare. Northumb, Cumb, Devon, Kent (Hyams, *Gdnrs' Chron.* **129**, 1951, 4–5).
In 1893 Cooke (*Gdnrs' Chron.* **14**, 1893, 33) said the disease was not known in Britain, but it was recorded soon after (ibid. **18**, 1895, 135). Various records in the older English horticultural papers to the occurrence of *Guignardia bidwelli* (Ellis) Viala & Ravaz (Bisby & Mason, *TBMS*, **24**, 1940, 145) in Britain probably refer to the conidial state of this species.

Elsinoe veneta (Burkh.) Jenkins, *J. agric. Res.* **44**, 1932, 696; Bisby & Mason, *TBMS*, **24**, 1940, 207; syn. *Plectodiscella veneta* Burkh., *Phytopathology*, **7**, 1917, 91; Sacc., *Syll.* xxiv, 1141.
The imperfect stage is usually listed as *Gloeosporium venetum*

Speg., *Mich.* i, 477; Sacc., *Syll.* iii, 706; Grove, *Coelomycetes,* ii, 225.

*Cane spot of raspberry and loganberry (*Rubus*). Known since 1925 and now widely distributed in Britain, but not a serious problem in raspberry cultivation. Also infects the leaves, leaf stalks and fruit. Sometimes very harmful to loganberry and other hybrid berries, but rarely troublesome on true blackberry. Occasional on lowberry.

Harris described the disease and its control on raspberry canes (*Rep. E. Malling Res. Sta. for 1926–7*, 57–63; *J. Pomol.* 9, 1931, 73–99) and fruits (*Rep. E. Malling Res. Sta. for 1932*, 86–9). Shaw (*J. Soc. chem. Ind., Lond.,* 58, 1939, 65) showed that copper oxychloride was an efficient substitute for Bordeaux Mixture, which leaves a disfiguring spray residue on the fruit.

Research on cane spot of loganberry was reviewed by Harris, Beakbane and Moore (*Rep. E. Malling Res. Sta. for 1939*, 68–9). Control by a system of weaving or fan training the canes was described by Beakbane and Bagenal (ibid. *for 1940*, 89–91) and by Beakbane (*J. Pomol.* 18, 1941, 379–93). The control of the disease in Scotland has been discussed by Cadman (*Scot. Agric.* 29, 1950, 220–6).

Endophyllum sempervivi (Alb. & Schw.) de Bary, *Morphol.* 1884, 304; Sacc., *Syll.* vii, 767; Grove, *Brit. Rust Fungi*, 335; Wilson & Bisby, *TBMS,* 37, 1954, 64.

*Sempervivum rust. Occasionally seen in England since 1879 (W. G. Smith, *Gdnrs' Chron.* 13, 1880, 660, 725 and 815), usually in the midlands. Occurs on *S. arachnoideum, S. calcareum, S. globiferum, S. montanum, S. tectorum,* etc. The life history of the fungus was studied by Ashworth (*TBMS,* 19, 1935, 240–58).

Entomosporium maculatum Lév. = stat. conid. of *Fabraea maculata.*

Entyloma calendulae (Oudem.) de Bary, *Bot. Ztg,* 32, 1874, 105; Sacc., *Syll.* vii, 492; Sampson, *TBMS,* 24, 1940, 301; Ainsworth & Sampson, *Brit. Smut Fungi,* 102.

Calendula smut. Common in south-west England. Occasional elsewhere in Britain. Hants, Kent, Norfolk, Suffolk, Forth, Dee, south-west Scotland.

Entyloma calendulae (Oudem.) de Bary f. **dahliae** (Syd.) Viégas, *Bragantia*, **4**, 1944, 748; Ainsworth & Sampson, *Brit. Smut Fungi*, 104; syn. *E. dahliae* Syd., *Ann. mycol.*, *Berl.*, **10**, 1912, 36; Sacc., *Syll.* xxiii, 624; Sampson, *TBMS*, **24**, 1940, 301.

Dahlia smut. Widely distributed and common throughout Britain, but rarely causes severe damage. First recognized at Worplesdon, Surrey (Pethybridge, *Gdnrs' Chron.* **84**, 1928, 393). Studied by Green (*J. R. hort. Soc.* **57**, 1932, 332–9).

Entyloma dahliae Syd. = *E. calendulae* f. *dahliae*.

Entyloma fergussoni (Berk. & Br.) Plowr., *Brit. Ured. Ustil.* 1889, 289; Sacc., *Syll.* vii (2), 488 (as *E. canescens* Schröt.); Sampson, *TBMS*, **24**, 1940, 301; Ainsworth & Sampson, *Brit. Smut Fungi*, 105.

Myosotis smut. Rare. Reported on *M. scorpioides*, *M. arvensis* and *M. caespitosa*, Norfolk (Ellis, *Trans. Norfolk Norw. Nat. Soc.* **16**, 1946, 172).

Epichloe typhina (Pers. ex Fr.) Tul., *Sel. Fung. Carp.* iii, 1865, 24; Sacc., *Syll.* ii, 578; Bisby & Mason, *TBMS*, **24**, 1940, 203.

*Choke of grasses. Widely distributed in England and Wales; common in hedgebank grasses and locally abundant in pastures, but not important except in some seed crops. The disease is best known on cocksfoot (*Dactylis*), but occurs on other grasses including timothy (*Phleum*), fescues (*Festuca*), meadow foxtail (*Alopecurus*), *Agrostis*, *Poa* and *Agropyron*.

Choke was described by Sampson (*Nature, Lond.*, **121**, 1928, 92; *Gdnrs' Chron.* **85**, 1929, 280–1), who also studied its method of systemic infection (*TBMS*, **18**, 1933, 30–47). Large (*Plant Path.* **1**, 1952, 23–8; **3**, 1954, 6–11) summarized the results of a special three years' survey (1951–3) of seed crops in England and Wales, which showed that the incidence of the disease increases with the age of stand. The overall annual loss of seed yield due to choke was about 3 %; there was negligible loss in first year crops, an average of 1·3 % in second year, 7 % in third year, 9 % in fourth year and 21 % in 5–7-year-old crops. No varietal, seasonal or regional differences were revealed.

Ingold (*TBMS*, **31**, 1948, 277–80) studied the water relations of spore discharge in *Epichloe*.

Erostrotheca multiformis Martin & Charles, see *Ramularia deusta*.

Erysiphe cichoracearum DC., *Flor. Fr.* ii, 1805, 274; Salmon, *Monogr. Erysiph*. 1900, 193; Bisby & Mason, *TBMS*, **24**, 1940, 135.

Powdery mildew of:

*Cucumber (*Cucumis*). Common; also on gherkins. Read (*Rep. exp. Res. Sta. Cheshunt for 1951*, 30–2) studied its control by chemicals.

*Salsify (*Tragopogon*). Herefs, 1929 (*Report*, vii, 64); Kent, 1943.

*Vegetable marrow (*Cucumis*). Common in August and September.

*Melon (*Cucumis*). Rare. Farnham, Surrey, 1931.

*Artemisia. Evesham, Worcs, 1953.

*Campanula. Bucks, 1937.

*Comfrey (*Symphytum*). On Russian comfrey, Mildenhall, Suffolk, Sept. 1956. The pathogen was parasitized by a species of *Cicinnobolus*.

*Michaelmas daisy (*Aster*). Widely distributed; common late in the season.

*Myosotis. Common in England and Wales, especially in the south-west, in the *Oidium* stage; Forth, Tay, Clyde. It is doubtful if perithecia have been seen in Britain on this host.

*Phlox. Uncommon. Ches, 1933.

*Scorzonera. Long Ashton, Som, 1918.

**Valeriana officinalis* (cultiv.). Herts, Aug. 1943.

*Verbascum. Occasional.

Erysiphe graminis DC., *Flor. Fr.* vi, 1815, 106; Salmon, *Monogr. Erysip.* 1900, 209; Sacc., *Syll.* i, 19; Bisby & Mason, *TBMS*, **24**, 1940, 135.

*Cereal mildew. Occurs throughout Britain on wheat, barley and oats, less commonly on rye; and on most grasses, including cocksfoot (*Dactylis*), ryegrass (*Lolium*), *Bromus mollis*, and occasionally meadow grass (*Poa*), fescue (*Festuca*) and couch (*Agropyron*). Seasonal and in some years severe, but economic importance still uncertain.

Mildew was recorded over a century ago (Graham, *Gdnrs' Chron.* **12**, 1852, 501) but has been given little attention until recently. Salmon (*Ann. mycol., Berl.*, **2**, 1904, 70–99) tested the susceptibility of different barley species and varieties, and varietal susceptibility has been studied in wheat by Ellerton (*Nature, Lond.*, **153**, 1944, 776) and in oats by Jones and Griffiths (*TBMS*, **35**, 1952, 71–80). Corner (*New Phytol.* **34**, 1935, 180–200) made observations on the stage of growth at which infection is checked on resistant cereal varieties, and Lupton (*TBMS*, **39**, 1956, 51–9) demonstrated at least three stages at which the parasite may be checked when a resistant variety is inoculated. Kent (*AAB*, **28**, 1941, 189) showed that the addition of lithium salts to the soil increased resistance in wheat. In some wheat varieties the presence of mildew increases susceptibility to *Puccinia triticina* (Manners & Gandy, ibid. **41**, 1954, 393–404).

The ecology of the fungus has been studied by Grainger (*TBMS*, **31**, 1947, 54–65) and by MacFarlan and Grainger (*Scot. J. Agric.* **26**, 1947, 211–15), and Last investigated the effect on mildew of temperature and nitrogen supply (*AAB*, **40**, 1953, 312–22), of time of application of nitrogenous fertilizers (ibid. **41**, 1954, 381–92), and of date of sowing (ibid. **45**, 1957, 1–10), as well as the effect of mildew on yield (*Plant Path.* **4**, 1955, 22–4). Turner (*TBMS*, **39**, 1956, 495–506) investigated sources of infection and concluded that ascospores, shed from July to late September, are the source for successive crops of winter wheat and barley, and conidia for winter oats. Overwintering cleistocarps play no part in spring outbreaks.

Erysiphe nitida (Wallr.) Rabenh., see *E. polygoni* (*Clematis*).

Erysiphe polygoni DC., *Flor. Fr.* 2, 1805, 273; Salmon, *Monogr. Erysip.* 1900, 174; Bisby & Mason, *TBMS*, 24, 1940, 135.

Powdery mildew in a wide range of crops, viz.:

ROOTS

*Turnip and swede (*Brassica*). Becomes conspicuous in August–September, and occurs everywhere, but is very seasonal in intensity. In hot, dry summers (e.g. 1921, 1933, 1949) usually very severe. Swedes suffer rather more than turnips. W. G. Smith (*Gdnrs' Chron.* 14, 1880, 392–3) referred to an epidemic in Dorset, and Searle (*TBMS*, 6, 1920, 274–93) studied host specialization. See also *Rep. econ. Mycol. Wye, 1913–14*, 487–93.
*Rape (*Brassica rapa*). Common in some districts.

PULSE

*Pea (*Pisum*). Occurs everywhere on garden pea, but injurious only on late-sown crops. Searle (*TBMS*, 6, 1920, 279) regarded it as a distinct biologic form.

PASTURE AND FORAGE CROPS

*Clover (*Trifolium*). Widely distributed, especially on crimson and red clover, and usually worst on the aftermath.
*Sainfoin (*Onobrychis*). Sometimes badly attacked.

VEGETABLES

*Brassicae. Widely distributed on cabbage, cauliflower, brussels sprouts and kale, but rarely serious. Occasional on marrow-stem kale.
*Parsley (*Petroselinum*). Tay; Kent, 1947.
*Parsnip (*Pastinaca*). Common in many districts; sometimes found on the seed.

ORNAMENTALS

*Aquilegia. Ross-shire, 1953.
*Clematis. Not uncommon in the south, occasional in Scotland. First seen 1903 and then called *Ovularia clematidis* Chittend. (*J. R. hort. Soc.* 28, 1903, clxxvi; *Gdnrs' Chron.* 34, 1903, 299).

Salmon (*J. Bot., Lond.*, **43**, 1905, 41–4) showed the true affinities of the fungus, which is sometimes listed as *Erysiphe nitida* (Wallr.) Rabenh.

*Delphinium. Widely distributed and very common, especially towards end of season (Langdon, *J. R. hort. Soc.* **55**, 1930, 119).

*Lupin. Bucks, Forth, north Scotland. Also on forage lupines, Suffolk, Norfolk.

*Ranunculus. On garden varieties, Dawlish, Devon, 1951.

*Scabious. Reported since 1938 in Corn, Som, Oxon, Cambs, Suffolk, Elgin, Dundee, Dee.

*Statice (*Limonium*). Seen most years in the *Oidium* stage on *Limonium latifolium* and *L. sinuatum* in the Badsey area of Worcs. Occasional elsewhere. F. Joan Moore (*Plant Path.* **1**, 1952, 53–5) found the perfect stage on *L. chiloensis* at St Albans, Herts, in 1950, and it has since been recorded in Middx, Beds, Cambs and Lincs.

*Sweet pea (*Lathyrus*). Very common everywhere late in the season.

*Wallflower (*Cheiranthus*). Northumb, 1931, 1933.

Erysiphe polyphaga Hammarl., *Bot. Notiser*, 1945, 108.

*On *Kalanchoe blossfeldiana* grown experimentally under glass, Harpenden, Herts, 1951 (F. Joan Moore, *Plant Path.* **1**, 1952, 54). Had existed for several years in the *Oidium* stage.

*For possible record on *Pyrethrum* see *Erysiphe* sp.

Erysiphe taurica Lév. = *Leveillula taurica*.

Erysiphe sp.

*Mildew of sugar beet (*Beta*). First British record at Bishop's Lydeard, Som, Oct. 1935; Hants, 1941; since 1943 common in parts of Suffolk, Norfolk and Lincs late in the autumn, or earlier during hot summers; East Lothian, 1951. Only the immature perfect stage has been seen in Britain; elsewhere it has been named *Erysiphe polygoni* DC. or *Microsphaera betae* Vañha.

*On China aster (*Callistephus*), Dur, 1955.

*On pyrethrum (*Chrysanthemum coccineum*), Luddington, War,

1953 (Taylor & F. Joan Moore, *Plant Path.* 3, 1954, 30) possibly *Erysiphe polyphaga* (q.v.); Norfolk, 1953; War, 1956; *Oidium* sp. at Edinburgh and in Perthshire, 1953; Jersey, 1955.

Exoascus, see *Taphrina.*

Exobasidium camelliae Shirai, *Bot. Mag., Tokyo,* **10**, 1896, 51; Sacc., *Syll.* xiv, 229.

*Camellia gall. On flowers, Handcross, Sussex, 1944 (Dennis & Wakefield, *TBMS,* **29**, 1946, 142).

Exobasidium vaccinii (Fuckel) Woron., *Verh. naturf. Ges. Freiburg,* iv, 1867, 397; Sacc., *Syll.* vi, 664.

*Azalea gall. Frequent in England and Wales on the small-leaved greenhouse species of *Rhododendron* formerly included in the genus *Azalea* and commonly imported. Also on *Rhododendron laetevirens* (= *wilsonii*), Northumb, 1934 (*Report,* viii, 88) and *R. ferrugineum,* Ches, Glam, 1931 (*Report,* vii, 100). Widely distributed out-of-doors in Scotland but rare there on indoor plants.

The disease was described by Chittenden (*J. R. hort. Soc.* **34**, 1908, 45–6).

Exosporium depazeoides Desm. = *Cercospora depazeoides.*

Fabraea fragariae Kleb. = *Diplocarpon earlianum.*

Fabraea maculata Atk., *Science,* **30**, 1909, 452; Sacc., *Syll.* xxii, 148 (as *Stigmatea mespili* Sor.); stat. conid. *Entomosporium maculatum* Lév., in Mougeot, *Stirpes Cryptogamae Vogeso-Rhenanae,* no. 1458; Sacc., *Syll.* iii, 657; Grove, *Coelomycetes,* ii, 191.

The perfect stage has not yet been seen in Britain.

**Leaf blight and fruit spot of quince (*Cydonia*) and pear (*Pyrus*).* Fairly frequent on fruiting trees and layer rows of quince in the south and west of England; occasional on pear, notably when near affected quinces. Reported from Cupar, Fife.

PARASITES

*Leaf blight of medlar (*Mespilus*). Rare. Maidstone, Kent (Grove, *Coelomycetes*, ii, 191) and West Malling, Kent (M. H. Moore, *Gdnrs' Chron.* **104**, 1938, 440); Jersey, 1956 (Phillips, *Rep. States Exp. Sta. Jersey for 1956*, 43).
*On loquat (*Eriobotrya*). On *E.* (*Photinia*) *japonica*, Corn (Grove, *Coelomycetes*, ii, 191).

Fistulina hepatica Fr., *Syst. Myc.* i, 396; Sacc., *Syll.* vi, 54; Cooke, *Fung. Pests Cult. Pl.* 1906, 208; Rea, *Brit. Basid.* 629.
Beef-steak fungus.
*Brown chestnut (*Castanea*). Frequent.
*'Brown oak' of *Quercus*. Frequent on old parkland trees; rarely fruits on young trees. Braid (*TBMS*, **9**, 1924, 210–13) published notes on the fungus on pollarded stag-headed oaks in Richmond Park, and Cartwright (ibid. **21**, 1937, 68–83) reinvestigated the cause of 'brown oak'.

Fomes annosus (Fr.) Cooke, *Grevillea*, **14**, 1885, 20; Sacc., *Syll.* vi, 197; Rea, *Brit. Basid.* 595; syn. *Trametes radiciperda* Hartig, *Wichtige Krankheiten der Waldbäume*, 1874, 62. Bakshi, *TBMS*, **35**, 1952, 195 considered *Oedocephalum lineatum* Bakshi, *TBMS*, **33**, 1950, 111 to be the conidial stage.
Wilson (*TBMS*, **12**, 1927, 147–9) enumerated the host plants of this parasite; Peace (*Quart. J. For.* **32**, 1938, 81–104) dealt with the prevalence of butt rot and the various causal agents of it, including *F. annosus*; and Day (*Quart. J. For.* **42**, 1948, 99–101) described the method of infection of roots by this parasite.

*Heart rot or butt rot of conifers. The fungus is a common saprophyte in conifer woodland, but also attacks the roots parasitically. Trees more than ten years old are rarely killed, but they become heart-rotted. The larches (*Larix*) and spruces (*Picea*) are most often attacked, but Douglas fir (*Pseudotsuga*), western red cedar (*Thuja plicata*), Lawson's cypress (*Chamaecyparis lawsoniana*) and western hemlock (*Tsuga heterophylla*) are also susceptible. Scots pine (*Pinus sylvestris*) and Corsican pine (*P. nigra* var. *poiretiana*) are more resistant to decay, but may be killed, even when thirty years old, on the dry sands of east

Scotland and of East Anglia, where Rishbeth (*Forestry*, **22**, 1948, 174–83; **24**, 1951, 114–20; **25**, 1952, 41–50; *Ann. Bot., Lond.*, **14**, 1950, 365–83; **15**, 1951, 1–21 and 221–46) made a careful study of the disease and showed that treating the stumps of thinnings with a mixture of tar and creosote prevents them from becoming infected and so acting as sources of infection for surrounding trees.

Silver fir (*Abies*) is sometimes attacked (Day & Peace, *Forestry*, **9**, 1935, 60–1), and it has been found on *Juniperus communis* in Scotland. Anderson (*Trans. R. Scot. arb. Soc.* **35**, 1921, 112–17; **38**, 1924, 37–45) studied the effect of soil conditions on prevalence, and McHardy (*Scot. For. J.* **43**, 1929, 18–19) described a severe attack on larch in Fife.

FRUIT

*Apple (*Malus*). Associated with die-back of crab apple, Roxburgh, 1950; on apple, Lanark, 1955.

*Corylus. On *C. avellana*, Scotland.

*Peach (*Prunus*). Killing trees in a walled garden over many years, Corn, 1947.

TREES AND SHRUBS

*Alder (*Alnus*). On *A. glutinosa*, Scotland.

*Beech (*Fagus*). Scotland; on young trees, Thetford Chase, Norfolk, 1952. This host is relatively resistant.

*Birch (*Betula*). On *B. pendula*, Scotland.

*Hawthorn (*Crataegus*). On *C. oxyacantha* in hedgerow, East Lothian, 1949; on *C. monogyna*, East Anglia.

*Heather (*Calluna*). Dee (*Scot. For. J.* **41**, 1927, 225).

*Poplar (*Populus*). On *P. nigra italica*, East Anglia (Rishbeth, *Ann. Bot., Lond.*, **14**, 1950, 365).

*Prunus. On *P. padus*, Scotland.

*Rhododendron. On *R. ponticum*, Scotland.

*Sorbus. On *S.* (*Pyrus*) *aria*, Scotland; on *S. aucuparia*, Perth, 1953.

Fomes applanatus (Pers.) Wallr. = *Ganoderma applanatum*.

Fomes fomentarius (Fr.) Kickx, *Fl. Cr. Fl.* ii, 237; Sacc., *Syll.* vi, 179; Rea, *Brit. Basid.* 592.

*Heart rot of birch (*Betula*). Rare except in parts of the Scottish highlands (Macdonald, *Trans. bot. Soc. Edinb.* **32**, 1938, 396–408).

Fomes fraxineus (Bull.) Fr., *Syst. Myc.* i, 374; Sacc., *Syll.* vi, 199; Rea, *Brit. Basid.* 595.

*The fructifications are uncommon in Britain but occur at times on ash (*Fraxinus*) and laburnum. Montgomery (*AAB*, **23**, 1936, 465–86), using a culture isolated originally from a living tree of *Robinia pseudacacia* (locality not stated), studied its morphological and physiological characters in pure culture and on blocks of ashwood.

Fomes igniarius (Fr.) Kickx, *Fl. Cr. Fl.* ii, 237; Sacc., *Syll.* vi, 180; Rea, *Brit. Basid.* 593.

*On plum (*Prunus*). Occasional, but at most a weak parasite.
*Heart rot of willow (*Salix*). Common in cricket-bat willow (*S. alba* var. *coerulea*).

Fomes pomaceus (Pers. ex S. F. Gray) Lloyd, *Mycol. Notes*, **35**, 1910, 469; Rea, *Brit. Basid.* 594.

*Fomes heart rot of plum (*Prunus*). Frequent as a weak and slow-growing wound parasite on old plum and cherry trees in neglected orchards. Fisher (*TBMS*, **19**, 1935, 102–13) made observations on the disease.

Fomes ribis Fr., *Syst. Myc.* i, 375; Sacc., *Syll.* vi, 184; Rea, *Brit. Basid.* 594.

*Fomes collar rot of currant (*Ribes*). Frequent on old red currant bushes in southern districts; less frequent on gooseberry; occasional on black currant. A weak parasite only.

Fomes robustus (Karst.) Bres., *Bot. Contr.* 1890, 388; Sacc., *Syll.* ix, 173; Rea, *Brit. Basid.* 593.

*Yellow trunk rot of oak (*Quercus*). Rare.

FOMES

Fomes ulmarius (Sow. ex Fr.) Gillet, *Les Hyménomycètes*, Alençon, 1874, 683; Sacc., *Syll.* vi, 166; Rea, *Brit. Basid.* 595.
*Butt rot of elm (*Ulmus*). Common. See Plowright, *Gdnrs' Chron.* **25**, 1899, 392–3.

Fuckelia conspicua Marchal = stat. conid. of *Phacidiella discolor*.

Fuligo septica Gmel., in Linné, *Systema Naturae*, ii, 1791, 1466; Sacc., *Syll.* vii, 353.
Flowers of tan or 'fairy butter'.
*On grasses, in large clumps on pasture grasses, Dee, 1949 (*Fuligo* sp.).
*On cabbage, in abundance over half-acre field, Berks, 1940.
*On cucumber (*Cucumis*), commonly infesting the stems and leaves.

Fusarium album Sacc. = *Cylindrocarpon album*.

Fusarium arcuatum Berk. & Curt. = *F. avenaceum*.

Fusarium arthrosporioides Sherb., *Mem. Cornell agric. Exp. Sta.* no. 6, 1915, 175; Sacc., *Syll.* xxv, 972; Wakefield & Bisby, *TBMS*, **25**, 1941, 64.
*Found occasionally as a cause of dry rot of potato tubers (McKee, *AAB*, **39**, 1952, 38–43). See also *Fusarium culmorum* (cereals).

Fusarium avenaceum (Fr.) Sacc., *Syll.* iv, 713; Wakefield & Bisby, *TBMS*, **25**, 1941, 64.
*Brown foot rot of cereals, see *Fusarium culmorum*.
*Dry rot of potato. This species is not uncommon as a cause of dry rot (McKee, *AAB*, **39**, 1952, 38–43) and is found on tubers from all the seed-growing areas in Scotland. It was first reported as a potato pathogen in Britain by F. Joan Moore (ibid. **32**, 1945, 304–9), who found it causing wastage in stored tubers. McKee has studied its pathogenicity (ibid. **41**, 1954, 417–34) and its host-parasite relations (ibid. **43**, 1955, 147–8).

*On seed samples of ryegrass (*Lolium*), Edinburgh, 1950.
*Foot rot of shallot (*Allium*). Inveresk, 1949. (See also *Fusarium* sp.)
*As a cause of rot in apple fruits. Rare (Kidd & Beaumont, *TBMS*, **10**, 1924, 114; also as *F. arcuatum* Berk. & Curt. and *F. viticola* Thüm.).
*Killing raspberry buds. Scotland.
*Root rot of sweet pea (*Lathyrus*). Angus, 1952, associated with *Fusarium culmorum* (q.v.).
*Leaf spot of tulip. Occasional under glass (Beaumont & Buddin, *TBMS*, **22**, 1938, 113–15). Reported as *Fusarium* sp. in Corn and Middx. The fungus probably spreads from straw. *Fusarium* sp. has been found associated with a bulb rot of tulips, Forth.

Fusarium blackmani Brown & Horne = *F. lateritium.*

Fusarium bulbigenum Cooke & Massee = *F. oxysporum* f. *narcissi.*

Fusarium bulbigenum var. **lycopersici** (Brushi) Wr. & Reinking = *F. oxysporum* f. *lycopersici.*

Fusarium caeruleum (Lib.) Sacc., *Syll.* iv, 705; Wakefield & Bisby, *TBMS*, **25**, 1941, 64. The perfect stage has been described as *Hypomyces asclepiadis* Zerova, *J. Bot. Acad. Sci. Ukr.* **11**, 1937, 101, but it is not known in Britain. Snyder and Hansen (*Amer. J. Bot.* **28**, 1941, 740) named this species *Fusarium solani* (Mart.) Appel & Wr. f. *radicicola* (Wr.) Snyder & Hansen, with its perfect stage *Hypomyces solani* Reinke & Berth. em. Snyder & Hansen.

*Dry rot of potato. Mainly a storage disease which becomes apparent in December or earlier, and reaches its peak in March or April. Of considerable economic importance, in some seasons destroying many hundreds of tons of early or early maincrop varieties.

Pethybridge and his colleagues (*Econ. Proc. R. Dublin Soc.* **1**, 1908, 547–58; *Sci. Proc. R. Dublin Soc.* **15**, 1917, 193–222) were the first clearly to distinguish this disease in the British Isles, though it was probably vaguely known in England long before (see, for example, *J. R. agric. Soc.* **6**, 1845, 161–74; *Gdnrs' Chron.* **5**, 1876, 656; **22**, 1884, 40). It is only comparatively recently that careful investigations have been made of it, first in Scotland and then in England.

BIOLOGY. Foister (*Scot. J. Agric.* **23**, 1940, 63–7) published a general account of the disease, which has been shown in Scotland to be soil-borne (Foister, Wilson & Boyd, *Nature, Lond.*, **155**, 1945, 793) and to be worse in potatoes that have been chilled (McIntosh, *Gdnrs' Chron.* **116**, 1944, 87).

Working in England, Small investigated the origin of infection (*AAB*, **31**, 1944, 290–5), the effect of dipping and bruising the seed on the incidence of dry rot (ibid. **32**, 1945, 310–18; **33**, 1946, 211–19), and the effect of planting infected and contaminated sets on plant establishment (ibid. **33**, 1946, 219–21). Later, Foister, Wilson and Boyd (ibid. **39**, 1952, 29–37) followed the effects of commercial handling methods on the incidence of the disease. McKee and Boyd (ibid. **39**, 1952, 44–53) devised a method of assessing soil infectivity; and McKee studied pathogenicity (ibid. **41**, 1954, 417–34) and host-parasite relations (ibid. **43**, 1955, 147–8).

Bennett and Munro (*Agriculture, Lond.*, **52**, 1946, 500–3) had carried out varietal resistance trials in the north of England, and later Boyd (*AAB*, **39**, 1952, 322–57), having devised methods for assessing tuber susceptibility, used them to study varietal, seasonal and local differences in tuber susceptibility, including the effect of storage temperatures.

OTHER PATHOGENS. F. Joan Moore (*AAB*, **32**, 1945, 304–9) compared the role of *F. caeruleum* and *F. avenaceum* in causing wastage in stored tubers, and later McKee (ibid. **39**, 1952, 38–43) showed that in Britain two other species (*F. arthrosporioides* and *F. tricinctum*) may also occasionally be involved, though 90 % of dry rot in this country is caused by *F. caeruleum*.

CONTROL. Foister and Wilson (*Agriculture, Lond.*, **50**, 1943, 300–3) summarized the results of their early tuber-dipping experiments with mercury compounds, one of which was used on a large scale by a conveyor-belt method (Wakely & Mellor, *Nature, Lond.*, **150**, 1942, 769). Practical difficulties about dipping tubers led to tests with dusts, including thymol, by Foister, Wilson and Boyd (ibid. **156**, 1945, 394). It was commercial enterprise, however, which led to the development of an efficient treatment at lifting time with dusts containing tetrachlornitrobenzene (TCNB or tecnazene). These dusts were tried by Foister and Wilson (*Agriculture, Lond.*, **57**, 1950, 229–33), who showed that the disease may cause serious losses if riddling is carried out only a month or more before planting. TCNB proved also to have a marked sprout depressant effect, and unless care is taken yields may be affected (Brown & Reavill, *AAB*, **41**, 1954, 435–47 and 448–60), though the 2.3.4.5 isomer seems to give good control without retarding sprouting (Brook & Chesters, *AAB*, **45**, 1957, 623–34). Wilson and Dawson (*J. Sci. Fd Agric.* 1953, 305–10) examined residues and found only negligible amounts of TCNB on ware tubers that had been dusted 4 to 5 months before. Nevertheless, treated tubers may yield dehydrated products having a persistent taint (Gooding, Tucker & Harries, *J. Sci. Fd Agric.* **7**, 1956, 411–16). Resistant mutants may develop in cultures of the fungus treated with TCNB (McKee, *Nature, Lond.*, **167**, 1951, 611).

Fusarium conglutinans Wr. var. **callistephi** Beach = *F. oxysporum* f. *callistephi*.

Fusarium culmorum (W. G. Sm.) Sacc., *Syll.* xi, 651; Wakefield & Bisby, *TBMS*, **25**, 1941, 64.

*Brown foot rot and ear blight of cereals. Attacks wheat, barley, oats and, to a less extent, rye. Widely distributed in heavy, poorly drained acid soils but rarely serious except in Scotland and northern England.

Bennett (*AAB*, **15**, 1928, 213–44) showed that this species and *F. avenaceum*, acting together or separately, were principally responsible in northern England, though other species were sometimes involved, such as *F. equiseti* (Corda) Sacc. syn. *F. scirpi* Lamb. & Fautr. (ibid. **19**, 1932, 21–34), *F. nivale* (ibid. **20**, 1933, 272–90) and *F. sambucinum*, etc. (ibid. **22**, 1935, 479–507). *F. nivale* is a common cause in Scotland and is seed-borne

(Noble & Montgomerie, *Scot. Agric.* **34**, 1954, 51–2; *TBMS*, **39**, 1956, 449–59). Bennett (*J. Minist. Agric.* **45**, 1939, 1115–18) also published a general article on these diseases. Observations on them were made by Russell (*TBMS*, **16**, 1932, 253–69) in the Cambridge area, and by Dennis (*AAB*, **31**, 1944, 370–4) in Scotland.

Species of *Fusarium* are also responsible for the so-called red mould of barley grain and malt (F. A. M[ason], *Contrib. Lab. Murphy and Son, Ltd. Bulls.* 11 and 12, vol. ii, 1924, 78–86). *F. culmorum* is a common soil inhabitant which soon colonizes decaying wheat straw (Sadasivan, *AAB*, **26**, 1939, 497; Walker, ibid. **28**, 1941, 333–50), and its parasitic activity is largely determined by soil conditions (Shen, ibid. **27**, 1940, 323–9). It can often be isolated from healthy plants (Samuel & Greaney, *TBMS*, **21**, 1937, 114–17). Tveit and Wood (*AAB*, **43**, 1955, 538–52) showed that control of seedling blight caused mainly by *F. nivale*, comparable with that obtained by organo-mercury seed treatment, is possible when living material of certain species of *Chaetomium* is present on the seed.

Snyder and Hansen (*Amer. J. Bot.* **32**, 1945, 657–65) reduced all species of *Fusarium* in the sections Roseum, Arthrosporiella, Gibbosum and Discolor to one species, *F. roseum* Link em. Snyder & Hansen. They distinguished the members of these sections pathogenic to cereals as *F. roseum* f. *cerealis* (Cooke) Snyder & Hansen; those non-pathogenic to cereals simply as *F. roseum.* Under this proposal the species listed here as *F. arthrosporioides*, *F. avenaceum*, *F. culmorum*, *F. equiseti*, *F. graminearum* and *F. sambucinum* are all synonyms of *F. roseum* f. *cerealis*, the perfect stage of which is named by Snyder and Hansen as *Gibberella roseum* (Link) Snyder & Hansen f. *cerealis* (Cooke) Snyder & Hansen. As most of these species have also been reported in Britain on hosts other than cereals, however, I prefer to retain the distinctive names until the position is clearer.

*Cob and stem rot of maize (*Zea*). Som, 1936–7; on sweet corn, Herts, 1945 (*Fusarium* sp.). Species of *Fusarium* have also been found causing foot rot of maize in Sussex and Herts.

*On potato (*Solanum*). May play a part in rotting potato tubers in storage (Mon, 1941; Fife) and has been found associated with a haulm wilt (Worcs, 1941).

*Fusarium rot of sugar beet (*Beta*). Occasional as cause of rot of the crown and upper part of root, and of scurfy root. East Midlands, eastern England, Moray.

*On pea (*Pisum*). Sometimes associated with *Fusarium* foot rot (see *F. solani* f. *pisi*).

*On grasses, damaging *Agrostis* in a small grass plot (Drew Smith, *J. Sports Turf Res. Inst.* **8**, 1954, 449); also on *Festuca rubra*, Harrogate, Yorks, 1954; probably occurs on other turf grasses (Drew Smith, ibid. **9**, 1955, 68). Seed treatment protects against it (Drew Smith, ibid. **9**, 1956, 244–50).

*Foot rot of leek (*Allium*). Regularly seen in the Vale of Evesham since about 1932. Not uncommon elsewhere, but the species not always determined. Sussex, Norfolk, Lincs, Ches, Northumb, Dur, Forth.

*Rotting apple fruit (Kidd & Beaumont, *TBMS*, **10**, 1924, 116).

*On raspberry canes, following midge attack (see *Leptosphaeria coniothyrium*).

*Stem rot and die-back of carnation (*Dianthus*). Widely distributed and fairly common, but usually develops on plants already attacked by *Verticillium cinerescens* (q.v.) or *Fusarium oxysporum* f. *dianthi* (q.v.). Also on *Dianthus floribunda*, Caerns, 1936.

*On *Galtonia candicans*, rotting stored, imported bulbs (Ghamrawy, *TBMS*, **18**, 1933, 249–52).

*Foot rot of *Lampranthus* (*Mesembryanthemum*). Kirkcaldy, 1951; Aberdeen and Kincardine, 1955.

*Damping-off of *Lobelia*. Staffs, 1938.

*On lupin, associated with rotting of crowns, Beds, I. of Ely, Lancs, Carlisle.

*Root rot of sweet pea (*Lathyrus*). Occasional, especially in the north. Species not always determined. See Taubenhaus and Manns (*Gdnrs' Chron.* **54**, 1913, 23).

Fusarium dianthi Prill. & Delacr. = *F. oxysporum* f. *dianthi*.

Fusarium equiseti (Corda) Sacc., *Syll.* iv, 707; syn. *F. scirpi*
Lamb. & Fautr., *Rev. Mycol.* **16**, 1894, 111; Sacc., *Syll.* xi, 651;
Wakefield & Bisby, *TBMS*, **25**, 1941, 66.
*Brown foot rot of cereals. (See *F. culmorum.*)

Fusarium graminearum Schwabe = stat. conid. of *Gibberella zeae*.

Fusarium lateritium Nees ex Fr. = stat. conid. of *Gibberella
baccata*.

Fusarium lateritium var. **fructigenum** Fr., see *Gibberella baccata*.

Fusarium lateritium var. **mori** Desm. = stat. conid. of *Gibberella
moricola*.

Fusarium lini Bolley = *F. oxysporum* f. *lini*.

Fusarium moniliforme Sheldon = stat. conid. of *Gibberella fujikuroi*.

Fusarium nivale (Fr.) Ces. = stat. conid. of *Griphosphaeria nivalis*.

Fusarium orthoceras Appel & Wr. = *Fusarium oxysporum*.

Fusarium oxysporum Fr., *Syst. Myc.* 3, 471; Sacc., *Syll.* iv, 705;
Wakefield & Bisby, *TBMS*, **25**, 1941, 65; syn. *Fusarium orthoceras* Appel & Wr., *Arb. Biol. Land.- u. Forstw.* **8**, 1910, 155;
Sacc., *Syll.* xxii, 1477; Wakefield & Bisby, *TBMS*, **25**, 1941, 65;
syn. *Fusarium vasinfectum* Atkins. var. *lutulatum* (Sherb.) Wr.,
Z. Parasitenk. **3**, 1931, 424; *Fusarium autographice delineata*,
no. 1019.

Buxton and Richards (*J. gen. Microbiol.* **13**, 1955, 99–102)
provided a method for distinguishing strains of *F. oxysporum* by
their reaction to soil actinomycetes in culture, and Buxton (ibid.
15, 1956, 133–9) investigated heterokaryosis and parasexual
recombination in pathogenic strains of the species.
*Fusarium wilt of potato (*Solanum*). Wye, Kent, 1928 (Chona,
TBMS, **17**, 1932, 229–35). The only authenticated record, but
this species has also been isolated from potatoes affected with dry
rot, Fife, 1950.

*Fusarium wilt of dwarf and runner bean (*Phaseolus*). Local in the Vale of Evesham since 1932 (Davies, *TBMS*, **22**, 1939, 309; **25**, 1942, 418–26). Also Beds, Staffs, Bucks, Herts, Kent. Occasionally seen under glass.

*Associated with root rot of strawberry (Berkeley & Lauder-Thomsen, *J. Pomol.* **12**, 1934, 222), but doubtfully parasitic.

*Causing damping-off of mushroom (*Agaricus*) along with *Fusarium solani* and *Fusarium* spp. (Wood, *Gdnrs' Chron.* **97**, 1935, 243; *Phytopathology*, **27**, 1937, 85–94; **29**, 1939, 728–39).

*Corm and root rot of *Cyclamen*. Ches, 1946.

*Chalky dry rot of *Freesia* corms. East Anglia, 1953. Buxton (*Plant Path.* **4**, 1955, 69–70) proved pathogenicity and showed that *F. oxysporum* f. *gladioli* (q.v.) from *Gladiolus* will also rot freesias.

Fusarium oxysporum Fr. f. **callistephi** (Beach) Snyder & Hansen, *Amer. J. Bot.* **27**, 1940, 66; syn. *F. conglutinans* Wr. var. *callistephi* Beach, *Rep. Mich. Acad. Sci.* **20**, 1918, 281; Wakefield & Bisby, *TBMS*, **25**, 1941, 64.

*Wilt of China aster (*Callistephus*). Common in many districts of Britain, but not always distinguished from foot rot caused by *Phytophthora cryptogea* (q.v.). Not studied in Britain until about 1933 (Ogilvie & Mulligan, *Gdnrs' Chron.* **95**, 1934, 215–16).

Fusarium oxysporum Fr. f. **dianthi** (Prill. & Delacr.) Snyder & Hansen, *Amer. J. Bot.* **27**, 1940, 66; syn. *F. dianthi* Prill. & Delacr., *Ann. Inst. nat. agron., Paris*, **16**, 1900, 161; Sacc., *Syll.* xvi, 1100; Wakefield & Bisby, *TBMS*, **25**, 1941, 65.

*Fusarium wilt of carnation (*Dianthus*). Not uncommon but less important than Verticillium wilt. (See *Verticillium cinerescens* and Wickens, *AAB*, **22**, 1935, 630–83.)

Fusarium oxysporum Fr. f. **gladioli** (Massey) Snyder & Hansen, *Amer. J. Bot.* **27**, 1940, 66.

*Fusarium yellows and corm rot of *Gladiolus*. Widespread and common from Corn to Ayrshire. Frequent in imported corms from Holland.

From 1926 a *Fusarium* corm rot was observed from time to time on early-flowering gladioli (e.g. *Gladiolus colvillei*) in south-west England and the Isles of Scilly, as well as in imported corms, but it was not specifically identified (W. C. Moore, *Diseases of Bulbs*, 1939, 117–19). By 1953 Buxton and Robertson (*Plant Path.* 2, 1953, 61–4) were able to show that a yellows disease was present in many stocks throughout the country, notably in Cambs, Corn, Surrey, Ayrshire and Midlothian. They also found it frequently in imported corms.

Later, Buxton (*TBMS*, **38**, 1955, 193–201) expressed the view that the rot of stored corms attributed in America to *Fusarium oxysporum* var. *gladioli*, and long known in Britain, was caused by the same organism as the disease of growing plants called in America Fusarium yellows, and attributed there to *F. orthoceras* var. *gladioli*. He named the pathogen of both diseases *F. oxysporum* Fr. f. *gladioli* (Massey) Snyder & Hansen, and gave reasons for this in a separate article (*TBMS*, **38**, 1955, 202–12).
*On *Freesia*. This or *Fusarium* sp. causing wilt and corm rot of *Freesia*, I. of Ely, Lincs, Oxon, Corn, Devon, Ches. (See also *F. oxysporum*.)

Fusarium oxysporum Fr. f. **lini** (Bolley) Snyder & Hansen, *Amer. J. Bot.* **26**, 1940, 66; syn. *F. lini* Bolley, *Proc. Soc. Prom. Agric. Sci.* **22**, 1901, 1; *Bot. Gaz.* 1902, 150; Sacc., *Syll.* xviii, 670; Wakefield & Bisby, *TBMS*, **25**, 1941, 65.

*Flax wilt. Unidentified species of *Fusarium* have from time to time been associated with yellowing or wilting of flax in small patches in several districts. *F. lini* has so far been confirmed only at Aberystwyth (Wilson, *Nature, Lond.*, **154**, 1944, 709; *TBMS*, **29**, 1946, 221–31). It was seen in Northern Ireland in 1902 (Cooke, *TBMS*, **2**, 1902, 15) and was studied in Eire by Pethybridge and others (*J. Dep. Agric. Ire.* **20**, 1920, 325–42; **21**, 1921, 167–87; **22**, 1922, 103–20). Booer (*AAB*, **38**, 1951, 334–7) demonstrated experimentally that the addition of 0·01 % mercury to contaminated soil will greatly reduce seedling losses.

Fusarium oxysporum Fr. f. **lycopersici** (Sacc.) Snyder & Hansen, *Amer. J. Bot.* **27**, 1940, 66; Sacc., *Syll.* iv, 705; syn. *F. bulbigenum* Cooke & Massee var. *lycopersici* (Brushi) Wr. & Reinking, *Die Fusarien*, 1935, 114; Wakefield & Bisby, *TBMS*, **25**, 1941, 64.

*Fusarium wilt of tomato (*Lycopersicon*). Much less common than Verticillium wilt and probably only occasional in hot summers in both England and Scotland. Common in Guernsey. It is not usually distinguished properly from a root rot caused by *Fusarium* spp., and the earlier literature on the disease (Massee, *Gdnrs' Chron.* **17**, 1895, 707; *J. R. hort. Soc.* **19**, 1895, 20–4; Collenette, ibid. **19**, 1895, 13–19; *J. Bd Agric.* **5**, 1898, 192–7) probably concerned this root rot, or perhaps *Verticillium* wilt.

Fusarium oxysporum Fr. f. **narcissi** Snyder & Hansen, *Amer. J. Bot.* **27**, 1940, 66; syn. *F. bulbigenum* Cooke & Massee, *Grevillea*, **16**, 1887, 49; Sacc., *Syll.* x, 725; Wakefield & Bisby, *TBMS*, **25**, 1941, 64.

*Corm rot of *Crocus*. On *C. maesiacus (aureus)*, Forth. In England a corm rot caused by *Fusarium* sp. has been seen on *Crocus versicolor picturatus* (1931), on *C. chrysanthus* (1945) and frequently on imported corms (W. C. Moore, *Diseases of Bulbs*, 1939, 148).

*Root rot and bulb-scale rot of lily. On seedlings of *Lilium auratum*, *L. sargentiae* and *L. nobilissimum* in the south of England; on mature plants of *L. regale*, Slough, Bucks (Hawker & Singh, *TBMS*, **26**, 1943, 116–26). *Fusarium* sp. has been found associated with a stem rot of *Lilium hansonii*, Clyde.

*Basal rot of *Narcissus*. Not now common except after hot summers.

Between 1889 and 1900 there was much concern about an unknown 'basal rot' of daffodils (see, for example, Dod, *Gdnrs' Chron.* **15**, 1894, 379; Crawford, *J. Hort.* **102**, 1900, 347), and a similar trouble, this time associated with *Fusarium*, flared up during the hot summer of 1911 (Massee, *Kew Bull.* 1913, 307–9; *J. Bd Agric.* **20**, 1914, 1091–3). From 1926 onwards *Fusarium* basal rot became very troublesome, and remained so until after

Gregory (*AAB*, **19**, 1932, 475–514) had thoroughly investigated the disease, and until he, and later Hawker (ibid. **22**, 1935, 684–708), had confirmed experience in America and Holland of the value of adding fungicides to the hot-water bath when controlling eelworm disease. As soon as this and other control measures became established practice, the disease rapidly declined. W. C. Moore (*Diseases of Bulbs*, 1939, 72) gave a full account of the disease. More recent studies on the use of disinfectants were published by Wallace (*Rep. Bulb Exp. Kirton agric. Inst.* **7**, 1940, 43) and by Hawker (*AAB*, **27**, 1940, 205–17; **30**, 1943, 323–4; **31**, 1944, 31–3), who also studied summer infection through dying roots (ibid. **30**, 1943, 325–6) and provided a general summary of her work (*Daffodil Yearb. 1946*, 78–83).

Fusarium oxysporum Fr. f. **pisi** (Lindf.) Snyder & Hansen, *Amer. J. Bot.* **27**, 1940, 66.

*Fusarium wilt of pea (*Pisum*). The main cause of crop failures up to mid-June, after which the pathogen is commonly found associated for a few weeks with foot rot caused by *F. solani* f. *pisi* (q.v.). The presence of wilt in Britain, though suspected since 1945, was first confirmed by Buxton and Storey (*Plant Path.* **3**, 1954, 13–16), who found it widely distributed, particularly in the eastern counties, in 1953. Buxton (*TBMS*, **38**, 1955, 309–16) has shown that different biologic races of the pathogen occur in Britain, and that *F. oxysporum* var. *redolens* sometimes causes wilt. He has also shown (ibid. **40**, 1957, 145–54, 305–17) that root exudates from different pea cultivars affect the biologic races in different ways.

Fusarium oxysporum Fr. var. **redolens** Gordon, *Canad. J. Bot.* **30**, 1952, 238.

*Fusarium wilt of pea, see *F. oxysporum* f. *pisi*.

Fusarium pelargonii Cooke, *Gdnrs' Chron.* **20**, 1896, 92; *Fung. Pests Cult. Pl.* 1906, 40; Wakefield & Bisby, *TBMS*, **25**, 1941, 66.

*Stem rot of *Pelargonium*. Sutherland.

Fusarium poae (Peck) Wr. = *F. tricinctum* f. *poae*.

Fusarium roseum Link em. Snyder & Hansen, see *F. culmorum*.

Fusarium sambucinum Fuckel = stat. conid. of *Gibberella pulicaris*.

Fusarium scirpi Lamb. & Fautr. = *F. equiseti*.

Fusarium solani (Mart.) Appel & Wr., *Arb. biol. Anst. Landw. Forstw.* **8**, 1910, 77, em. Snyder & Hansen, *Amer. J. Bot.* **28**, 1941, 740; Sacc., *Syll.* iv, 705; syn. *F. solani* var. *martii* (Appel & Wr.) Wr., *Z. Parasitenk.* **3**, 1931, 270 and 516; Wakefield & Bisby, *TBMS*, **25**, 1941, 66.

Its perfect stage, and that of its various varieties and forms, has been named *Hypomyces solani* Reinke & Berth. em. Snyder & Hansen, *Amer. J. Bot.* **28**, 1941, 741, but it is unknown in Britain.

*On potato (*Solanum*). This species was reported, under the name *Fusarium viride* (Lechm.) Wr., as a wound parasite of potato tubers by Mitra (*Nature, Lond.*, **133**, 1934, 67).

*Rotting apple fruits (Kidd & Beaumont, *TBMS*, **10**, 1924, 116).

*Associated with root rot of strawberry (*Fragaria*). Cumb, 1931.

*One cause of damping-off of mushroom (*Agaricus*). (See *Fusarium oxysporum*.)

Fusarium solani f. **phaseoli** (Burkh.) Snyder & Hansen, *Amer. J. Bot.* **28**, 1941, 740; syn. *F. solani* var. *martii* (Appel & Wr.) Wr. f. 3 Snyder, *Zbl. Bakt.* II, **91**, 1934, 174; Wakefield & Bisby, *TBMS*, **25**, 1941, 66.

*Foot rot of dwarf bean (*Phaseolus*). Of regular occurrence since 1928 in Worcs (Ogilvie & Mulligan, *Rep. Long Ashton Res. Sta. for 1930*, 128), and occasionally seen elsewhere under glass and in the open. Runner beans are attacked by the same or a very closely related species; Jersey, on runner bean, 1952, on French bean, 1954. White (*Rep. exp. Res. Sta. Cheshunt for 1940*, 42–3) described a method of control in pot-grown plants.

Fusarium solani f. **pisi** (Jones) Snyder & Hansen, *Amer. J. Bot.* **28**, 1941, 740; syn. *F. solani* var. *martii* (Appel & Wr.) Wr. *f. 2* Snyder, *Zbl. Bakt.* II, **91**, 1934, 165; Wakefield & Bisby, *TBMS*, **25**, 1941, 66.

*Fusarium foot rot of pea (*Pisum*). Occurs throughout Britain, usually from mid-June onwards, and is the main cause of crop failures after mid-July. Early-season failures are more often caused by the wilt-producing *F. oxysporum* f. *pisi* (q.v.), and in June the two species are often found together, producing the wilt-foot rot complex called 'St John's Disease'.

Other species of *Fusarium*, including *F. culmorum* and *F. sambucinum*, sometimes play a part in causing foot rot, which is particularly common in gardens and allotments where peas are grown year after year, and which often occurs on plants also infested with *Heterodera göttingiana* (formerly called *H. schachtii*). See Walton, Ogilvie and Mulligan, *Rep. Long Ashton Res. Sta. for 1933*, 74–85. The disease was given much attention by Ogilvie and Mulligan (ibid. *for 1930–2*) and by Davies (ibid. *for 1943*, 103–7). Buxton (*TBMS*, **38**, 1955, 309–16) clearly distinguished foot rot from wilt, and showed that some strains of *Fusarium solani* from peas are non-pathogenic.

Fusarium solani f. **radicicola** (Wr.) Snyder & Hansen, see *F. caeruleum*.

Fusarium solani var. **martii** (Appel & Wr.) Wr. = *F. solani*.

Fusarium solani var. **martii** f. **2** Snyder = *F. solani* f. *pisi*.

Fusarium solani var. **martii** f. **3** Snyder = *F. solani* f. *phaseoli*.

Fusarium tricinctum (Corda) Sacc., *Syll.* iv, 700; Wakefield & Bisby, *TBMS*, **25**, 1941, 66.

Dry rot of potato tubers. Occasional (McKee, *AAB*, **39, 1952, 38–43).

Fusarium tricinctum f. poae (Peck) Snyder & Hansen, *Amer. J. Bot.* **32**, 1945, 663; syn. *F. poae* (Peck) Wr., *Die Fusarien*, 1935, 47; *Sporotrichum poae* Peck, *Rep. Stat. Botanist*, 1902, 29; Sacc., *Syll.* xviii, 525.

*On pink-stained oats. Aberdeen. Parasitism doubtful.

*On flower head of cocksfoot (*Dactylis*), Corstorphine, Edinburgh (*TBMS*, **29**, 1946, 164); Staffs, 1950. Parasitism doubtful.

*Associated with silver top of timothy (*Phleum*). Dorset, Wilts, Staffs. The primary cause of silver top is unknown; the mite *Pediculopsis graminum* is sometimes but by no means always present.

*Bud rot of carnation (*Dianthus*). Bethersden, Kent, 1956. The mite *Pediculopsis graminum* was also present.

Fusarium urticearum (Corda) Sacc., see *Gibberella moricola*.

Fusarium vasinfectum Atkins. var. **lutulatum** (Sherb.) Wr. = *F. oxysporum*.

Fusarium viride (Lechm.) Wr. = *F. solani*.

Fusarium viticola Thüm. = *F. avenaceum*.

Fusarium sp.

*Broad and field bean (*Vicia*). Causing foot and root rot. Yorks, Salop, east midlands, Herts, East Anglia.

*Mustard (*Brassica*). Associated with yellowing and death of seed crops, Essex, 1951; and with failure of white mustard, Norfolk.

VEGETABLES

Asparagus. Associated with root rot; doubtfully parasitic. Herts Kent, Middx, east midlands, Essex, Worcs, Anglesey. Also on *A. sprengeri*, Suffolk, 1937, and *A. plumosus*, Herts, 1940.

Brassicae. Associated with cut tops of Brussels sprouts, Beds, 1951.

*Cucumber (*Cucumis*). Commonly associated with a root rot and wilt, but the primary cause is probably adverse cultural conditions, such as lack of drainage and poor aeration (Williams, *Reps. exp. Res. Sta. Cheshunt for 1937 and 1938*).

*Mint (*Mentha*). Rhizome rot of forced mint, Glos, 1930.
*Onion (*Allium*). Root rot of onion and shallot, with symptoms resembling those of pink root (*Phoma*) in America. Not uncommon and widely distributed. Worcs, Glam, Bristol, Derby, Yorks, Wester Ross.
*Tomato (*Lycopersicon*). A frequent cause of root rot distinct but not normally differentiated from Fusarium wilt.

FRUIT

*Melon (*Cucumis*). Fairly frequently associated with a wilt or root rot. In Herts, 1956, the species was identified as *Fusarium oxysporum* f. *melonis*.
*Raspberry (*Rubus*). Associated with a damaging crown rot widely distributed in east Scotland since 1951. Frost may be a predisposing cause.
*Ring rot of walnut (*Juglans*) fruits (W. C. Moore, *TBMS*, **28**, 1945, 128).

ORNAMENTAL AND MISCELLANEOUS

*Damping-off of mushroom (*Agaricus*), see *F. oxysporum*.
**Alyssum*. Associated with dying plants, east midlands, 1931.
**Antirrhinum* wilt. Hants, Clyde, Dee.
*Stem rot of carnation (*Dianthus*). Several species may be involved (see *Verticillium cinerescens*).
*Root rot or wilt of *Chrysanthemum*. Common.
*Wilt of *Cineraria*.
*Associated with death of coriander (*Coriandrum sativum* L.), Chelmsford, Essex, 1943.
*Root rot of *Erysimum* (*Cheiranthus*) *allionii*. Herts, 1935.
*Foot and root rot of *Gypsophila*. Dumfries, 1955.
**Iris*. Associated with leaf rot of *I. japonica*, Kent, 1938; with root rot, Dorset, 1951; and with bulb rot, Jersey, 1954 (probably a form of *Fusarium oxysporum*).
*Bulb rot of *Lachenalia* (Green & Hewlett, *J. R. hort. Soc.* **74**, 1949, 211–15).
*Associated with foot rot of sweet blue lupin, grown for forage, Beds, Cambs, Suffolk.
*Foot rot of *Primula*. Clyde, 1949.

*Associated with death of pyrethrum plants, Cambs, 1946.
*Associated with root injury to rose (Bewley, *Sci. Hort.* **6**, 1938, 97).
*Killing mature sweet william (*Dianthus*) plants. Leics, 1930.
*Associated with root rot of *Viola*.

Fusicladium carpophilum (Thüm.) Oudem., *Kon. Akad. Wet. Amst.* 1900, 394; syn. *Cladosporium carpophilum* Thüm., *Öst. bot. Z.* **27**, 1877, 12; Sacc., *Syll.* iv, 353; Wakefield & Bisby, *TBMS*, **25**, 1941, 84.

*Scab of peach and nectarine (*Prunus*). Occasional on fruits and shoots, notably in Scotland. Described by Green (*Gdnrs' Chron.* **89**, 1931, 151–2).
*Plum scab. Widely distributed throughout Britain. Occasionally seen on damsons in the open (Westmorland, 1951) and under glass (Shetland, 1952).
*On *Prunus mirabella floribunda*. Herts, 1951.

Fusicladium cerasi (Rabenh.) Sacc. = stat. conid. of *Venturia cerasi*.

Fusicladium dendriticum (Wallr.) Fuckel = stat. conid. of *Venturia inaequalis*.

Fusicladium lini Sor., *Z. PflKrankh.* **5**, 1895, 103.

*On flax (*Linum*), associated with wilting and yellowing of seedlings. Norfolk, Wilts (W. C. Moore, *TBMS*, **31**, 1947, 90). Parasitism uncertain.

Fusicladium pirinum (Lib.) Fuckel = stat. conid. of *Venturia pirinum*.

Fusicladium pirinum var. **pyracanthae** Thüm. = *Fusicladium pyracanthae*.

Fusicladium pyracanthae (Otth) Rostr., *Plantepatologi*, 1902, 467; syn. *F. pirinum* (Lib.) Fuckel var. *pyracanthae* Thüm., *Mycotheca universalis*, no. 874; Sacc., *Syll.* iv, 346; Wakefield & Bisby, *TBMS*, **25**, 1941, 86.

*Scab of firethorn (*Pyracanthus coccinea*). Little reported but is common in southern and eastern districts. Listed from Clyde and Tay. A species of *Fusicladium* has also been recorded on *Crataegus orientalis* in Glos, 1933.

This disease was recognized in England many years ago (M.J.B., *Gdnrs' Chron.* 8, 1848, 716). It was described by McKay (*J. R. hort. Soc.* 69, 1944, 204–7), who discussed its control with lime sulphur.

Fusicladium saliciperdum (Allesch. & Tub.) Lind = stat. conid. of *Venturia chlorospora*.

Fusicoccum castaneum Sacc. = stat. conid. of *Cryptodiaporthe castanea*.

Fusoma triseptatum Sacc. = *Septogloeum oxysporum*.

Gaeumannomyces graminis (Sacc.) Arx & Olivier = *Ophiobolus graminis*.

Ganoderma applanatum (Pers. ex Wallr.) Pat., *Hyménomycètes de l'Europa*, Paris, 1887, 143; Sacc., *Syll.* vi, 176 (as *Fomes applanatus* (Pers.) Wallr.); Rea, *Brit. Basid.* 597.

*Heart rot of beech (*Fagus*). Frequent (Cartwright & Findlay, *Decay of Timber and its Prevention*, 1946, 112–14). Occurs also on many other leafy trees, notably poplar, willow and oak.

Geotrichum roseum Grove, see *Corticium fuciforme*.

Gibberella baccata (Wallr.) Sacc., *Syll.* ii, 553; stat. conid. *Fusarium lateritium* Nees ex Fr., *Syst. Myc.* iii, 470; Sacc., *Syll.* iv, 694; Wakefield & Bisby, *TBMS*, 25, 1941, 65; syn. *F. lateritium* var. *fructigenum* Fr.

*Bud rot of apple (*Malus*) and pear (*Pyrus*). Frequent but seasonal. Affects both wood and fruit buds, especially of Grenadier and Early Victoria. Occasional on pear. Sometimes rots apple fruits in storage.

Bud rot was described by Salmon and Ware (*J. S.-E. agric. Coll. Wye*, **21**, 1912, 392–3) and by Dillon Weston (*TBMS*, **12**, 1927, 170–2), and the parasitism of the fungus was proved by M. H. Moore (*Rep. E. Malling Res. Sta. for 1938*, 86). The fruit rot was reported by Kidd and Beaumont (*TBMS*, **10**, 1924, 115). It was this fruit rotting species which Brown and his colleagues (see, for example, *Ann. Bot., Lond.*, **38**, 1924, 379–83; **42**, 1928, 285–304) studied under the name *Fusarium blackmani* Brown & Horne.

*On black currant (*Ribes*), associated with die-back, Cumb, 1933.

*Wilt of *Dimorphotheca barberiae*. Botanic Garden, Bristol, 1949 (Nicholls, *TBMS*, **34**, 1951, 220–2).

*On *Corylus avellana* as a minor wound parasite of the nuts (M. H. Moore, *J. hort. Sci.* **25**, 1950, 213–24).

Gibberella fujikuroi (Saw.) Wr., *Z. Parasitenk.* 3, 1931, 514; stat. conid. *Fusarium moniliforme* Sheldon, *Rep. Neb. agric. Exp. Sta.* **17**, 1904, 23; Sacc., *Syll.* xxii, 1485; Wakefield & Bisby, *TBMS*, **25**, 1941, 65.

The perfect stage has not been found in Britain.

*On maize (*Zea*), causing a foot and root rot of seedlings raised, for use as cattle fodder, by sprouting the grain in cabinets containing nutrient solutions, London, 1937 (W. C. Moore, *TBMS*, **26**, 1943, 20). Singh and Wood (*Ann. Bot., Lond.*, n.s., **20**, 1956, 89) have described the production and properties of pectic enzymes secreted by *Fusarium moniliforme*.

Gibberella moricola (de Not.) Sacc., *Mich.* i. 347; *Syll.* ii, 553; Bisby & Mason, *TBMS*, **24**, 1940, 197; stat. conid. *Fusarium lateritium* Nees ex Fr. var. *mori* Desm., *Ann. Sci. nat.* 2 Ser. **8**, 1837, 10; Sacc., *Syll.* iv, 698 as *F. urticearum* (Corda) Sacc.; Wakefield & Bisby, *TBMS*, **25**, 1941, 65.

*Mulberry (*Morus*) canker. Uncommon. Sussex, 1914; Kent, 1916; Hants, 1926; Cambs, 1945.

The disease was described by Salmon and Wormald (*Gdnrs' Chron.* **60**, 1916, 95–6). The perfect stage has been found associated with the disease on mulberry at Cambridge (Brooks, *Plant Diseases*, 1953, 188) and on a dead fig twig in Dorset, 1954 (W. C. Moore, *TBMS*, **31**, 1947, 86).

Gibberella pulicaris (Fr.) Sacc., *Mich.* i, 1877, 43; *Syll.* ii, 553; Bisby & Mason, *TBMS*, **24**, 1940, 197; stat. conid. *Fusarium sambucinum* Fuckel, *Symb. myc.* 1869, 167; Sacc., *Syll.* iv, 695; Wakefield & Bisby, *TBMS*, **25**, 1941, 66.

*Hop canker. Common in the hop districts, especially under moist conditions and where drainage is defective. The disease was briefly described by Percival (*J. S.-E. agric. Coll. Wye*, **11**, 1902, 87–9) and more fully by Salmon and Wormald (*J. Minist. Agric.* **29**, 1922, 354–9; *Misc. Publ. Min. Agric. Fish. Lond.* **42**, 1925, 49–53). The perfect stage has been known in Britain since 1924 but was not determined specifically for many years (Davies, *Rep. Long Ashton Res. Sta. for 1938*, 115–23). It may be abundant on cut-back crowns in grassed down plots (Cheal & Taylor, *Plant Path.* **5**, 1956, 75). Though usually a rootstock canker it can cause bine canker (Salmon & Ware, *J. S.-E. agric. Coll. Wye*, **28**, 1931, 62–4).

*The conidial stage is sometimes associated with Fusarium foot rot of pea (see *F. solani* f. *pisi*); with root rot of sweet pea (see *F. culmorum* and *Fusarium* sp.) and viola; and is a minor agent of brown foot rot of cereals (see *F. culmorum*).

Gibberella roseum (Link) Snyder & Hansen f. **cerealis** Cooke, see *Fusarium culmorum*.

Gibberella saubinetii (Mont.) Sacc., see *G. zeae*.

Gibberella zeae (Schw.) Petch, *Ann. mycol., Berl.*, **34**, 1936, 260; Bisby & Mason, *TBMS*, **24**, 1940, 197; stat. conid. *Fusarium graminearum* Schwabe, *Fl. Anhaltina*, **2**, 1838, 285; Sacc., *Syll.* xxii, 1984; Wakefield & Bisby, *TBMS*, **25**, 1941, 65.

The name *Gibberella saubinetii* (Mont.) Sacc. was formerly used incorrectly for this species.

*Scab or ear blight of cereals. First recognized in Britain on wheat in 1928 (*Report*, vii, 19) and subsequently seen also on barley and oats, chiefly in Wales, northern England and Scotland; and on rye in Corn, 1947, and ryegrass (*Lolium*) in west Scotland, 1944. Slight except for an epidemic on oats in mid-Wales in 1942.

Bennett summarized his investigations of this fungus in a series of three papers (*AAB*, **17**, 1930, 43–58; **18**, 1931, 158–77; **20**, 1933, 377–80). For epidemics in Eire see McKay (*Sci. Proc. R. Dublin Soc.* **23**, 1943, 111–29; *J. Dep. Agric. Eire*, **43**, 1946, 31–4) and McKay and Loughnane (*Sci. Proc. R. Dublin Soc.* **24**, 1945, 9–18). McKay (ibid. **26**, 1952, 55–62) also described failures in oats due to a root rot simulating Take-all (*Ophiobolus graminis*), but caused by *Gibberella zeae*.

Gibellina cerealis Passer., *Rev. mycol.* **8**, 1886, 177; Sacc., *Syll.* ix, 740; Bisby & Mason, *TBMS*, **24**, 1940, 174.

*White foot rot of wheat (*Triticum*). Regularly seen at Rothamsted, Herts, since 1935, but not found elsewhere (Glynne, *TBMS*, **20**, 1936, 121).

Gloeodes pomigena (Schw.) Colby, *Trans. Ill. Acad. Sci.* **13**, 1920, 158; Sacc., *Syll.* xxv, 501.

*Sooty blotch of apple (*Malus*), pear (*Pyrus*) and plum (*Prunus*). Common in south and south-east England on apple in wet and sunless summers. Otherwise not prominent. Not known in Scotland. Occasional on pear, notably the variety Catillac; common on Warwickshire Drooper plum and occasional on other varieties.

Described on apple by Salmon (*Gdnrs' Chron.* **48**, 1910, 443) and on pear by Salmon and Wormald (ibid. **59**, 1916, 58–9). Reported more recently by Wormald (*Rep. E. Malling Res. Sta.* **14–15**, II Suppl. 1928, 114; **16–18**, II Suppl. 1931, 127) who also described methods of dealing with it (*J. Minist. Agric.* **43**, 1937, 923–5; *Rep. E. Malling Res. Sta. for 1936*, 194–7). *Leptothyrium pomi* (q.v.) is not infrequently associated with the *Gloeodes* on all three hosts.

Gloeosporium affine Sacc., *Mich.* i, 129; *Syll.* iii, 709; Grove, *Coelomycetes*, ii, 208; Cooke, *Fung. Pests Cult. Pl.* 166; *Grevillea*, **19**, 1890, 42.

*On leaves of *Hoya* and *Aeschynanthus* in hot-houses, Glasgow.

Gloeosporium album Osterw., *Zbl. Bakt.* II, **18**, 1907, 825; Sacc., *Syll.* xxii, 1180; Grove, *Coelomycetes*, ii, 221.

*Bitter rot of apple (*Malus*). Common in stored apples for many years but now less so than *Gloeosporium perennans* (q.v.). First recognized in 1924 (Kidd & Beaumont, *TBMS*, **10**, 1924, 107) and probably mistaken previously for *G. fructigenum*. Disease described by Wilkinson (*Rep. Long Ashton Res. Sta. for 1943*, 81–9; *AAB*, **41**, 1954, 354–7). The fungus occurs in the orchard growing saprophytically on dead wood (Edney, ibid. **44**, 1956, 113–28) and is not parasitic on wood (Corke, *J. hort. Sci.* **31**, 1956, 272–83) except that it may cause small cankers on the spurs (Ogilvie, *J. Pomol.* **13**, 1935, 140).

*Fruit rot of pear (*Pyrus*). Not uncommon in storage.
*Fruit rot of quince (*Cydonia*). Herts, 1931.

Gloeosporium ampelophagum (Passer.) Sacc., see *Elsinoe ampelina*.

Gloeosporium aquilegiae Thüm. = *Actinonema aquilegiae*.

Gloeosporium aristoteliae Smith & Ramsb., *TBMS*, **5**, 1917, 429; Sacc., *Syll.* xxv, 538; Grove, *Coelomycetes*, ii, 210.

*On leaves of *Aristotelia macqui*, Torquay, Devon, 1916.

Gloeosporium begoniae Magnaghi, *Atti Ist. bot. Univ. Pavia*, **8**, 1902, 11; Sacc., *Syll.* xviii, 451.

*Leaf spot of *Begonia*. Forth.

Gloeosporium bidgoodi Cooke, *TBMS*, **2**, 1902, 15; Sacc., *Syll.* xviii, 457; *J. R. hort. Soc.* **26**, 1901, cxxxix and cxli; Grove, *Coelomycetes*, ii, 218.

*On leaves of *Odontoglossum* in hot-houses. *Gloeosporium* sp. was noted on *Cymbidium* leaves in East Anglia, 1938.

169

Gloeosporium carpini Desm., *Ann. Sci. nat.* **20**, 1853, 214; Sacc., *Syll.* iii, 712; Grove, *Coelomycetes*, ii, 212.

*On leaves of hornbeam (*Carpinus*), Hampstead, Highgate.

Gloeosporium caulivorum Kirchn. = *Kabatiella caulivora.*

Gloeosporium concentricum (Fr.) Berk. & Br., *Ann. Mag. nat. Hist.* **5**, 1850, 455; Sacc., *Syll.* iii, 701; Grove, *Coelomycetes*, ii, 211.

*Light leaf spot of *Brassicae.* Common on cabbage, cauliflower and broccoli in northern England; sporadic elsewhere, including Scotland. Largely restricted to outer, older leaves but sometimes causes wastage in cabbage; occasional on marrow-stem kale, thousand-headed kale and swede (Stranraer, Lawes, 1946). Also occurs on seed heads, and on buttons of Brussels sprouts, especially autumn-sown ones.

Berkeley (*J. hort. Soc.* **6**, 1851, 117–21) published some notes on the fungus, and Thomson (*TBMS*, **20**, 1926, 123), who studied it, considered it was not a typical *Gloeosporium* and should be named *Cylindrosporium concentricum* Grev. Hickman, Schofield and Taylor (*Plant Path.* **4**, 1952, 129–31) obtained the perfect stage in culture. It is not yet identified but belongs to the Dermateaceae.

Gloeosporium crotalariae Massee, *Kew Bull.* 1913, 198; Sacc., *Syll.* xxv, 547; Grove, *Coelomycetes*, ii, 214.

*On shoots of *Crotalaria juncea*, Kew Gardens.

Gloeosporium cytisi Berk. & Br., *Ann. Nat. Hist.* **7**, 1881, 129; Sacc., *Syll.* iii, 705; Grove, *Coelomycetes*, ii, 215.

*Leaf spot of laburnum. War, Tay.

Gloeosporium dianthi Cooke, *Gdnrs' Chron.* **31**, 1902, 193; *J. R. hort. Soc.* **27**, 1902, 31; *Fung. Pests Cult. Pl.* 1906, 31.

*On leaves of carnation (*Dianthus*). Occasional.
*On leaves of sweet william (*Dianthus*), Ches, 1933.

Gloeosporium diervillae Grove, *J. Bot., Lond.,* **60,** 1922, 145; *Coelomycetes,* ii, 215.

*On leaves of *Weigelia florida,* Ayrshire.

Gloeosporium fagi Westend. *Notice,* **7,** 1853, 12; Sacc., *Syll.* iii, 713; Grove, *Coelomycetes,* ii, 215.

*On leaves of beech (*Fagus*), Essex, Bucks, Kent, Surrey, Yorks, widely distributed in Scotland. The species on this host is sometimes named *Gloeosporium fagicolum* Passer., *Rev. mycol.* **8,** 1886, 206; Sacc., *Syll.* x, 454, which has smaller conidia than *G. fagi.*

Gloeosporium filicinum Rostr. = *Herpobasidium filicinum.*

Gloeosporium fructigenum Berk. = stat. conid. of *Glomerella cingulata.*

Gloeosporium laeticolor Berk., see *Glomerella cingulata.*

Gloeosporium mezerei Cooke, see *Marssonina daphnes.*

Gloeosporium nervisequum (Fuckel) Sacc., *Mich.* ii, 381; *Syll.* iii, 711; Grove, *Coelomycetes,* ii, 219.

*Leaf disease of plane (*Platanus*). Not uncommon in the south and midlands. Day (*Quart. J. For.* **41,** 1947, 22) discussed the control of the disease.

Gnomonia veneta (Sacc. & Speg.) Kleb., *Jb. wiss. Bot.* **41,** 1905, 557, was described as the perfect stage of this fungus but the name is invalid (Bisby & Mason, *TBMS,* **24,** 1940, 167). *G. platani* Kleb., *Vortr. Gesamt. Bot. dtsch. bot. Ges.* **1,** 1914, 28, is a valid name but von Arx (*Antonie van Leeuwenhoek J. Microb. Serol.* **17,** 1951, 259) called it *Apiognomonia errabunda* (Rob.) Höhnel.

Gloeosporium nymphaearum Allesch. = *Ovularia nymphaearum.*

Gloeosporium orbiculare Berk. & Mart., *Cryptogamic Plants Collected in Portugal by Dr F. Welwitsch,* 1853, 7; Sacc., *Syll.* iii, 720; Grove, *Coelomycetes,* ii, 214.

*On fruits of vegetable marrow, melon, etc., when nearly ripe, King's Cliffe, Benefield, Northants, etc. Not reported for many years.

Gloeosporium orchidearum Karst. & Har., see *Hypodermium orchidearum.*

Gloeosporium paradoxum (de Not.) Fuckel, *Symb. Myc.* 277; Sacc., *Syll.* iii, 707; Grove, *Coelomycetes*, ii, 217.

*On leaves of ivy (*Hedera*). Widely distributed and fairly common throughout Britain.

Gloeosporium perennans Zeller & Childs = stat. conid. of *Neofabraea perennans.*

Gloeosporium phacidiellum Grove = stat. conid. of *Trochila laurocerasi.*

Gloeosporium phomoides Sacc. = *Colletotrichum phomoides.*

Gloeosporium platani Kleb., see *G. nervisequum.*

Gloeosporium quercinum Westend. *Exs.* no. 981; Sacc., *Syll.* iii, 714; Grove, *Coelomycetes*, ii, 222.

*Leaf spot of oak (*Quercus*). Widely distributed and common in Britain on *Q. robur*. Known for many years (Berkeley & Broome, *Ann. Mag. nat. Hist.* 3 Ser. **18**, 1866, 121).

Gloeosporium rhododendri Bri. & Cav., *Fungh. Parass.* no. 198, 1892; Sacc., *Syll.* xi, 565; Grove, *Coelomycetes*, ii, 223.

*On *Azalea* and *Rhododendron*, Devon, Corn.

Gloeosporium ribis (Lib.) Mont. & Desm. = stat. conid. of *Pseudopeziza ribis.*

Gloeosporium rosarum (Passer.) Grove = *Sphaceloma rosarum.*

Gloeosporium rufomaculans (Berk.) Thüm. = stat. conid. of *Glomerella cingulata.*

Gloeosporium salicis Westend., *Exs*. no. 1269; Sacc., *Syll*. iii, 711; Grove, *Coelomycetes*, ii, 225.

*On leaves of willow (*Salix*). Common on *S. alba*, *S. caprea* and especially *S. fragilis*. The lesions are often barren.

Gloeosporium tiliae Oudem., *Arch. néerl. Sci.* **2**, 1867, 31; Sacc., *Syll*. iii, 701; Grove, *Coelomycetes*, ii, 226.

*Leaf spot of lime (*Tilia*). Common in Scotland; less so in England. Described by A. L. Smith (*TBMS*, **2**, 1904, 55–6). Salmon and Wormald (*Gdnrs' Chron.* **58**, 1915, 193–4) attributed it to *Gloeosporium tiliaecolum* Allesch.

Gloeosporium venetum Speg., see *Elsinoe veneta*.

Gloeosporium sp.

*On leaves of arum lily (*Zantedeschia*), Glam, 1938.

Gloeotinia temulenta (Prill. & Delacr.) Wilson, Noble & Gray, *TBMS*, **37**, 1954, 31; Dennis, *Mycol. Pap.* **62**, 1956, 135; syn. *Phialea temulenta* Prill. & Delacr., *Bull. Soc. mycol. Fr.* **8**, 1892, 22; Sacc., *Syll*. xi, 404; *P. mucosa* Gray, *TBMS*, **25**, 1942, 329.

*Blind seed disease of ryegrass (*Lolium*). Common and widespread throughout Britain wherever the crop is grown, and often reduces germination.

Neill and Hyde (*N.Z. J. Sci. Tech.* **20**, 1939, 281 A–301 A) reported finding the fungus in English seed samples of ryegrass (*Lolium*), but Noble (*AAB*, **26**, 1939, 630–3) was the first to study it in Britain. Shortly afterwards Glasscock (*Nature, Lond.*, **146**, 1940, 368) observed the disease in Kent, and Gemmell (*Bull. W. Scot. agric. Coll.* **136**, 1940, 16 pp.) found it in several other districts, notably south-west Scotland. At first there was some confusion about the cause, for the common saprophyte *Pullularia pullulans* (de Bary) Berkh. (Wakefield & Bisby, *TBMS*, **25**, 1941, 53) was frequently associated with the parasite, but this was soon cleared up (Wilson, Noble & Gray, *Nature, Lond.*, **145**, 1940, 783; Muskett & Calvert, ibid. **146**, 1940, 200) and Gray (*TBMS*, **25**, 1942, 329–33) provisionally named the true parasite *Phialea mucosa*. Its relationship with *P. temulenta* described by Prillieux and Delacroix (*Bull. Soc. mycol. Fr.* **8**, 1892, 22) was meanwhile confirmed (Wilson, Noble & Gray, *Nature, Lond.*, **146**, 1940, 492; Calvert & Muskett, ibid. **153**, 1944, 287). A full account of the disease was published in 1945 (Wilson, Noble & Gray, *Trans. roy. Soc. Edinb.* **61**, 1945, 327–40), and this included comparison with the endophyte of *Lolium perenne* described in Britain by Sampson (*TBMS*, **19**, 1935, 337–43; **21**, 1937, 84–97; **23**, 1939, 316–19). Later, Wilson, Noble and Gray (ibid. **37**, 1954, 29–32) concluded that the parasite differed from other genera of the Sclerotiniaceae and from the Phialeoideae, and erected the genus *Gloeotinia* to accommodate it.

Certain aspects of blind seed disease, including its control by hot-water seed treatment, have been dealt with by Calvert and Muskett (*AAB*, **32**, 1945, 329–43) and by Noble and Gray (*Scot. J. Agric.* **25**, 1945, 94–7). Wright (*Rep. Exp. Rec. Min. Agric. N. Ire. for 1956*, **6**, 1–18) has studied the techniques for evaluating resistance in the field. Lafferty (*J. Dep. Agric. Eire*, **45**, 1948, 192–201) gave a useful summary and recorded the disease in Eire.

Glomerella cincta (Stonem.) Spauld. & Schrenk, *Science*, **17**, 1903, 750; Sacc., *Syll.* xvii, 573 and xvi, 453; stat. conid. *Colletotrichum cinctum* Stonem., *Bot. Gaz.* **26**, 1898, 106; Grove, *Coelomycetes*, ii, 230.

The perfect stage has not been seen in Britain.

*On leaves of orchid in a conservatory. Glasgow.

Glomerella cingulata (Stonem.) Spauld. & Schrenk, *Science*, **17**, 1903, 750; Sacc., *Syll.* xvii, 573 and xvi, 453; Bisby & Mason, *TBMS*, **24**, 1940, 144; stat. conid. *Gloeosporium fructigenum* Berk., *Gdnrs' Chron.* **16**, 1856, 245; Sacc., *Syll.* iii, 718; Grove, *Coelomycetes*, ii, 221.

*Bitter rot of apple (*Malus*) fruit. Perhaps fairly common on the tree but infrequent in stored fruit. It is probable that the fruit rot caused by *Gloeosporium album* (q.v.) has frequently been recorded as due to *G. fructigenum*. A useful historical account was given by M. C. C[ooke] in *Gdnrs' Chron.* **36**, 1904, 249–51. (See also Wilkinson, *Rep. Long Ashton Res. Sta. for 1938*, 84–90.)

*Bitter rot of cherry (*Prunus*). Clyde.

*On fig (*Ficus*). Reported by M. J. B[erkeley] (*Gdnrs' Chron.* **24**, 1864, 818) from Sussex. There is a further record in *Gdnrs' Chron.* **11**, 1892, 666, and in 1926 *Gloeosporium* sp. was found on fig shoots under glass in Northumb.

*Ripe rot of grape (*Vitis*). Not infrequent under glass. Described by Wormald (*Gdnrs' Chron.* **88**, 1930, 498–500). The imperfect stage on this host has usually been called *Gloeosporium rufo-maculans* (Berk.) Thüm.; it was first described by Berkeley (ibid. **14**, 1854, 676) as *Septoria rufomaculans*. See also W. G. Smith (ibid. **8**, 1890, 657) and Grove (*Coelomycetes*, ii, 209).

*On peach and nectarine (*Prunus*). Rare. Reported by M. J. B[erkeley] (*Gdnrs' Chron.* **19**, 1859, 603–4) and W. G. Smith (ibid. **25**, 1886, 796) under the name *Gloeosporium laeticolor* Berk. (loc. cit.), Sacc., *Syll.* iii, 718.

*Fruit rot of pear (*Pyrus*) and quince (*Cydonia*). Occasional.

*On privet (*Ligustrum*). Brooks (*Plant Diseases*, 1953, 230) reported recently seeing in England a disease similar to that described by Atkinson (*Bull. Cornell agric. Exp. Sta.* no. 49, 1892). It was reported in a privet hedge in Berks in 1956 (Roberts, *Plant Path.* **6**, 1957, 76).

Glomerella lagenarium (Passer.) Stevens, see *Colletotrichum lagenarium.*

Glomerella lindemuthianum Shear, see *Colletotrichum lindemuthianum.*

Glomerella phacidiomorpha (Ces.) Petrak, *Ann. mycol., Berl.*, **25**, 1927, 251; stat. conid. *Colletotrichum rhodocyclum* (Mont.) Petrak, loc. cit.

*Leaf blotch of *Phormium*. On leaves of *P. tenax*, Devon, 1934 (Kinghorn, *AAB*, **23**, 1936, 30–44).

Glomerella phomoides Swank, see *Colletotrichum phomoides.*

Gnomonia erythrostoma (Fr.) Auersw., in Rabenhorst, *Mycologia europaea*, 1869, 26; Sacc., *Syll.* i, 566; Bisby & Mason, *TBMS*, **24**, 1940, 166.

*Leaf scorch of cherry (*Prunus*). Common in Kent and epidemic there in some years; occasional elsewhere, but not known in Scotland. Less common than formerly.

The disease was described by Carruthers (*J. R. hort. Soc.* **25**, 1900–1, 313–16; *J. R. agric. Soc.* **62**, 1901, 241–6), by Salmon (*J. S.-E. agric. Coll. Wye,* **15**, 1906, 221–9; **16**, 1907, 286–91; *J. Bd Agric.* **14**, 1907, 334–44) and by Brooks (*Ann. Bot., Lond.,* **24**, 1910, 585–605). Goodwin, Salmon and Ware (*J. S.-E. agric. Coll. Wye,* **25**, 1928, 147–51) carried out spraying trials against it.

Gnomonia fructicola (Arnaud) Fall., *Canad. J. Bot.* **29**, 1951, 310; stat. conid. *Zythia fragariae* Laib., *Arb. Kais. Biol. Anst.* **6**, 1908, 76; Sacc., *Syll.* xxv, 485; syn. *Phyllosticta grandimaculans* Bub. & Krieg., *Ann. mycol., Berl.,* **10**, 1912, 46; Sacc., *Syll.* xxv, 66; *P. gei* Bres., *Hedwigia,* **39**, 1900, 325; Sacc., *Syll.* xvi, 831; Grove, *Coelomycetes,* i, 39.

*Leaf blotch of strawberry (*Fragaria*). Local in Kent and Worcs (Wormald & Montgomery, *Gdnrs' Chron.* **110**, 1941, 180; *Rep. E. Malling Res. Sta. for 1941,* 44). Occasional elsewhere. Flint, 1950; Fife and the Lothians, 1946; Cheddar, Som, 1951; Devon and Hants, 1954; Argyll, 1955. Wormald (*Gdnrs' Chron.* **116**, 1944, 160–1) showed the perfect stage was a *Gnomonia* but did not name it. It occurs naturally in Britain along with the imperfect stage.

*Leaf blotch of *Geum*. Rare. On *Geum × borisii*, Harpenden, Herts, 1942, in the imperfect state only (W. C. Moore, *TBMS,* **32**, 1949, 95).

Gnomonia leptostyla (Fr.) Ces. & de Not., *Schema,* Sfer., 1863, 232; Sacc., *Syll.* i, 568; Bisby & Mason, *TBMS,* **24**, 1940, 166; stat. conid. *Marssonina juglandis* (Lib.) Magn., *Hedwigia,* **45**, 1906, 88; Sacc., *Syll.* iii, 768; Grove, *Coelomycetes,* ii, 275.

*Leaf blotch of walnut (*Juglans*). Common and widely distributed in England and Wales; also occurs in Scotland (Solway); occasionally found on the fruit.

What appeared to be a distinct species of *Marssonina* was found on this host in Som in 1928.

GNOMONIA

Gnomonia padicola (Lib.) Kleb., *Z. PflKrankh.* **18**, 1908, 137; stat. conid. *Actinonema padi* Fr., *Summ. Veg. Scand.* no. 424; Sacc., *Syll.* iii, 409; Grove, *Coelomycetes*, ii, 271.
The perfect stage has not been seen in Britain.
*Leaf spot of bird cherry. Not uncommon on *Prunus padus* in Scotland.

Gnomonia rubi (Rehm) Wint., *Hedwigia*, **26**, 1887, 62; Sacc., *Syll.* ix, 673; Bisby & Mason, *TBMS*, **24**, 1940, 166.
*Die-back of rose. Wisley, Surrey, 1921. Described on rambler roses by Dowson (*Gdnrs' Chron.* **76**, 1924, 374; *J. R. hort. Soc.* **50**, 1925, 55–72), who also recorded it on brambles and wild roses at Keswick, Cumb, 1922, and on loganberry in Kent. The disease has not been reported in recent years.

Gnomonia veneta (Sacc. & Speg.) Kleb., see *Gloeosporium nervisequum*.

Graphiola phoenicis Poit., *Ann. Sci. nat.* **3**, 1824, 473; Sacc., *Syll.* vii, 522; Sampson, *TBMS*, **24**, 1940, 305; Ainsworth & Sampson, *Brit. Smut Fungi*, 112.
*On leaves of the date palm (*Phoenix dactylifera*) and other members of the Palmae in greenhouses. Sheffield, Glasgow, Devon.

Graphium ulmi Schwarz = stat. conid. of *Ceratostomella ulmi*.

Griphosphaeria corticola (Fuckel) Höhnel, *Ann. mycol., Berl.*, **16**, 1918, 87; Sacc., *Syll.* xxiv, 1024; stat. conid. *Coryneopsis microsticta* (Berk. & Br.) Grove, *J. Bot., Lond.*, **70**, 1932, 34; *Coelomycetes*, ii, 328; Sacc., *Syll.* iii, 775 (as *Coryneum microstictum* Berk. & Br.).
*Canker and die-back of rose. Not uncommon in England and Wales, especially on hybrid tea varieties; also reported from Scotland. The fungus is usually found in the imperfect stage. The disease was fully investigated by Brooks and Alaily (*AAB*, **26**, 1939, 213–26).

*Associated with an internal rot of apple fruits, Mon, 1945. Kidd and Beaumont (*TBMS*, **10**, 1924, 108) listed *Coryneum microstictum* Berk. & Br. var. *mali* Kidd & Beaumont among the fungi that rot stored apples.

Griphosphaeria nivalis (Schaffnit) Müller & von Arx, *Phytopath. Z.* **24**, 1955, 356; syn. *Calonectria nivalis* Schaffnit, *Landw. Jb.* **43**, 1912, 521; *Mykol. Zbl.* ii, 1913, 257; Sacc., *Syll.* xxiv, 681; stat. conid. *Fusarium nivale* (Fr.) Ces. in Klotzsch, *Herbarium vivum mycologicum*, no. 1439; Sacc., *Syll.* x, 726; Wakefield & Bisby, *TBMS*, **25**, 1941, 65. See Petch, ibid. **27**, 1945, 150.

*Snow mould of turf. The commonest and most harmful turf disease in Britain. Occurs in many districts on bowling and golf greens and lawns. The most susceptible grasses are *Agrostis stolonifera*, *Poa annua*, *Festuca rubra* and *F. ovina*. *Lolium perenne* is said to be markedly resistant, but this is evidently not true for North Scotland (Gray & Nicholson, *Trans. bot. Soc. Edinb.* **37**, 1957, 123–8). Descriptions of the disease are given by Sampson (*J. Bd Greenk. Res.* **2**, 1931, 116–18) and by Bennett (ibid. **3**, 1933, 79–86), who also dealt with its control with Bordeaux Mixture containing malachite green. A full report on its causes, prevention and cure was issued by the Board of Greenkeeping Research in 1936, 16 pp., and later by Drew Smith (*J. Sports Turf Res. Inst.* **8**, 1953, 230–52), who also tested some modern fungicides against it (ibid. **8**, 1954, 445–7), including the antibiotic griseofulvin (ibid. **9**, 1956, 203–9; see also *AAB*, **45**, 1957, 206).

*Brown foot rot and seedling blight of cereals. *Fusarium nivale* was one of a number of species of *Fusarium* found by Bennett (*AAB*, **20**, 1933, 272–90) to cause brown foot rot of cereals in the north of England (see also *F. culmorum*). Noble and Mont-gomerie (*Scot. Agric.* **34**, 1954, 51–2; *TBMS*, **39**, 1956, 449–59) showed that it is common on oats in Scotland and is often seed-borne there in wheat, barley and oats (Noble, *Scot. Agric.* **36**, 1956, 86–90). The perithecia develop freely on the culms and leaf sheaths of affected plants in Scotland, northern England and occasionally elsewhere (Hants, 1953).

GUIGNARDIA

Guignardia aesculi (Peck) Stewart, *Phytopathology*, **6**, 1916, 9.

*On horse chestnut (*Aesculus*), causing leaf blotching and premature leaf-fall in nursery stock, Glos, 1953.

See also *Phyllosticta paviae*, which is probably the imperfect stage.

Guignardia bidwelli (Ellis) Viala & Ravaz, see *Elsinoe ampelina*.

Gymnosporangium clavariiforme (Pers.) DC., *Flor. Fr.* ii, 217; Sacc., *Syll.* vii, 737 p.p.; Grove, *Brit. Rust Fungi*, 304; Wilson & Bisby, *TBMS*, **37**, 1954, 64.

On hawthorn and juniper.

*Rust of hawthorn (*Crataegus*). Aecidia on *C. monogyna* and *C. oxyacantha*. Infrequent. Salop, Hants, Westmorland, Sussex, west Scotland, Moray.

*Branch rust of juniper (*Juniperus*). Teleutospores fairly common on *J. communis*; also on *J. nana* and other garden species.

(See *G. fuscum* for occurrence on pear.)

Gymnosporangium confusum Plowr., *Brit. Ured. Ustil.* 1889, 232; Sacc., *Syll.* xii, 284; Grove, *Brit. Rust Fungi*, 306; Wilson & Bisby, *TBMS*, **37**, 1954, 64.

*Cluster cup rust of medlar (*Mespilus*). Aecidial stage. Rare. East Malling, Kent, 1931 (M. H. Moore, *TBMS*, **28**, 1945, 13–15) and, from 1943, seen most years on the same tree until about 1955, with teleutospores on adjacent plants of *Juniperus sabina*.

The aecidial stage was also listed by Wilson and Bisby (loc. cit.) on *Crataegus monogyna*, *C. oxyacantha* and quince (*Cydonia oblonga*).

In 1921 (*Report*, iv, 84) this rust in its teleuto-stage was said to be frequent annually on junipers in the chalk districts near Worthing, Sussex, and to have been found at Sevenoaks, Kent.

Gymnosporangium fuscum DC., *Flor. fr.* ii, 217; Wilson & Bisby, *TBMS*, **37**, 1954, 64; syn. *G. sabinae* Wint., *Pilze Deutschl.*, 232; Sacc., *Syll.* vii, 739; Grove, *Brit. Rust Fungi*, 308; Cooke, *Fung. Pests Cult. Pl.* 1906, 120.

On pear and juniper.

*Pear rust. Aecidia on *Pyrus communis*. Now scarce, but has been seen from time to time since 1850 and earlier thought to be not uncommon. Was formerly listed sometimes as *Gymnosporangium clavariiforme* (q.v.) and sometimes as *G. sabinae*. Most records have been from southern England (Berkeley, *Gdnrs' Chron.* **16**, 1856, 693; **24**, 1864, 1009; Güssow, ibid. **40**, 1906, 134).

The teleuto-stage occurs on *Juniperus sabina*. A teleuto-stage was formerly found frequently on the chalk downs of Sussex, where wild juniper was common, but its precise identity is now uncertain (cf. *Gymnosporangium confusum*).

Gymnosporangium juniperi Link, *Observationes in Ordines plantarum naturales*, i, 1809, 7; Sydow, *Monogr. Uredin.* iii, 27; Grove, *Brit. Rust Fungi*, 307; Wilson & Bisby, *TBMS*, **37**, 1954, 64.

*Rowan rust. The aecidia occur on rowan (*Sorbus aucuparia*) and the teleuto-stage on *Juniperus communis*. Scotland; rare in England.

Gymnosporangium juniperi-virginianae Schw., *Schr. naturw. Ges. Leipzig*, i, 1822, 74; Sydow, *Monogr. Uredin.* iii, 73; Sacc., *Syll.* vii, 740 (as *G. macropus* Link).

*Apple rust. Has occasionally been intercepted on imported fruits, 1934, 1938. In 1933 a species of *Gymnosporangium* was found in small amount on apple in Hants.

Gymnosporangium sabinae Wint. = *G. fuscum.*

Hadrotrichum virescens Sacc. & Roum., *Mich.* ii, 640; Sacc., *Syll.* iv, 301; Wakefield & Bisby, *TBMS*, **25**, 1941, 73.

*On leaves of *Agrostis*, producing dark olive-green spots. Not uncommon, but unimportant.

Hainesia lythri (Desm.) Höhnel, *S.B. Akad. Wiss. Wien*, **115**, 1906, 687.

*Associated with root rot of strawberry (*Fragaria*) in Britain (Berkeley & Lauder-Thomsen, *J. Pomol.* **12**, 1934, 222) but doubtfully parasitic. Clyde, Kent.

Shear and Dodge (*Mycologia*, **13**, 1921, 162) have described this species as one of the imperfect stages of *Pezizella lythri* (Desm.) Shear & Dodge.

Hapalosphaeria deformans Syd., *Ann. mycol., Berl.*, **6**, 1908, 305; Sacc., *Syll.* xxii, 868; Grove, *Coelomycetes*, i, 130.

*Stamen blight of bramble (*Rubus*). First recorded in the British Isles on *Rubus fruticosus* at Aberlady, East Lothian, in 1904 (Wilson, *TBMS*, **7**, 1921, 84), but there is a specimen on *R. latifolius* from Perth, 1871, in the Perth Museum. Not seen again until 1947 and then at the same place as well as in Perthshire and near Edinburgh (Noble & Gray, *Gdnrs' Chron.* **122**, 1947, 92). Now known to be rather common in the Lothians. Found at Penrith, Cumb, and in Hants in 1954.

*Stamen blight of raspberry. First seen on cultivated raspberry in 1952 at a number of places in Scotland (Dee, Angus, Perth) with infected brambles invariably in the neighbourhood. 1953, Inverness-shire; Forfar; on wild raspberry, Kinross, Aberdeen.

Haplobasidium pavoninum Höhnel, *Ann. mycol., Berl.*, **3**, 1905, 407; Sacc., *Syll.* xxii, 1350; Wakefield & Bisby, *TBMS*, **25**, 1941, 73.

*Peacock spot of *Aquilegia*. Forming 'peacock spots' on leaves of *Aquilegia* sp., St Keverne, Corn, May 1938 (W. C. Moore, *TBMS*, **22**, 1939, 266).

Heleococcum aurantiacum Jørgensen, *Bot. Tidsskr.* **37**, 1922, 417.

*In mushroom compost. Very rare. Fordingbridge, Hants, Nov. 1934; in compost samples, Worthing, Sussex, Feb. 1957 (Wood, *Nature, Lond.*, **180**, 1957, 283).

Helicobasidium purpureum Pat., *Bull. Soc. bot. Fr.* **32**, 1885, 171; Sacc., *Syll.* vi, 666; Rea, *Brit. Basid.* 726; stat. mycel. *Rhizoctonia crocorum* Fr., *Syst. Myc.* ii, 265; Sacc., *Syll.* xiv, 1175 (as *R. violacea* Tul.); Wakefield & Bisby, *TBMS*, **25**, 1941, 102.

*Violet root rot. Common and widely distributed on a wide range of crops in England and Wales. Rare in Scotland.

Buddin and Wakefield (*AAB*, **11**, 1924, 292–309) successfully grew *Rhizoctonia crocorum* in pure culture, and ultimately (*TBMS*, **12**, 1927, 116–40; **14**, 1929, 97–9) proved the connexion between it and *Helicobasidium purpureum*. Garrett (ibid. **29**, 1946, 114–27) followed this up with a study of some of the factors affecting the production and growth of mycelial strands, and (ibid. **32**, 1950, 217–23) the effect of substratum on its survival in the soil.

*POTATO (*Solanum*). Reported off and on since 1847 (*Gdnrs' Chron.* **7**, 1847, 603; **13**, 1853, 116), and probably more common and widely distributed now than at any time in England and Wales, but rarely causes significant loss; Forth and Tay. (See *J. Bd Agric.* **2**, 1896, 437–9; **12**, 1906, 667–8.)

ROOTS

*Sugar beet (*Beta*). Known at least since 1902 and now the most prevalent and damaging root disease of sugar beet. Occurs mainly in eastern districts, Yorks and Salop, especially on light, sandy and limestone soils, but often has no adverse effect in the field, though obvious as soon as the roots are lifted. In Scotland, Moray and Dumfries.

Less common on mangolds; occasional on fodder beet (Dumfries, 1953).

The disease was described by Stirrup (*Brit. Sug. Beet Rev.* **13**, 1939, 232–4), by Hull and Wilson (*AAB*, **33**, 1946, 420–33), and by Hull (*Bull. Minist. Agric., Lond.*, no. 142, 1950, 38).

*Turnip and swede (*Brassica*). Not uncommon.

PULSE

*On roots of broad bean (*Vicia*), Glam, Lancs, Kent, Suffolk.

PASTURE AND FORAGE CROPS

*On red and white clover, alsike, trefoil and lucerne. Occasional since 1922 (Ware, *J. Minist. Agric.* 30, 1923, 48–52; *TBMS,* 14, 1929, 94–5). Erect violet stromata found among grass and white clover in old turf, Salop, 1954.

*Meadow fescue (*Festuca*). Colesbourne, Glos, 1948.

Poa spp. (See Ware, *J. Minist. Agric.* 30, 1923, 51.)

VEGETABLES

Asparagus. Fairly common in Worcs. Widely distributed elsewhere in England, mainly on older plants. Known since 1853 (*Gdnrs' Chron.* 13, 1853, 116). Barker and Gimingham (*Rep. Long Ashton Res. Sta. for 1916,* 39; *for 1917,* 28–32) experimented with various soil treatments.

*Carrot (*Daucus*). Frequent throughout Britain, sometimes rendering the roots worthless.

*Celery (*Apium*). Kent, 1930 (*Report,* vii, 56); I. of Ely, 1945; east and south-east England, a few records in 1947–55.

*Chicory (*Cichorium*). Hunts, 1933; Suffolk, 1939; Norfolk and I. of Ely, 1947.

*Cucumber (*Cucumis*). Worcs, 1954.

*Parsley (*Petroselinum*). Oxon, 1941; east midlands, 1949; Sussex, 1950; Rutland, 1953.

*Parsnip (*Pastinaca*). Wye, Kent, 1906; Worcs, 1948.

*Rhubarb (*Rheum*). Portobello, N.B., 1907.

*Salsify (*Tragopogon*). Wye, Kent, 1906; Oxon, 1941.

*Seakale (*Crambe*). Occasional. Described by Salmon and Crompton (*J. S.-E. agric. Coll. Wye,* 17, 1908, 348–53; *Gdnrs' Chron.* 44, 1908, 1–3).

FRUIT

*Apple (*Malus*). Dartford, Kent, 1932 (Salmon & Ware, *J. S.-E. agric. Coll. Wye,* 31, 1933, 13); Surrey, 1953.

*Black currant (*Ribes*). Kent (*Report,* v, 80); Bristol, 1922.

*Gooseberry (*Ribes*). Dartford, Kent, 1932; East Malling, Kent, 1948.

*Raspberry (*Rubus*). Kent, Worcs (*Report,* v, 80).

*Strawberry (*Fragaria*). Kent, 1931.

PARASITES

ORNAMENTAL AND OTHER PLANTS

*Antirrhinum. Romney Marsh, Kent, 1946.

*Arum lily (*Zantedeschia*). War, 1951.

*Carnation (*Dianthus*). Berks, Essex; on pink, Norfolk, 1950.

*Crocus. Rare. 1860 (Berkeley, *Outl. Brit. Fung.* 410); Hereford, 1926 (*Report*, vi, 68).

*Holly (*Ilex*). Yorks, 1934.

*Iris. On *I. germanica*, Harpenden, 1929 (W. C. Moore, *Diseases of Bulbs*, 1939, 143); Buntingford, Herts, 1949.

*Pansy (*Viola*). Wirksworth (*J. Bd Agric.* **16**, 1909, 390).

*Phlox. Killing the plant. Berks, 1932.

*Poppy (*Papaver*). On oriental poppy. Hants, 1928.

*Primula. Bricket Wood, Herts, 1952 (Jenkins, *Plant Path.* **2**, 1953, 106).

*Solomon's seal (*Polygonatum*). Honiton, Devon, 1956.

*Teazle (*Dipsacus*). Som, 1949, 1955.

*Tulip. Manchester, 1936.

TREES

*Ash (*Fraxinus*). Perfect state observed on bark at Alresford, Hants, 1924, and Ipplepen, Devon.

*Sitka spruce (*Picea*). On *P. sitchensis* in two areas in forest nursery, Beauly, Inverness-shire, 1928 (Watson, *Scot. For. J.* **42**, 1928, 58–61; *TBMS*, **14**, 1929, 95–6).

WEEDS

Among weed hosts in Britain may be mentioned *Alliaria officinalis*, wild carrot (*Daucus*), *Chenopodium album*, *Convolvulus arvensis*, *Mentha arvensis*, *Mercurialis perennis*, *Plantago major*, *Polygonum aviculare*, *P. persicaria*, *Ranunculus repens*, *Rumex acetosella*, *R. obtusifolius*, *Taraxacum officinale*, *Urtica dioica* and *Veronica agrestis*.

Helminthosporium allii Campanile, *Nuovi Ann. Minist. Agric.* **4**, 1924, 87.

*Bulb canker of garlic (*Allium*). Slight in a consignment of bulbs imported from the Argentine, and about to be planted in Worcs, 1942 (W. C. Moore, *TBMS*, **26**, 1943, 22–3).

HELMINTHOSPORIUM

Helminthosporium atrovirens (Harz) Mason & Hughes = *Spondylocladium atrovirens*.

Helminthosporium avenae Eidam = stat. conid. of *Pyrenophora avenae*.

Helminthosporium dematioideum Bubák & Wróblewski, *Hedwigia*, **57**, 1916, 337; Sacc., *Syll.* xxv, 820.
*On *Anthoxanthum odoratum*, Dorking, Surrey, June 1945 (*TBMS*, **29**, 1946, 160).

Helminthosporium dictyoides Drechsler, *J. agric. Res.* **24**, 1923, 679.
*Net blotch of *Lolium*. Kew (*TBMS*, **29**, 1946, 159); Scotland. (See also *Helminthosporium siccans*.)

Helminthosporium dictyoides var. **phlei** Graham, *Phytopathology*, **45**, 1955, 228.
*Leaf blight of timothy (*Phleum*). Noted occasionally on seed and growing plants at Edinburgh. Fairly common in Carse of Stirling and Midlothian. The first record of a *Helminthosporium* on *Phleum* in England appears to be one from Cambridge in 1955 (*Rep. nat. Inst. agric. Bot. for 1955*, 22).

Helminthosporium gramineum Rabenh. = stat. conid. of *Pyrenophora graminea*.

Helminthosporium papaveris Hennig = stat. conid. of *Pleospora calvescens*.

Helminthosporium sativum Pamm., King & Bakke = stat. conid. of *Ophiobolus sativus*.

Helminthosporium siccans Drechsl., *J. agric. Res.* **24**, 1923, 682.
*Leaf spot of ryegrass (*Lolium*). Occurs on perennial and Italian ryegrass. First recorded by Sampson and Western (*TBMS*, **24**, 1940, 255–63), but disease known in Wales since 1921. Also occurs in many English counties and relatively common in Scotland. Occasional on meadow fescue (*Festuca pratensis*).

Dovaston (*TBMS*, **31**, 1948, 249–53) reported both spot and net lesions on perennial and Italian ryegrass at Auchincruive, Scotland. He obtained an ascigerous stage which produced net lesions on Italian ryegrass. He named it *Pyrenophora lolii* n.sp. and said the conidial stage corresponded to *Helminthosporium siccans*.

Helminthosporium stenacrum Drechsl., *J. agric. Res.* **24**, 1923, 683.
*On *Agrostis* sp., Kew Gardens, 1945 (*TBMS*, **29**, 1946, 161); on *A. stolonifera* in golf green, Corn (Drew Smith, *J. Sports Turf Res. Inst.* **9**, 1955, 73). *Helminthosporium* sp. also recorded on *Agrostis tenuis* in turf on a cricket pitch, London.

Helminthosporium teres Sacc. = stat. conid. of *Pyrenophora teres*.

Helminthosporium triseptatum Drechsl., *J. agric. Res.* **24**, 1923, 686.
*On *Holcus lanatus*. Hackhurst Downs, Surrey, 1945 (*TBMS*, **29**, 1946, 160).

Helminthosporium tritici-repentis Died. ex Drechsl., *J. agric. Res.* **24**, 1923, 667.
*On *Agropyron repens*, Kew, 1945 (*TBMS*, **29**, 1946, 159).

Helminthosporium vagans Drechsl., *J. agric. Res.* **24**, 1923, 688.
*On *Poa pratensis*, Aberystwyth, Cards, Cambs, Forth, Clyde, Shetland. First recorded by Sampson and Western (*TBMS*, **24**, 1940, 255–63). Common in turf, causing leaf spot and lesions on the stem bases and rhizomes (Drew Smith, *J. Sports Turf Res. Inst.* **9**, 1955, 71–2).

Hemeleia americana Massee, *Gdnrs' Chron.* **38**, 1905, 153; Sacc., *Syll.* xxi, 599; Grove, *Brit. Rust Fungi*, 381; Wilson & Bisby, *TBMS*, **37**, 1954, 77.
*Rust of *Cattleya*. Massee (loc. cit.) originally described this rust on *Oncidium cavendishianum* recently imported, and on *Oncidium* sp. growing in England for some time, but later (*Kew*

Bull. 1906, 40) listed the host as *Cattleya dowiana* imported from Costa Rica, 1899.

Hemileia oncidii Griff. & Maubl., *Bull. Soc. mycol. Fr.* **25**, 1909, 138; Sacc., *Syll.* xxi, 599; Grove, *Brit. Rust Fungi*, 383; as *Uredo behnickiana* P. Henn. in Wilson & Bisby, *TBMS*, **37**, 1954, 77.

*Uredo-stage on *Oncidium varicosum* imported from Brazil, 1909.

Hemileia phaji Syd., see *Uredo phaji*.

Hendersonia acicola Münch. & Tubeuf, *Nat. Z. Landw. Forstw.* **8**, 1910, 44; Sacc., *Syll.* xxii, 1069.

*On *Pinus*. Common throughout Britain. It causes whitish grey discoloration and defoliation of *P. sylvestris* and *P. mugo* (Laing, *Scot. For. J.* **43**, 1929, 48–52). Also on Corsican pine (*P. nigra* var. *poiretiana*). It has been regarded as the imperfect stage of *Hypodermella sulcigena*, but this is still doubtful.

Hendersonia rubi Sacc., *Syll.* iii, 424; Grove, *Coelomycetes*, ii, 83. On p. 330, however, Grove transfers this species to his genus *Coryneopsis* as *C. rubi* n.c. on bramble (*Rubus fruticosus*) and loganberry, and *C. rubi* forma *rubi-idaei* Brun., *Rev. mycol.* 1886, 141; Sacc., *Syll.* x, 321 on cultivated raspberry and loganberry (*Kew Bull.* 1913, 198).

*On raspberry canes (*Rubus*). Common in Britain but doubtfully parasitic. Has also been reported on loganberry (Corn, Worcs, Carns) and blackberry (Som).

Herpobasidium filicinum (Rostr.) Lind, *Ark. Bot.* **7**, 1908, 5; Sacc., *Syll.* xxi, 444 and xvi, 198; syn. *Gloeosporium filicinum* Rostr., in Thüm. *Mycotheca universalis*, no. 2083, 1881.

*Leaf spot of ferns. On *Dryopteris filix-mas*, Harpenden, Herts, 1931; Montgomery, 1942; on *D. phegopteris*, Scotland, 1932 (*TBMS*, **29**, 1946, 144).

Herpotrichia nigra Hartig, *Hedwigia*, **27**, 1888, 13; Sacc., *Syll.* ix, 858; Bisby & Mason, *TBMS*, **24**, 1940, 177.

*On leaves of seedling spruce (*Picea*). Yorks, 1905 (*J. Bd Agric.* **12**, 1905, 177–9).

Heteropatella antirrhini Buddin & Wakef., *TBMS*, **11**, 1928, 188; Grove, *Coelomycetes*, ii, 156.

*Shot hole of antirrhinum. Fairly common in southern England; Perthshire.

The disease was first recognized in 1917 (Wakefield, *Kew Bull.* 1918, 233), and the pathogen, then regarded as a hyphomycete, was named *Cercosporella antirrhini* Wakef. (Sacc., *Syll.* xxv, 747). Seen again by Cayley (*Gdnrs' Chron.* **68**, 1920, 158). Buddin and Wakefield (ibid. **76**, 1924, 150–2; *TBMS*, **11**, 1926, 169–86) studied the fungus in pure culture and obtained a pycnidial stage which they named *Heteropatella antirrhini*. Later (*TBMS*, **14**, 1929, 215–21) they concluded that the *Cercosporella* stage was an acervulus and renamed it *Pseudodiscosia antirrhini* (Wakef.) Buddin & Wakef.

Heteropatella dianthi Buddin & Wakef. = *H. valtellinensis*.

Heteropatella valtellinensis (Trav.) Wr., *Z. Parasitenk.* **3**, 1931, 499; Grove, *Coelomycetes*, ii, 156; syn. *H. dianthi* Buddin & Wakef., *TBMS*, **14**, 1929, 220.

*Leaf rot of carnation (*Dianthus*). Frequent in south and west England and eastern Scotland. Also occurs on border pinks.

This fungus was first recognized in Jan. 1927 at Brighton (Salmon & Ware, *Gdnrs' Chron.* **81**, 1927, 196–7, 216) and was identified with *Pseudodiscosia dianthi* Hösterm. & Laub. (*Gartenwelt*, **25**, 1921, 65). Later, Buddin and Wakefield (*TBMS*, **14**, 1929, 215–21) described a pycnidial stage and named it *Heteropatella dianthi*.

Heterosporium allii Ell. & Mart., *J. Mycol.* **1**, 1885, 100; Sacc., *Syll.* iv, 480.

*On leaves of chives (*Allium schoenoprasum*), Crowborough, Sussex, 1934 (W. C. Moore, *TBMS*, **29**, 1946, 90–2).
*On moribund leaves of leek (*Allium porrum*), Sussex, Herts, 1950.

Heterosporium allii Ell. & Mart. var. **cepivorum** Nich. & Aggery, *Rev. Path. veg.* **14**, 1927, 195–8; Wakefield & Bisby, *TBMS*, **25**, 1941, 92.

*Leaf blotch of onion (*Allium cepa*). Not uncommon in southwest England. Occasional elsewhere in the south and Wales (W. C. Moore, *TBMS*, **29**, 1946, 90–2).

Heterosporium auriculae Cooke, *J. R. hort. Soc.* **27**, 1902, 380; Sacc., *Syll.* xxii, 1384; Wakefield & Bisby, *TBMS*, **25**, 1941, 92.

*On leaves of auricula (*Primula*). First seen about 1888 and little or nothing heard of it since.

Heterosporium echinulatum (Berk.) Cooke = stat. conid. of *Didymellina dianthi*.

Heterosporium gracile Sacc. = stat. conid. of *Didymellina macrospora*.

Heterosporium ornithogali Klotzsch. ex Cooke, *Grevillea*, **5**, 1877, 123; Sacc., *Syll.* iv, 480; Wakefield & Bisby, *TBMS*, **25**, 1941, 93.

*On *Ornithogalum*. Occasional (W. G. Smith, *Gdnrs' Chron.* **3**, 1888, 658).

Heterosporium phlei C. T. Gregory, *Phytopathology*, **9**, 1919, 580; Sacc., *Syll.* xxv, 815.

*Leaf spot of timothy (*Phleum*). On seed crop, Axminster, Devon, 1950, first British record; Dee, Edinburgh, 1955 (*Heterosporium* sp.).

Heterosporium syringae Oudem., *Ned. kruidk. Arch.* 3 Ser. **1**, 1898, 529; Sacc., *Syll.* xvi, 1065; Wakefield & Bisby, *TBMS*, **25**, 1941, 93. The attribution of the epithet to Klebahn

(*Krankheiten des Flieders*, 1909, 11), given in Sacc., *Syll.* xxii, 1386, was an error.

*Leaf blotch of lilac (*Syringa*). Occasional in England since 1911 (Massee, *Kew Bull.* 1911, 81–2); Devon, Glos. A species of *Heterosporium* was recorded on lilac in west Scotland in 1950.

Heterosporium variabile Cooke, *Grevillea*, **5**, 1877, 123; Sacc., *Syll.* iv, 480; Wakefield & Bisby, *TBMS*, **25**, 1941, 93.

*Leaf spot of spinach (*Spinacia*). Three records only, but probably more widespread. Welshpool, 1877; Appledore, Kent, 1939; Wimborne, Dorset, 1944 (Glasscock & Ware, *Gdnrs' Chron.* **106**, 1939, 100–2; W. C. Moore, *TBMS*, **28**, 1945, 129).

Hormiactis sp.

*Cap spot of mushroom (*Agaricus*). Everingham, Yorks, 1949; Oulton, Yorks, 1950. Parasitism doubtful.

Hyalodendron album (Dowson) Diddens = *Ramularia deusta*.

Hyalopsora polypodii (Diet.) Magn., *Ber. dtsch. bot. Ges.* **19**, 1901, 582; Sacc., *Syll.* xvii, 268 and vii, 857; Grove, *Brit. Rust Fungi*, 375; Wilson & Bisby, *TBMS*, **37**, 1954, 64.

*On *Cystopteris fragilis* and var. *dentata*. Not uncommon in gardens and greenhouses.

Hydnum diversidens Fr., *Syst. Myc.* i, 411; Sacc., *Syll.* vi, 451; Rea, *Brit. Basid.* 637.

*On beech (*Fagus*). Acting as wound parasite, Epping Forest (Massee, *Dis. Cult. Pl. Trees*, 1915, 389).

Hypholoma fasciculare (Huds.) Fr., *Syst. Myc.* i, 288; Sacc., *Syll.* v, 1029; Rea, *Brit. Basid.* 262.

*Killing crown of rhubarb (*Rheum*). Middx, 1928 (*Report*, vii, 64).

*Causing a soft brown rot of European larch (*Larix*), Sitka spruce (*Picea*) and *Thuja plicata* (Peace, *Quart. J. For.* **32**, 1938, 81–104).

Hypochnus solani Prill. & Delacr. = *Corticium solani*.

HYPODERMA

Hypoderma brachysporum (Rostr.) Tubeuf, *Beiträge zur Kenntnis der Baumkrankheiten*, 1888, 30; Sacc., *Syll.* ix, 1125 (as *Lophodermium brachysporum* Rostr.); Bisby & Mason, *TBMS*, 24, 1940, 209. Darker (*Contr. Arnold Arbor.* 1, 1932, 25) places it as a synonym of *H. desmazierii* Duby.

*On leaves of Weymouth pine (*Pinus strobus*). Not uncommon on this species and its var. *nana*, sometimes causing serious defoliation of young trees. First reported by Wilson (*TBMS*, 7, 1922, 81) at Murthly, Perthshire, on var. *nana*. Occurs also on other five-needled pines, including *P. cembra*, *P. monticola* and *P. parviflora*, and Dr Batko has found it on *P. radiata*.

Hypoderma desmazierii Duby, see *H. brachysporum*.

Hypoderma pinicola, see *Hypodermella conjuncta*.

Hypodermella conjuncta Darker, *Contr. Arnold Arbor.* 1, 1932, 25. It is not clear how far this species is distinct from *H. sulcigena* (Rostr.) Tubeuf, *Bot. Zbl.* 41, 1895, 48; Sacc., *Syll.* xi, 385 and iii, 729; Bisby & Mason, *TBMS*, 24, 1940, 210; syn. *Hypoderma pinicola* Brunch., *Z. Parasitenk.* 1894, 242; Sacc., *Syll.* xi, 389. (See also *Hendersonia acicola.*)

*Needle cast of Scots pine (*Pinus sylvestris*). Frequent. Described by Wilson (*TBMS*, 7, 1922, 79). Occasional on Corsican pine (*P. nigra* var. *poiretiana*) in northern England.

Hypodermella sulcigena (Rostr.) Tubeuf, see *H. conjuncta*.

Hypodermium orchidearum Cooke & Mass., *Grevillea*, 16, 1887, 48; Sacc., *Syll.* x, 466; Grove, *Coelomycetes*, ii, 263. Grove (loc. cit. 218) suggests that *Gloeosporium orchidearum* Karst. & Har., *J. Bot., Paris*, 4, 1890, 360; Sacc., *Syll.* x, 462 belongs here.

*On orchid leaves. Originally reported on *Cymbidium eburneum*, locality unstated; causing death of leaf tips of *Cymbidium*, Anglesey, 1924 (*Report*, v, 89); on leaves of several orchids, including *Thunia* and *Dendrobium*, Botanic Garden, Cambridge (Brooks, *Gdnrs' Chron.* 50, 1911, 27).

Hypomyces asclepiadis Zerova, see *Fusarium caeruleum.*

Hypomyces solani Reinke & Berth. em. Snyder & Hansen, see *Fusarium solani* and *F. caeruleum.*

Hysterium juniperi Grev. = *Lophodermium juniperinum.*

Isaria fuciformis Berk. = *Corticium fuciforme.*

Itersonilia perplexans Derx, *Buitenz. Jard. Bot. Bull.* 3 Ser., **17,** 1948, 465.

*Parsnip canker. War, Bucks, Kent and perhaps widespread as the cause of one form of parsnip (*Pastinaca*) canker (Channon, *Nature, Lond.,* **178,** 1956, 217; *Rep. nat. Veg. Res. Sta. Warwick for 1956,* 63).

*Flower scorch of *Chrysanthemum.* Frequent in some districts under glass where humidity is high. First seen in Som, 1936, later in Dorset, Hants, Devon, Corn, Sussex, Herts, Glam, Yorks, Lancs. Occurs outdoors when autumn damp. The pathogen was first thought closely to resemble *Entyloma calendulae* (Robertson, *Plant Path.* **4,** 1955, 33), but Dosdall (*Phytopathology,* **46,** 1956, 232) suggested it was *Itersonilia perplexans,* which causes a similar disease in U.S.A.

Itersonilia sp.

*Petal blight of *Anemone.* Corn, 1956, associated with *Botrytis cinerea.*

Kabatiella caulivora (Kirchn.) Karak., *Bot. mater. Inst. crypt. pl., Chief Bot. Gdn Russian Repub.* ii, 1923, 101; syn. *Gloeosporium caulivorum* Kirchn., *Z. PflKrankh.* **12,** 1902, 13; Sacc., *Syll.* xviii, 449; Grove, *Coelomycetes,* ii, 226.

*Scorch of red clover (*Trifolium*). Widely distributed in Great Britain and in wet seasons does much damage in western seed-growing districts.

First recognized near Cambridge, 1918 (*Report,* ii, 49), and soon after at Aberystwyth (Sampson, *Bull. Welsh Pl. Breed. Sta.*

Ser. H 1, 1922, 83–7) and in south-east England (Ware, *J. Minist. Agric.* **30**, 1923, 833–6). Fully described by Sampson (*TBMS*, **13**, 1928, 103–42) who showed that alsike, white clover and shaftal clover (*Trifolium suaveolens*) can be infected artificially.

Kabatiella microsticta Bubák, *Hedwigia*, **46**, 1907, 297; Sacc., *Syll.* xxii, 1297.
*On leaf spots on *Lilium umbellatum*, Sussex, 1930.

Keithia tetraspora Phill. & Keith = *Didymascella tetraspora*.

Keithia thujina Durand = *Didymascella thujina*.

Keithia tsugae (Farl.) Durand = *Didymascella tsugae*.

Kuehneola albida (Kühn) Magn. = *K. uredinis*.

Kuehneola uredinis (Link) Arth., *N. Amer. Flora*, **7**, 1912, 186; Wilson & Bisby, *TBMS*, **37**, 1954, 64; syn. *K. albida* (Kühn) Magn., *Bot. Zbl.* **74**, 1898, 169; Sacc., *Syll.* vii, 761 (as *Chrysomyxa albida* Kühn); Grove, *Brit. Rust Fungi*, 300.
*Stem rust of blackberry (*Rubus*). Fairly common. Norfolk, Oxon, Herefs, Glos, Sussex, etc.

Kunkelia nitens (Schw.) Arth., *Bot. Gaz.* **63**, 1917, 501.
*Orange rust of *Rubus*. Very rare. In 1931 three out of 15,000 plants of Lucretia dewberry imported a few months previously from U.S.A. were found infected with this rust in Hants. The affected plants and their immediate neighbours were burnt and the rust has not been seen again in Britain (*Report*, vii, 84).

Lasiobotrys lonicerae Kunze ex Fr., *Mycol. Heft* II, 88; Sacc., *Syll.* i, 30; Cooke, *Fung. Pests Cult. Pl.* 49; Bisby & Mason, *TBMS*, **24**, 1940, 136.
*On leaves of honeysuckle (*Lonicera*). Uncommon on garden forms.

Lembosia aulographoides Bomm., Rouss. & Sacc. = *Echidnodes aulographoides.*

Leptosphaeria avenaria Weber, *Phytopathology,* **12**, 1922, 454; stat. conid. *Septoria avenae* Frank, *Ber. dtsch. bot. Ges.* **13**, 1895, 64; Sacc., *Syll.* xi, 547; Grove, *Coelomycetes,* i, 426.

*Septoria leaf spot of oats. Formerly only occasionally seen, but since 1947 has become prevalent and damaging in oat-growing areas of Wales and the north of England. Also Hants, Dorset, War, Worcs. Widely distributed in Scotland; common and sometimes severe in the north and east.

Noble and Montgomerie (*Scot. Agric.* **34**, 1954, 51–2; *TBMS,* **39**, 1956, 449–59) called it the straw-breaking fungus, and described, in addition to leaf spot, the condition known as 'black stem' ('brown stem' would be better) in America, as well as spotting of the glumes, chaff and seeds. Perithecia, believed to be of this species, were found by Noble (ibid. **39**, 1956, 455) in Scotland in June 1956.

Leptosphaeria coniothyrium (Fuckel) Sacc., *N. Giorn. bot. ital.* **7**, 1875, 317; Sacc., *Syll.* ii, 29; Bisby & Mason, *TBMS,* **24**, 1940, 182; stat. conid. *Coniothyrium fuckelii* Sacc., *Mich.* i, 207; Sacc., *Syll.* iii, 306; Grove, *Coelomycetes,* ii, 2.

*On maiden apple trees, Nantwich, Ches, 1938.
*Cane blight of raspberry (*Rubus*). Common in north-east England and Kent, occasional elsewhere. Widely distributed in Scotland (Clyde, Tay, Inverness). It attacks and kills the fruiting canes in summer. The varieties Norfolk Giant, Reader's Perfection, Lloyd George and Red Cross are very susceptible.

The association of the disease with cane midge (*Thomasiniana theobaldi*) was considered by Fox-Wilson and Green (*J. R. hort. Soc.* **69**, 1944, 79–86) and by Pitcher (*Rep. E. Malling Res. Sta. for 1947,* 141–3) who, with Webb (*Nature, Lond.,* **163**, 1949, 574; *J. hort. Sci.* **27**, 1952, 95–100), also showed that other fungi, including *Didymella applanata* and *Fusarium culmorum,* may also follow midge attack.

*Associated with root rot of strawberry, in its pycnidial stage, but doubtfully parasitic (Berkeley & Lauder-Thomson, *J. Pomol.* **12**, 1934, 222–46).
*Stem canker of rose. Fairly common throughout Britain. Described by Güssow (*J. R. hort. Soc.* **34**, 1908, 222–30) on material from Northern Ireland. See also Green (ibid. **59**, 1934, 470).

Leptosphaeria culmicola (Fr.) Karst., see *L. herpotrichoides.*

Leptosphaeria culmorum Auersw., see *L. herpotrichoides.*

Leptosphaeria herpotrichoides de Not., *Sferiacei italici*, 1863, 80; Sacc., *Syll.* ii, 77.
*On wheat. This and other species of *Leptosphaeria*, including *L. culmorum* Auersw., *L. culmicola* (Fr.) Karst., and *L. tritici* (Gar.) Passer., have been recorded occasionally on wheat stubble or stem bases since 1915, but their parasitism, though sometimes assumed, has not been studied (*Report*, viii, 6).

Leptosphaeria heterospora (de Not.) Niessl, *Beiträge zur Kenntniss der Pilze*, 1872, 23; Sacc., *Syll.* ii, 67; Bisby & Mason, *TBMS*, **24**, 1940, 183.
*On roots and rhizomes of *Iris*. Occasionally seen, probably common; usually found on unthrifty and yellowing plants but parasitism uncertain.
On *I. germanica*, London, Surrey, Bucks; on *I. pumila formosa*, Kent (W. C. Moore, *TBMS*, **31**, 1947, 87).

Leptosphaeria nigrans (Desm.) Ces. & de Not., *Schema Sfer.* 1863, 61; Sacc., *Syll.* ii, 70; Bisby & Mason, *TBMS*, **24**, 1940, 184.
*On wheat stubble, associated with take-all (*Ophiobolus graminis*), Glam, 1944. Also occurs on grasses. Parasitism doubtful. The perithecia and pycnidia were described by Hughes (*TBMS*, **32**, 1949, 63–8).

13-2

Leptosphaeria nodorum Müller, *Phytopath. Z.* **19**, 1952, 409; stat. conid. *Septoria nodorum* Berk., *Gdnrs' Chron.* **5**, 1845, 601; Sacc., *Syll.* iii, 561; Grove, *Coelomycetes*, i, 422; syn. *S. glumarum* Passer., *Funghi parmensi enumerati,* n. 147; Sacc., *Syll.* iii, 561.

The perfect stage has not yet been found in Britain.

*Glume blotch of wheat (*Triticum*). Occurs in all parts of Britain in July and August, but is not harmful. Described by Grove (*Gdnrs' Chron.* **60**, 1916, 194 and 210). It is frequently seed-borne (Noble, *Scot. Agric.* **36**, 1956, 88).

Leptosphaeria tritici (Gar.) Passer., see *L. herpotrichoides.*

Leptothyrium anemones Grove, *Coelomycetes*, ii, 169, 363.

*On leaves of *Anemone*. On living foliage of cultivated plants of the var. St Brigid, Isles of Scilly.

Leptothyrium gentianaecolum Bauml. = *Pycnothyrium gentianicola.*

Leptothyrium periclymeni Sacc., *Syll.* iii, 626; Grove, *Coelomycetes*, ii, 172.

*Leaf spot of honeysuckle (*Lonicera*). Widely distributed on *L. periclymenum.*

Leptothyrium platanoidis Passer. apud Brun., *Champignons nouvellement observés aux environs de Saintes*, vi, 1887, 4; Sacc., *Syll.* x, 413; Grove, *Coelomycetes*, ii, 168.

*On leaves of seedling sycamore (*Acer*). Gower, south Wales; Himley Park, Staffs, etc. (See also *Phyllosticta platanoidis*.)

Leptothyrium pomi (Mont. & Fr.) Sacc., *Mich.* ii, 113; Sacc., *Syll.* iii, 632; Grove, *Coelomycetes*, ii, 173.

*Fly speck of apple and pear fruit. Not common and usually associated with *Gloeodes pomigena* (q.v.). Seen rarely on plums.

Leveillula taurica (Lév.) Arn., *Ann. Epiphyt.* **7**, 1921, 108; Sacc., *Syll.* i, 16 (as *Erysiphe taurica* Lév.); Salmon, *Monogr. Erysiph.* 1900, 215 (as *E. taurica*).

*Powdery mildew of rock rose (*Helianthemum*). Known only since the hot summer of 1947 and now probably widely distributed. On cultivated varieties of *H. vulgare*, Harpenden, Herts; Luton, Beds; Exmouth and Exeter, Devon; Sept. 1947: on *H. rhodanthe carnea*, Reading, Berks, Sept. 1947 (W. C. Moore & F. Joan Moore, *TBMS*, **32**, 1950, 273–4). In Harpenden the same and other seedling plants were subsequently attacked annually until 1958, though the mildew was severe only in the hot summers of 1949 and 1955.

Libertella ulcerata Massee = *Phomopsis cinerascens.*

Lilliputia insigne (Wint.) Dennis & Wakef., *TBMS*, **29**, 1946, 145.
*On mushroom (*Agaricus*) compost, Uxbridge, Middx, 1938.

Linocarpon cariceti (Berk. & Br.) Petrak = *Ophiobolus graminis.*

Lophodermellina macrospora (Hartig) Tehon = *Lophodermium macrosporum.*

Lophodermium brachysporum Rostr. = *Hypoderma brachysporum.*

Lophodermium juniperinum (Fr.) de Not., *Pirenomiceti Isterini* 1847, 40; Sacc., *Syll.* ii, 794; Bisby & Mason, *TBMS*, **24**, 1940, 210.
*On *Juniperus communis*, Clyde, Moray, Forth. First described by Greville (*Scottish Cryptogamic Flora*, i, 1823, 26) under the name *Hysterium juniperi* from the Pentland Hills and very abundant in the Cairngorm Mountains.

Lophodermium macrosporum (Hartig) Rehm, Rabenh., *Krypt. Fl.* 1 (3), 1887, 45; Sacc., *Syll.* ii, 786 (as *Hypoderma macrosporum* Hartig). Tehon (*Illinois biol. Monogr.* **13**, 1935, 76) transferred it to *Lophodermellina* as *L. macrospora* (Hartig) Tehon.
*On Norway spruce (*Picea abies*). Not common, but occurs in Scotland. Campbell and Vines (*New Phytol.* **37**, 1938, 358–68) studied the factors that prevent needle fall in the α form of this disease.

Lophodermium melaleucum (Fr.) de Not., *Pirenomiceti Isterini*, 1847, 40; Sacc., *Syll.* ii, 791; Bisby & Mason, *TBMS*, **24**, 1940, 211.

*Leaf blight and canker of rhododendron. Tweed. (See *L. vagulum.*)

Lophodermium pinastri (Schrad. ex Fr.) Chevall., *Flore générale des environs de Paris*, i, 1826, 430; Sacc., *Syll.* ii, 794; Bisby & Mason, *TBMS*, **24**, 1940, 211.

*Leaf cast of Scots pine (*Pinus*). Common in nurseries and young plantations in Britain and in some cases damaging. Jones (*Ann. Bot., Lond.*, **49**, 1935, 699–728) described the method and cytology of infection. Peace (*Scot. For.* **7**, 1953, 17–22) suggested its control by spraying with Bordeaux Mixture and spreader.

Lophodermium vagulum Wils. & Roberts., *Trans. roy. Soc. Edinb.* **61**, 1947, 524.

*On *Rhododendron*, affecting the leaves and twigs of Chinese species cultivated in Scotland.

Macrophoma draconis Allesch. in Rabenh., *Krypt. Fl.* vii, 836; Sacc., *Syll.* xiv, 864 (as *Phyllosticta draconis* Berk.); Grove, *Coelomycetes*, i, 126.

*Leaf spot of *Dracaena*. Frequent in glasshouses and conservatories where the plant is grown.

Macrophoma sp.

*Leaf and stem blight of gentian. On *Gentiana sino-ornata*. Solway.

Macrosporium parasiticum Thüm., see *Pleospora herbarum*.

Macrosporium porri Ell. = *Alternaria porri*.

Macrosporium sarcinaeforme Cav. = *Stemphylium sarciniiforme*.

Macrosporium sp.

*Hop drop. Seen occasionally in Kent since 1922 (Salmon & Wormald, *J. Minist. Agric.* **30**, 1923, 434–5).

Marasmius androsaceus Fr., *Epicr.*, 385; Sacc., *Syll.* v, 543.

*Die-back ('Rhizomorph fungus') of heather (*Calluna*). Found associated with dying-off of *C. vulgaris* in Scotland (Braid & Tervet, *Scot. J. Agric.* **20**, 1937, 365–72). Macdonald (ibid. **28**, 1948, 99–101) discussed the identity of the fungus, and studied it in culture (*Proc. roy. Soc. Edinb.* Sect. B, **63**, 1949, 230–41). On white heather, Argyll, 1952 (as *Marasmius* sp.).

Marasmius oreades (Fr.) Fr., *Epicr.*, 375; Sacc., *Syll.* v, 510; Rea, *Brit. Basid.* 519.

*Fairy rings. With other fungi the cause of fairy rings in turf, which occur everywhere. Ramsbottom (*J. Quekett micr. Cl.* **15**, 1927, 231–42) published an interesting historical account of research into fairy rings. See also Bayliss-Elliott (*AAB*, **13**, 1926, 277–88). Drew Smith (*J. Sports Turf Res. Inst.* **9**, 1955, 62–8) discussed their control by the use of chemicals.

Marasmius rotula (Scop.) Fr., *Epicr.*, 335; Sacc., *Syll.* v, 541.

*Associated with root rot in hop (*Humulus*), Kent, 1935 (Salmon & Ware, *J. S.-E. agric. Coll. Wye*, **37**, 1936, 26).

Marssonina daphnes Magn., *Hedwigia*, **45**, 1906, 88; Sacc., *Syll.* iii, 769 (as *Marsonia daphnes* (Desm. & Rob.) Sacc.); Grove, *Coelomycetes*, ii, 274.

*Leaf spot of *Daphne*. Widely distributed in Britain on *D. mezereum* and not uncommon. First reported in Scotland from Peebles-shire, Oct. 1908 (A. L. Smith, *TBMS*, **3**, 1909, 119), and in England from Hants (Green, *Gdnrs' Chron.* **96**, 1934, 305; *J. R. hort. Soc.* **60**, 1935, 156–8).

The fungus may not be distinct from *Gloeosporium mezerei* Cooke, *Grevillea*, xix, 8, known in Britain since 1890 but very rarely reported.

Marssonina fragariae (Sacc.) Kleb. = stat. conid. of *Diplocarpon earlianum*.

Marssonina juglandis (Lib.) Magn. = stat. conid. of *Gnomonia leptostyla*.

Marssonina panattoniana (Berl.) Magn., *Hedwigia*, **45**, 1906, 88; Sacc., *Syll*. xiv, 1021 (as *Marsonia panattoniana* Berl.); Grove, *Coelomycetes*, ii, 276.

*Ring spot of endive (*Cichorium*). Penzance, Corn, 1933.
*Ring spot of lettuce (*Lactuca*). Rare in glasshouses and frames, and not common on spring or summer lettuce, but often serious in southern England on plants that have stood through the winter. Markedly seasonal according to the weather. Now rare in Scotland.

The disease was first reported under glass in 1912 (Chittenden, *J. R. hort. Soc.* **37**, 1912, 541–3) and in the open in 1922 (*Report*, v, 53). It was described by Salmon and Wormald (*J. Minist. Agric.* **30**, 1923, 147–51). Stevenson (*J. Pomol.* **17**, 1939, 27–50) studied its occurrence, spread and control, and Ogilvie and Mulligan (*Rep. Long Ashton Res. Sta. for 1933*, 110) carried out variety resistance trials.

Marssonina populi (Sacc.) Magn., *Hedwigia*, **45**, 1906, 88; Sacc., *Syll*. iii, 767 (as *Marsonia populi* (Lib.) Sacc.); Grove, *Coelomycetes*, ii, 278.

*On leaves of poplar (*Populus*). Common and occasionally damaging.

Marssonina salicicola (Bres.) Magn., *Hedwigia*, **45**, 1906, 88; Sacc., *Syll*. xi, 574 (as *Marsonia salicicola* Bres.); Grove, *Coelomycetes*, ii, 280.

*Anthracnose of willow (*Salix*). Not uncommon in Som, occasional elsewhere, Yorks, Bucks, Surrey, Kent, Glos. First recognized in 1926, and described by Nattrass (*Bull. Minist. Agric. Egypt*, no. 99, 1930, 19 pp.).

MASTIGOSPORIUM

Mastigosporium album Riess, in Fresenius, *Beiträge zur Mykologie*, 1852, 56; Sacc., *Syll.* iv, 220; Wakefield & Bisby, *TBMS*, **25**, 1941, 67.

*Leaf fleck of meadow foxtail (*Alopecurus*). Locally common on *A. pratensis* in Britain, particularly in late autumn and early spring. Also occurs on *A. nigricans, Cynosurus cristatus* and *Deschampsia caespitosa*.

Bollard (*TBMS*, **33**, 1950, 265–75) investigated the mode of parasitism of this and other species of *Mastigosporium* on grasses. (See also note under *Dilophospora alopecuri*.)

Mastigosporium cylindricum Sprague, *Mycologia*, **32**, 1940, 43.

*Leaf spot of *Phleum*. On *P. nodosum*, Aberystwyth, 1947 (Bollard, *TBMS*, **33**, 1950, 250–64). It will also attack *P. pratense*.

Mastigosporium deschampsiae Jørstad, *K. norske vidensk. Selsk. Forh.* **19**, 1947, 25.

*Leaf spot of *Deschampsia*. On *D. caespitosa*, Wicken Fen, Cambs, 1948 (Austwick, *TBMS*, **37**, 1954, 161–5).

Mastigosporium rubricosum (Dearn. & Barth.) Sprague, *J. agric. Res.* **57**, 1938, 287; Wakefield & Bisby, *TBMS*, **25**, 1941, 67.

*Leaf fleck of cocksfoot (*Dactylis*). Widely distributed on *D. glomerata* in Britain. It was first recorded in 1918 as *Mastigosporium album* var. *muticum* Sacc. (Wakefield, *Kew Bull.* **7**, 1918, 233).

The fungus on *Agrostis stolonifera*, formerly ascribed to this species, was regarded by Bollard (*TBMS*, **33**, 1950, 250–64) as a distinct variety and named by him var. *agrostidis* (q.v.).

Mastigosporium rubricosum var. **agrostidis** Bollard, *TBMS*, **33**, 1950, 262.

*Leaf fleck of *Agrostis*. Occasional in Britain on *A. stolonifera*.

Melampsora allii-fragilis Kleb., *Jb. wiss. Bot.* **35**, 1901, 671; Sacc., *Syll.* xvii, 462; Grove, *Brit. Rust Fungi*, 344; Wilson & Bisby, *TBMS*, **37**, 1954, 64.

*On *Allium* and *Salix*. Caeomata on *Allium cepa* (onion) and other species of *Allium*: uredo- and teleutospores on *Salix fragilis* and *S. pentandra*.

Without knowledge of the teleuto-host, the caeomata of this species cannot be separated from those of *Melampsora allii-populina* Kleb. on *Allium* and *Populus*, or of *Melampsora allii-salicis-albae* Kleb. on *Allium* and *Salix*. The last-named is damaging to the osier (*S. alba* var. *vitellina*) because it makes the rods too brittle for basket making.

Melampsora allii-populina Kleb., see *M. allii-fragilis*.

Melampsora allii-salicis-albae Kleb., see *M. allii-fragilis*.

Melampsora amygdalinae Kleb., *Jahrb. wiss. Bot.* **24**, 1900, 352; Sacc., *Syll.* xvii, 463; Wilson & Bisby, *TBMS*, **37**, 1954, 65.

*Willow rust. Common in Som on the stems and leaves of *Salix triandra*; occasional elsewhere in England and Scotland. Cankered rods are useless for basket making.

Ogilvie and Hutchinson (*Rep. Long Ashton Res. Sta. for 1932*, 125–30) made observations on the disease, and Ogilvie (ibid. 131–8) carried out spore-germination experiments. For an early note on willow rusts see W. G. Smith (*Gdnrs' Chron.* **16**, 1881, 497).

Melampsora epitea Thüm., *Mitth. Versuch. Oest.* ii, 1879, 15; Sacc., *Syll.* vii, 588; Wilson & Bisby, *TBMS*, **37**, 1954, 65; syn. *M. larici-epitea* Kleb., *Z. PflKrankh.* **9**, 1899, 88; Sacc., *Syll.* xvii, 463; Grove, *Brit. Rust Fungi*, 340.

*Willow rust. The uredo- and teleutospores occur on a number of species of *Salix* in Britain. On the foliage, but not the stems of *S. viminalis*, Som (Ogilvie, *Rep. Long Ashton Res. Sta. for 1931*, 133).

The caeomata occur on larch (*Larix europaea*).

Melampsora euonymi-caprearum Kleb., *Jb. wiss. Bot.* **34**, 1900, 358; Sacc., *Syll.* xvii, 463; Grove, *Brit. Rust Fungi*, 339; Wilson & Bisby, *TBMS*, **37**, 1954, 65.
*Uredo- and teleutospores on *Salix aurita*, *S. caprea* and *S. cinerea*; caeomata on *Euonymus europaeus*, rare.

Melampsora hypericorum Wint., *Pilze*, 241; Sacc., *Syll.* vii, 591; Grove, *Brit. Rust Fungi*, 354; Wilson & Bisby, *TBMS*, **37**, 1954, 65.
*Hypericum rust. Occasional in the west on garden forms. On *Hypericum elatum*, Devon; on *H. calycinum*, Som, Glam, Herefs; on *H. androsaemum*, Devon.

Melampsora larici-caprearum Kleb., *Forstl.-nat. Z.* 1897, 469; Sacc., *Syll.* xvii, 463; Grove, *Brit. Rust Fungi*, 1913, 338; Wilson & Bisby, *TBMS*, **37**, 1954, 65.
*On willow (*Salix*) and *Larix*. Common. The uredo- and teleutospores occur on *Salix caprea*, etc., and the caeomata on *Larix europaea*.

Melampsora larici-epitea Kleb. = *M. epitea*.

Melampsora larici-populina Kleb., *Z. PflKrankh.* **12**, 1902, 43; Sacc., *Syll.* xvii, 463; Grove, *Brit. Rust Fungi*, 348; Wilson & Bisby, *TBMS*, **37**, 1954, 65.
*Poplar rust. The uredo- and teleuto-stages are not uncommon on *Populus* spp., including *P. nigra* and *P. balsamifera*, and sometimes they cause premature defoliation, especially on *P. generosa*.
The caeomata occur on *Larix europaea*.

Melampsora larici-tremulae Kleb. = *M. tremulae*.

Melampsora lini (Ehrenb.) Lév., *Ann. Sci. nat.* 3 Ser. **8**, 1847, 376; Sacc., *Syll.* vii, 588; Grove, *Brit. Rust Fungi*, 356; Wilson & Bisby, *TBMS*, **37**, 1954, 65.
*Flax rust. Found at Cobham more than a century ago (M.J.B., *Gdnrs' Chron.* **11**, 1851, 611), but rarely seen again until 1942.

During the next few years common in south-east England and Pembs, and seen in other widely scattered areas throughout Britain. At times severe (Ware & Glasscock, *Agriculture, Lond.*, **50**, 1943, 16–19). Since the 1939–45 War less common with the greatly decreased acreage devoted to flax. A different physiologic race occurs on the wild *Linum catharticum*.

Rust was studied in Ireland by Pethybridge *et al.* (*J. Dep. Agric. Ire.* **20**, 1920, 325–42; **21**, 1921, 167–87; **22**, 1922, 103–20). Colhoun (*Gdnrs' Chron.* **118**, 1945, 191) could not cure it by applications of boron. For varietal resistance see under *Colletotrichum linicola*.

Arif (*TBMS*, **37**, 1954, 353–61) distinguished fifteen physiologic races from collections made in Northern Ireland and six races from collections in Great Britain.

Melampsora pinitorqua Rostr. = *M. tremulae*.

Melampsora ribesii-purpureae Kleb., *Jb. wiss. Bot.* **35**, 1901, 667; Sacc., *Syll.* xvii, 463; Grove, *Brit. Rust Fungi*, 342; Wilson & Bisby, *TBMS*, **37**, 1954, 65.

*On *Salix* and *Ribes*. Uredo- and teleutospores on *Salix purpurea*, Herts. There is no definite British record of the caeomata on *Ribes*.

Melampsora ribesii-viminalis Kleb., *Jb. wiss. Bot.* **34**, 1900, 363; Sacc., *Syll.* xvii, 463; Grove, *Brit. Rust Fungi*, 342; Wilson & Bisby, *TBMS*, **37**, 1954, 65.

*On *Salix* and *Ribes*. Uredo- and teleutospores on *Salix viminalis*. Rare. The caeomata on *Ribes grossularia*, *R. nigrum* and *R. rubrum* have apparently not yet been seen in Britain.

Melampsora rostrupii Wagner = *M. tremulae*.

Melampsora tremulae Tul., *Ann. Sci. nat.* 4 Ser. **2**, 1854, 95; Sacc., *Syll.* vii, 589; Grove, *Brit. Rust Fungi*, 349; Wilson & Bisby, *TBMS*, **37**, 1954, 65; syn. *M. larici-tremulae* Kleb., *Forstl.-nat. Z.* 1897, 468; *M. pinitorqua* Rostr., *Tidsskr. Skovbrug*, **12**, 1889, 177; *M. rostrupii* Wagner, *Öst. bot. Z.* **46**, 1896, 273.

*Poplar rust. Widely distributed throughout Britain. Uredo- and teleutospores on *Populus tremula* and *P. alba*. Caeomata on *Larix europaea* and *Mercurialis perennis*.
*Branch rust of pine (*Pinus*). Limited to southern half of Britain. Uredo- and teleutospores on *Populus tremula*, *P. alba* and *P. canescens*, often very conspicuous on *P. tremula*.

Caeomata on young shoots of Scots pine (*Pinus sylvestris*) and occasionally other species of *Pinus*, often causing severe shoot distortion and die-back.

The occurrence in south-east England of this rust was discussed by Peace (*Forestry*, **18**, 1944, 47–8), who has also dealt (*Bull. For. Comm., Lond.*, **19**, 1952, 40) with the varietal resistance of poplars to this and other rusts.

Melampsorella caryophyllacearum Schröt., *Hedwigia*, **13**, 1874, 85; Grove, *Brit. Rust Fungi*, 360; Wilson & Bisby, *TBMS*, **37**, 1954, 66; syn. *M. cerastii* (Pers.) Schröt., *Kr. Fl. Schles.* 1887, 366; Sacc., *Syll.* vii, 596.

*'Witches brooms' of silver fir (*Abies*). Aecidia on *A. alba* and *A. nordmanniana*, causing witches brooms or cankers; not common. The uredo- and teleutospores occur on species of *Cerastium* and *Stellaria*.

Melampsorella cerastii (Pers.) Schröt. = *M. caryophyllacearum*.

Melampsorella symphyti (DC.) Bubák, *Dtsch. bot. Ges.* **21**, 1903, 356; Sacc., *Syll.* xvii, 464.

*On cultivated comfrey (*Symphytum*). West Scotland, 1950; Kinross, 1954; Midlothian, 1955; on Russian comfrey (*S. asperum*) grown for fodder, Yorks, 1952–3; Cambs, 1955; Sussex, 1956.

Melampsoridium betulinum (Desm.) Kleb., *Z. PflKrankh.* 1899, 21; Sacc., *Syll.* xvii, 464; Grove, *Brit. Rust Fungi*, 358; Wilson & Bisby, *TBMS*, **37**, 1954, 66.

*On *Betula* and *Larix*. Uredo- and teleutospores on *Betula alba*, *B. pendula* and *B. pubescens*; very common but seldom damaging except in nurseries. Aecidia on leaves of *Larix europaea*; very rare, King's Lynn, Norfolk.

Melanconium pandani Lév., *Ann. Sci. nat.* 3 Ser. 3, 1845, 66; Sacc., *Syll.* iii, 759; Grove, *Coelomycetes*, ii, 317.

*On screw-pine (*Pandanus*), Kew Gardens (Massee, *Dis. Cult. Pl. Trees*, 1915, 433); Scotland, 1947.

Melasmia acerina Lév. = stat. conid. of *Rhytisma acerinum*.

Melasmia salicina Lév. apud Tul. = stat. conid. of *Rhytisma salicinum*.

Meria laricis Vuill., *Bull. Soc. Sci. Nancy*, 2 Ser. 14, 1896, 13; Sacc., *Syll.* xiv, 431; Wakefield & Bisby, *TBMS*, 25, 1941, 74.

*Meria needle cast of larch (*Larix*). Common on European larch (*L. decidua*), but troublesome mainly in nurseries, and less prevalent than formerly. Also on *L. occidentalis* and rarely on *L. eurolepis* (Carns, 1954) and *L. leptolepis* (Corn, Carns, Kirkcudbright, Wigtownsh in 1954). (See Batko, *TBMS*, 39, 1956, 13–16.)

Descriptions are given by Hiley (*Quart. J. For.* 15, 1921, 57–62) and by Peace and Holmes (*Oxf. For. Mem.* 15, 1933, 29 pp.). Peace (*Forestry*, 10, 1936, 78–82) has dealt with its control by spraying with sulphur compounds. Biggs (*Rep. For. Res., Lond., 1955–6*, 89–90) recognized eight strains of the fungus.

Microsphaera alphitoides Griffin & Maubl., *Bull. Soc. mycol. Fr.* 28, 1912, 88; syn. *M. quercina* (Schw.) Burrill, *J. Mycol.* 4, 1887, 35; Sacc., *Syll.* i, 22 (as *Erysiphe quercina* Schw.); stat. conid. *Oidium quercinum* Thüm., *Contrib. flor. myc. lusit.* 6; Sacc., *Syll.* iv, 44.

*Oak mildew. Common everywhere, and a constant irritant in nurseries. Occurs equally on *Quercus petraea* and *Q. robur*; occasional on *Q. rubra*.

The literature was reviewed by Day (*Forestry*, 1, 1927, 108–12), and Woodward, Waldie and Steven (ibid. 3, 1929, 38–56) gave a full account of the disease and pointed out that the mildew on

beech (*Fagus*) was identical. Fresh foliage formed after defoliation by the oak roller moth (*Tortrix viridana*) is particularly
susceptible (Robinson, *Quart. J. For.* **21**, 1927, 25–7).

The perithecial stage was first found in England, in minimal
amount, by Robertson and Macfarlane (*TBMS*, **29**, 1947, 219–
20), and it was found frequently during the hot summer of 1947,
at Aberystwyth (Knoyle, *Nature, Lond.*, **61**, 1948, 938) and in
Herts, Staffs, Middx and Essex.

*Beech mildew. Uncommon. Sevenoaks, Kent, 1918 (Cotton,
TBMS, **6**, 1919, 198–200).

Microsphaera berberidis (DC.) Lév., *Ann. Sci. nat.* **15**, 1851,
    159; Sacc., *Syll.* i, 13; Bisby & Mason, *TBMS*, **24**, 1940, 135.
*Mildew of *Berberis*. Salop; on *B. hookeri*, Devon, 1937; Oxon,
1949; widely distributed on *B. vulgaris* in Scotland.
*Mildew of *Mahonia*. On *M. aquifolium*, Glam, Salop.

Microsphaera betae Vanha, see *Erysiphe* sp.

Microsphaera grossulariae (Wallr.) Lév., *Ann. Sci. nat.* **15**, 1851,
    160; Sacc., *Syll.* i, 12; Bisby & Mason, *TBMS*, **24**, 1940, 135.
*European gooseberry mildew. Widely distributed, but much less
common than *Sphaerotheca mors-uvae* (q.v.) and rarely harmful.
(See *J. Bd Agric.* **4**, 1897, 202–4.)

This mildew also occurs occasionally on red and black currant
in England and Wales. It was first recorded on red currant by
Salmon (*Gdnrs' Chron.* **40**, 1906, 294).

Microsphaera mougeotii Lév., *Ann. Sci. nat.* **15**, 1851, 158; Sacc.,
    Syll. i, 10; Bisby & Mason, *TBMS*, **24**, 1940, 135.
*On *Lycium*. The *Oidium* stage has been recorded on the leaves
of *Lycium barbarum* by Cooke (*Fung. Pests Cult. Pl.* 1906, 192),
under the name *Microsphaera lycii*, and on *Lycium halimifolium*,
East Wittering, Sussex, 1950.

Microsphaera polonica Siemaszko, *Rev. Path. vég.* **20**, 1933, 142;
    stat. conid. *Oidium hortensiae* Jørstad, *The Erysiphaceae of
    Norway*, 1925, 106.

The perfect stage has not been seen in Britain.

*Hydrangea mildew. Occasional, widely distributed in Britain, and perhaps of long standing, though not reported before 1930 (*Report*, vii, 98).

Microsphaera quercina (Schw.) Burrill = *M. alphitoides.*

Microthyriella rubi Petrak, *Ann. mycol., Berl.*, **21**, 1923, 15.

*Raspberry (*Rubus*). Present on fruiting canes of var. Reader's Perfection, Kent, 1928 (Harris, *Rep. E. Malling Res. Sta. for 1928–30*, II Suppl., 136).

Milesia blechni (Syd.) Arth., *Bot. Gaz.* **73**, 1922, 61; Sacc., *Syll.* vii, 680 (as *Uredo scolopendrii* (Fuckel) Schröt. p.p.); Grove, *Brit. Rust Fungi*, 377; Wilson, *Trans. bot. Soc. Edinb.* **31**, 1934, 435; Hunter, *TBMS*, **20**, 1936, 117; Wilson & Bisby, ibid. **37**, 1954, 66.

*On *Blechnum.* Uredo- and teleutospores on the native *B. spicant.* Widely distributed in Scotland; Devon, Hants, Surrey. Aecidia on *Abies alba.*

Milesia carpatica (Wrobl.) Faull, *Contr. Arn. Arbor.* **2**, 1932, 55; Sacc., *Syll.* xxiii, 846 (as *Milesina carpatica* Wrobl.); Hunter, *TBMS*, **20**, 1936, 117; Wilson & Bisby, ibid. **37**, 1954, 66.

*On *Dryopteris.* Uredo- and teleutospores on the native *D. filix-mas*, Newton Abbot, Devon.

Milesia kriegeriana (Magn.) Arth., *Mycologia*, **7**, 1915, 176; Wilson, *Trans. bot. Soc. Edinb.* **31**, 1934, 433; Hunter, *TBMS*, **20**, 1936, 117; Wilson & Bisby, ibid. **37**, 1954, 66; syn. *Aecidium pseudo-columnare* Kühn, *Hedwigia*, **23**, 1884, 168; Sacc., *Syll.* vii, 826.

*On *Dryopteris.* Uredo- and teleutospores on the native *D. filix-mas, D. spinulosa* and *D. austriaca.* Widely distributed in Scotland; Corn, Devon, Hants, Norfolk. The aecidia occur on *Abies alba* and other species of *Abies.*

Milesia murariae (Syd.) Faull, *Contr. Arn. Arbor.* **2**, 1932, 34; Wilson, *Trans. bot. Soc. Edinb.* **31**, 1934, 434; Hunter, *TBMS*, **20**, 1936, 118; Wilson & Bisby, ibid. **37**, 1954, 66.

*On *Asplenium*. Uredospores on the native *A. ruta-muraria*, Forth, Clyde, Argyll.

Milesia polypodii White, *Scot. Nat.* **4**, 162; Sacc., *Syll.* vii, 768; Grove, *Brit. Rust Fungi*, 376 (as *Milesina dieteliana* Magn.); Wilson, *Trans. bot. Soc. Edinb.* **31**, 1934, 436; Hunter, *TBMS*, **20**, 1936, 118; Wilson & Bisby, ibid. **37**, 1954, 66.

*On *Polypodium*. Uredo- and teleutospores on the native *P. vulgare*, Devon, Hants, Surrey, Merioneth, Argyll, Perthsh. Aecidia on *Abies alba* and *A. concolor* by inoculation (Hunter, *J. Arnold Arbor.* **17**, 1936, 26–37).

Milesia scolopendrii (Fuckel) Arth., *Bull. Torrey bot. Cl.* **51**, 1924, 52; Sacc., *Syll.* xxiii, 846 (as *Milesina scolopendrii* Jaap); Grove, *Brit. Rust Fungi*, 378 (as *Uredo scolopendrii* Schröt.); Wilson, *Trans. bot. Soc. Edinb.* **31**, 1934, 436; Hunter, *TBMS*, **20**, 1936, 118; Wilson & Bisby, ibid. **37**, 1954, 66.

*On *Phyllitis* (*Scolopendrium*). Uredo- and teleutospores on the native *P. scolopendrium* (*Scolopendrium vulgare*), Corn, Devon, Hants, Lancs, Clyde. Aecidia on *Abies alba* and *A. concolor* by inoculation (Hunter, *J. Arnold Arbor.* **17**, 1936, 26–37.

Milesia whitei Faull, *Contr. Arn. Arbor.* **2**, 1932, 111; Hunter, *TBMS*, **20**, 1936, 118; Wilson & Bisby, ibid. **37**, 1954, 66.

*On *Polystichum*. Uredo- and teleutospores on the native *P. seti-ferum*, Devon, Scotland.

Milesina = *Milesia*.

The name *Milesia* is accepted on the basis of arguments put forward by Rogers (*Mycologia*, **40**, 1948, 251).

Monilia cinerea Bon. = stat. conid. of *Sclerotinia laxa*.

Monilia fimicola Cost. & Matr. = *Scopulariopsis fimicola*.

Monilia fructigena Pers. = stat. conid. of *Sclerotinia fructigena.*

Monilinia cydoniae (Schellenb.) Whetz. = *Sclerotinia cydoniae.*

Monilinia fructigena (Aderh. & Ruhl.) Honey = *Sclerotinia fructigena.*

Monilinia johnsonii (Ell. & Ev.) Honey = *Sclerotinia crataegi.*

Monilinia laxa (Aderh. & Ruhl.) Honey = *Sclerotinia laxa.*

Monilinia linhartiana (Prill. & Delacr.) Dennis = *Sclerotinia cydoniae.*

Monilinia mespili (Schellenb.) Whetz. = *Sclerotinia mespili.*

Mortierella sp.
*Associated with the death of lawn grasses. Kent, 1934.

Mucilago spongiosa Morgan = *Spumaria alba.*

Mucor sp.
*On cress (*Lepidium*), causing considerable loss by reason of abundant growth among seedlings in boxes, Aylesbury, Bucks, 1946–7.

Myceliophthora lutea Cost., *Rev. gén. bot.* **6**, 1894, 289; Sacc., *Syll.* xi, 587.
*Yellow mould of mushroom (*Agaricus*) beds. Contaminates the compost in mushroom beds as a mould, white at first, then yellow (Wood, *Mushroom News*, **4**, 1953, 64–8). Not uncommon in recent years. Sussex, Kent, Salop, Yorks, Lancs.

Edwards and Gandy (*Rep. Mushr. Res. Sta. Yaxley for 1953*, 38) suggested the name yellow mould for this and other yellowish fungi in mushroom composts, to replace the older names 'mat disease' and 'vert-de-gris'.

Mycogone perniciosa Magn., *Bot. Zbl.* **34**, 1888, 394; Sacc., *Syll.* xvi, 1040; Wakefield & Bisby, *TBMS*, **25**, 1941, 86.

*White mould of mushroom (*Agaricus*). Widely distributed in mushroom beds and often serious where cleanliness, fumigation and sterilization of casing soil is not regularly practised. It has been seen (Devon, 1938) on *A. arvensis* in a pasture.

The disease was referred to by Berkeley (*Gdnrs' Chron.* **23**, 1863, 1227), Cooke (ibid. **5**, 1889, 434–5) and others (e.g. *J. Bd Agric.* **12**, 1905, 47–9), but Smith (*TBMS*, **10**, 1924, 81–97) was the first to study it carefully. Ware (*Ann. Bot., Lond.*, **47**, 1933, 763–85) also made observations on the fungus.

Mycogone rosea Link ex Chev., *Flore générale des environs de Paris*, i, 1826, 53; Sacc., *Syll.* iv, 183; Wakefield & Bisby, *TBMS*, **25**, 1941, 86.

*On mushroom (*Agaricus*). Rare. Arborfield, Berks, 1938 (Williams, *Gdnrs' Chron.* **105**, 1939, 236; *Rep. exp. Res. Sta. Cheshunt for 1938*, 43); Som, 1949 and 1956.

Mycosphaerella brassicicola (Duby) Oudem. *Révision des champignons dans les Pays Bas*, **2**, 1897, 210; Sacc., *Syll.* i, 502 (as *Sphaerella brassicicola* (Duby) Ces. & de Not.); Bisby & Mason, *TBMS*, **24**, 1940, 168; stat. conid. *Phyllosticta brassicicola* McAlp., *Bull. Dep. Agric. Vict.* 1901, 27; Grove, *Coelomycetes*, i, 9.

*Ring spot of *Brassicae*. Occurs regularly on cabbage, brussels sprouts, and broccoli in south and south-west England and in south and mid-Wales, especially between October and March; epidemic in some years. Less frequent in northern districts and Scotland. Found occasionally on marrow-stem kale, ox cabbage, and horse-radish. Known in Britain since 1841 and described by Grove in *J. R. hort. Soc.* **40**, 1914, 76–7.

A special survey carried out in 1949 showed that ring spot was of economic importance only in south-west England, especially in Devon, Corn, and the market-garden areas around Bristol, and in the Gower and Pembroke peninsulas of Wales. There was no evidence that it is becoming more serious or spreading eastwards.

Mycosphaerella carinthiaca Jaap, *Ann. mycol.*, *Berl.*, **6**, 1908, 210; Sacc., *Syll.* xxii, 128; Bisby & Mason, *TBMS*, **24**, 1940, 168.

*Mid-vein spot of red clover (*Trifolium*). Local and unimportant. Hants and other southern and east midland counties, Glam, Aberystwyth.

Mycosphaerella citrullina (C. O. Smith) Grossenb., *Tech. Bull. N. Y. St. agric. Exp. Sta.* no. 9, 1909, 226; Sacc., *Syll.* xxii, 124.

*Stem rot of cucumber (*Cucumis*). Fairly common in the Lea Valley (Essex, Herts); occasional elsewhere, Hants, Glos, Wilts, Dorset, Yorks. Also causes die-back of the shoots (Williams, *Rep. exp. Res. Sta. Cheshunt for 1952*, 25–6) and a fruit rot.

Massee (*Kew Bull.* 1909, 292–3) recorded *M. citrullina* on tomato from Waltham Cross and on cucumber from Glos, and his view was accepted for a time, but later Brooks and Searle (*TBMS*, **7**, 1921, 173–97) concluded that this species had not been found in England, an opinion that held sway until it was identified on melon in 1947 (see below). It is possible that some of the early records of *Didymella lycopersici* (q.v.) on cucumber belong here. Sheard (*J. hort. Sci.* **28**, 1953, 278–94) showed that the host ranges of the two fungi are different.

*Stem rot of melon (*Cucumis*). Hockley, Essex, July 1947 (W. C. Moore & F. Joan Moore, *TBMS*, **32**, 1950, 277–9). (See also *Ascochyta cucumeris*.)

Mycosphaerella cucumis (Fautr. & Roum.) Chiu & Walker, see *Ascochyta cucumeris*.

Mycosphaerella fragariae (Tul.) Lindau, Rabenh., *Krypt. Fl.* i, 8, 458; Sacc., *Syll.* i, 505 (as *Sphaerella fragariae* (Tul.) Sacc.); Bisby & Mason, *TBMS*, **24**, 1940, 169; stat. conid. *Ramularia tulasnei* Sacc., *Mich.* i, 536; Sacc., *Syll.* iv, 203; Wakefield & Bisby, *TBMS*, **25**, 1941, 96.

*Leaf spot of strawberry (*Fragaria*). Occurs everywhere and is sometimes locally serious. It has been found causing brown lesions on the fruits in Som (Ogilvie, *Plant Path.* **5**, 1956, 37).

Mycosphaerella hedericola (Desm.) Lindau, in Engler's *Pflanzenfamilien*, I. Abt. i, 1897, 424; Sacc., *Syll.* i, 481 (as *Sphaerella hedericola* (Desm.) Cooke); Bisby & Mason, *TBMS*, **24**, 1940, 169; stat. conid. *Septoria hederae* Desm., *Ann. Sci. nat.* **19**, 1843, 340; Sacc., *Syll.* iii, 490; Grove, *Coelomycetes*, i, 385.
*On leaves of ivy (*Hedera*). Very common.

Mycosphaerella ligustri (Rob. in Desm.) Lindau, in Engler's *Pflanzenfamilien*, I. Abt. i, 1897, 424; Sacc., *Syll.* i, 480 (as *Sphaerella ligustri* (Desm.) Cooke); Bisby & Mason, *TBMS*, **24**, 1940, 169; stat. conid. *Phyllosticta ligustri* Sacc., *Mich.* i, 134; *Syll.* iii, 21; Grove, *Coelomycetes*, i, 25.
*Leaf spot of privet (*Ligustrum*). Widely distributed and common.

Mycosphaerella pinodes (Berk. & Blox.) Vestergr., *Bih. svensk VetenskAkad. Handl.* **22**, 1896, 15; Sacc., *Syll.* i, 514 (as *Sphaerella pinodes* (Berk. & Blox.) Niessl); Bisby & Mason, *TBMS*, **24**, 1940, 169.
*Foot rot and leaf and pod spot of pea (*Pisum*). Widely distributed in Britain, but prevalence uncertain. (See *Ascochyta pisi*.)

Mycosphaerella ribis (Fuckel) Kleb., *Haupt u. Nebenfruchtf. Askom.* 1918, 66; Sacc., *Syll.* i, 486 (as *Sphaerella ribis* Fuckel); stat. conid. *Septoria ribis* Desm., *Ann. Sci. nat.* **17**, 1842, 111; Sacc., *Syll.* iii, 491; Grove, *Coelomycetes*, i, 403.
The perfect stage has apparently not been seen in Britain.
*Septoria leaf spot of black currant (*Ribes*). North Wales, Som, and perhaps elsewhere in England; widely distributed in Scotland. It is comparatively harmless. Found occasionally on red currant.
The disease was described by Collinge (*Rep. econ. Biol.* **2**, 1912, 45–6) and Wormald (*Gdnrs' Chron.* **104**, 1938, 424–5).

Mycosphaerella sentina (Fr.) Schröt., in Cohn, *Kryptogamenflora von Schlesien*, 1894, 334; Sacc., *Syll.* i, 482 (as *Sphaerella pyri* Auersw.); Bisby & Mason, *TBMS*, **24**, 1940, 170; stat. conid.

Septoria piricola Desm., *Ann. Sci. nat.* 14, 1850, 114; Sacc., *Syll.* iii, 487; Grove, *Coelomycetes*, i, 400.

*Leaf fleck of pear (*Pyrus*). Not uncommon in southern districts of England and Wales. Listed from Clyde.

Myrothecium roridum Tode ex Fr., *Syst. Myc.* iii, 217; Sacc., *Syll.* iv, 750; Wakefield & Bisby, *TBMS,* 25, 1941, 57.

Preston (*TBMS,* 26, 1943, 158–68) gave an amended description of *Myrothecium* and its three classic species, including *M. roridum.*

*On old pea plants, Oxon, 1950.

*On stem bases of tomato (*Lycopersicon*), Glam, Salop, south-east England. Doubtfully parasitic.

*On *Antirrhinum*, Cambridge, 1939.

*On *Delphinium*, along with *Diplodina* sp., Botley, Hants, 1946.

*On lupin, Cambridge (Brooks, *TBMS*, 27, 1945, 155), Clyde.

*Stem rot of *Viola*. First seen on pansy and viola in the west midlands in 1932 and now known to be fairly widely distributed on these hosts in England. Seen rarely on violet, Salop, 1953 (Preston, *Plant Path.* 3, 1954, 30).

The parasitism of this fungus has been proved by Preston (*TBMS,* 20, 1936, 242–51) and by Brooks (ibid. 27, 1945, 155–7). Brooks also obtained a non-pathogenic strain from potato. (See also under *Pythium violae.*)

Mystrosporium adustum Massee, *Gdnrs' Chron.* 25, 1899, 412; Wakefield & Bisby, *TBMS*, 25, 1941, 99.

*Ink disease of *Iris*. Widely distributed on bulbous irises in England and Wales, but rarely severe except in the extreme south-west of England, the Isles of Scilly and parts of Wales. Listed from Moray and Perthshire. *I. reticulata* bulbs are particularly susceptible.

After its early appearances as a bulb disease of *I. reticulata* between 1887 and 1900 (Dod, *Gdnrs' Chron.* 2, 1887, 313; 26, 1899, 14) the 'ink mildew' was lost sight of until 1927 (*Report*, vi, 68), and since then it has become prominent mainly as a destructive leaf disease. Green (*Gdnrs' Chron.* **89**, 1931, 55; *J. R. hort. Soc.* **61**, 1936, 167–75) proved the pathogenicity of the fungus. The disease was described by W. C. Moore (*Diseases of Bulbs*, 1939, 133).

*Leaf spot of *Lachenalia*. Rare. On *L. glaucina* (Green & Hewlett, *J. R. hort. Soc.* **74**, 1949, 211–15); on *Lachenalia* × *boundii* in glasshouse, Harpenden, Herts, 1940.

*On *Tritonia* (*Montbretia*). (See *J. Bd Agric.* **15**, 1908, 509.)

Myxocyclus cenangioides (Ell. & Rothr.) Petrak, *Ann. mycol. Berl.*, **25**, 1927, 305; Sacc., *Syll.* x, 508 (as *Steganosporium cenangioides* Ell. & Rothr.); syn. *Camarosporium abietis* Wilson & Anders., *TBMS*, **9**, 1924, 150; *Camarographium abietis* Grove, *Coelomycetes*, ii, 107.

*On branches of *Abies*. On *A. concolor* var. *lowiana*, Arniston, Midlothian, 1923. Parasitism doubtful. Described by Wilson and Anderson (*TBMS*, **9**, 1923, 144–51). (See also Wilson, ibid. **23**, 1939, 206–8.) On *A. pinsapo*, Droitwich.

Myxosporium corticola Edgert. = stat. conid. of *Pezicula corticola*.

Myxosporium rosae Fuckel, *Symb. Myc.* 399; Sacc., *Syll.* iii, 723; Grove, *Coelomycetes*, ii, 255.

*On shoots and stems of rose, Harpenden, Herts; Edgbaston, War.

In May 1947 a species of *Myxosporium* was seen on a cankered rose shoot in Ches. The spores ($19–25 \times 5–8\,\mu$, av. $22 \times 6\,\mu$) were too large for this species or *M. vogelii* Laub.

Naemospora crocea (Bon.) Sacc., *Syll.* iii, 747; Grove, *Coelomycetes*, ii, 261.

*On peach (*Prunus*). Massee (*Kew Bull.* 1908, 269–71) described the fungus as causing a die-back of peach shoots, but this requires confirmation.

Nectria cinnabarina (Fr.) Fr., *Summ. Veg. Scand.* 1849, 388; Sacc., *Syll.* ii, 479; Bisby & Mason, *TBMS*, **24**, 1940, 199.

Coral spot. The fungus usually occurs as a saprophyte on many hosts, but sometimes becomes parasitic after entering through dead snags or wounds, and may kill branches or whole bushes.

*Apple (*Malus*) and pear (*Pyrus*). Unimportant.

*Apricot (*Prunus*). Yorks.

*Red currant (*Ribes*). Common and sometimes damaging. Line (*TBMS*, **8**, 1922, 22–8) carried out field observations and experiments.

*Fig (*Ficus*). Fairly common.

*Gooseberry (*Ribes*). Unimportant.

*Beech (*Fagus*). Edinburgh. Severe in five-year-old hedge.

*Sycamore (*Acer*) and Elm (*Ulmus*). Boyd (*Quart. J. For.* **13**, 1919, 139) discussed parasitism.

Nectria coccinea (Fr.) Fr., *Summ. Veg. Scand.* 1849, 368; Sacc., *Syll.* iv, 481; Bisby & Mason, *TBMS*, **24**, 1940, 199.

*Bark disease of beech (*Fagus*). This species is consistently associated with the death of patches of bark on older, and particularly on old beech trees. Mr T. R. Peace of the Forestry Commission tells me that the cankers which occur sporadically in young beech woods, notably in the southern Cotswolds and Sussex, are symptomatic of a different disease which appears to be constantly associated with *Nectria punicea* (Kunze & Schm.) Fr. (*Summ. Veg. Scand.* 387; Sacc., *Syll.* ii, 480), though proof of pathogenicity is still lacking. At one time the cankers were attributed variously to *N. galligena* Bres., *N. ditissima* Tul. and *N. coccinea*.

Nectria cucurbitula (Tode) Fr., *Summ. Veg. Scand.* 1849, 388; Sacc., *Syll.* ii, 484; Bisby & Mason, *TBMS*, **24**, 1940, 199.

*On spruce (*Picea*). A common saprophyte on dead coniferous branches, but occasionally associated with stem canker of Norway spruce (*Picea abies*) and Sitka spruce (*P. sitchensis*). See Laing, *Forestry*, **21**, 1947, 217–20.

Nectria galligena Bres., in Strasser, *Pilzfl. Sonntagbl.* 4, 1901, 413; Sacc., *Syll.* xvii, 788; as *Dialonectria galligena* (Bres.) Petch in Mason & Grainger, *Cat. Yorks Fungi*, 1937, 32; Bisby & Mason, *TBMS*, 24, 1940, 196; stat. conid. *Cylindrocarpon mali* (Allesch.) Wr., *Phytopathology*, 3, 1913, 225; Sacc., *Syll.* xxv, 982.

*Canker and eye rot of apple (*Malus*) and pear (*Pyrus*). Very common on apple trees throughout Britain; often severe in low-lying places on heavy soils. Less common on pear. The eye-rot stage on the fruit very prominent in some years, especially on Worcester Pearmain.

BIOLOGY. An early account of apple canker was that by Plow-right (*Gdnrs' Chron.* 21, 1884, 509–10), though the disease was known long before. Serious investigation of it began when Cayley (*Ann. Bot., Lond.*, 35, 1921, 79–92) made observations on the life history of the fungus, and Wiltshire showed that infection could take place through leaf scars (*AAB*, 8, 1921, 182–92) and scab wounds (ibid. 9, 1922, 275–81). The occurrence of conidia and perithecia on apple fruits was mentioned by Salmon and Wormald (*Gdnrs' Chron.* 58, 1915, 289), by Kidd and Beaumont (*TBMS*, 10, 1924, 116) and by Dillon-Weston (*AAB*, 12, 1925, 398–400; *TBMS*, 12, 1927, 5–12). M. H. Moore (*Rep. E. Malling Res. Sta. for 1933*, 166–75) made field observations on the disease, Umpleby and Swarbrick (*Rep. Long Ashton Res. Sta. for 1935*, 98–103) dealt with its incidence in cider-apple orchards, and more recently Crowdy (*AAB*, 36, 1949, 483–95) described the anatomy of the stem canker, and returned to the question of leaf-scar infection (ibid. 39, 1952, 569–80), while M. H. Moore and Bennett (ibid. 39, 1952, 588–98) dealt with the effect of manuring on the disease.

Pear canker was dealt with specifically by Harris (*Rep. E. Malling Res. Sta. for 1924*, 135–6) and Wormald (*J. Minist. Agric.* 34, 1927, 162–5) on the shoots and branches; and by Dillon-Weston (*Gdnrs' Chron.* 80, 1926, 373) on the fruit.

CONTROL. Munson (*AAB*, 26, 1939, 440–57) and Marsh (ibid. 26, 1939, 458–69) discussed control of canker in the light of information obtained about the times of spore dispersal. Crowdy (*Nature, Lond.*, **161**, 1948, 320; *Rep. Long Ashton Res. Sta. for 1947*, 158–63; *AAB*, **40**, 1953, 197–207) tried the effect of treating canker lesions with plant-growth substances, whilst the role of eradicant sprays has been studied by Byrde, Crowdy and Roach (*Rep. Long Ashton Res. Sta. for 1951*, 134–7; *AAB*, **39**, 1952, 581–7) and by Byrde and Corke (*Rep. Long Ashton Res. Sta. for 1953*, 159–62). Taylor and Byrde (*Plant Path.* **3**, 1954, 72) showed the value of phenyl mercury chloride as an eradicant, if expensive, spray.

*Canker of ash (*Fraxinus*). Great Missenden, Bucks, and doubt-less elsewhere.

Nectria mammoidea Phill. & Plowr. var. **rubi** (Osterw.) Weese, *Z. Garungsphys.* **1**, 1912, 128; Petch, *TBMS*, **21**, 1938, 260; Bisby & Mason, ibid. **24**, 1940, 200; syn. *N. rubi* Osterw., *Ber. dtsch. bot. Ges.* **29**, 1911, 620; Sacc., *Syll.* xxiv, 675 (as *Hypomyces rubi* (Osterw.) Wr.).

*Crown rot of raspberry (*Rubus*). Occasional in England and Wales, and may be damaging. Widely distributed in Scotland.

Pethybridge and Nattrass (*TBMS*, **12**, 1927, 20–7) published notes on the disease in Ireland and the west of England, and Alcock (*Trans. bot. Soc. Edinb.* **29**, 1925, 197–8) on its occurrence in Scotland. There is some doubt about the parasitism of the fungus.

Nectria peziza (Tode) Fr. = *Dialonectria peziza*.

Nectria punicea (Kunze & Schm.) Fr., see *N. coccinea*.

Nectria rubi Osterw. = *N. mammoidea* var. *rubi*.

Neofabraea perennans Kienholz, *J. agric. Res.* **59**, 1939, 635; stat. conid. *Gloeosporium perennans* Zeller & Childs, *Bull. Ore. agric. Exp. Sta.* no. 217, 1925, 1–17.

*Perennial canker and fruit rot of apple (*Malus*). Widely distri-buted in Britain on the shoots and branches, and now the chief species of *Gloeosporium* causing wastage of fruit (especially of

Cox's Orange Pippin) in storage. First recorded in England in 1942 and in Scotland in 1950.

The disease was first described in England by Wilkinson (*Gdnrs' Chron.* **111**, 1942, 269; **114**, 1943, 159), who later concluded (*J. Pomol.* **21**, 1945, 180–5) that the most serious phase of its attack was die-back in summer-pruned trees. He also (*TBMS*, **28**, 1945, 77–85) studied the cultural characters and pathogenicity of the fungus, and compared its attack on apple fruits with that of *G. album* and *G. fructigenum*. Perennial canker subsequently increased markedly in intensity (Wilkinson, *AAB*, **41**, 1954, 354–7). Corke (*Rep. Long Ashton Res. Sta. for 1954*, 164–8; *J. hort. Sci.* **31**, 1956, 272–83) studied the rate of growth of cankers, and succeeded in inducing infection all the year round by inoculation with mycelium, but only in November–December with spore suspensions.

Edney (*AAB*, **44**, 1956, 113–28) showed that winter- as well as summer-pruning cuts became infected. He followed the course of spore production in the orchard and studied the factors influencing fruit rotting in storage.

Trials by Marsh *et al.* (*Plant Path.* **6**, 1957, 30–41; see also *Rep. Long Ashton Res. Sta. for 1956*, 109–15) showed that some reduction of rotting in storage can be obtained by late summer spraying, and Hamer & Hunnam (*Grower*, **47**, 1957, 746–7) found spraying with captan beneficial.

*Fruit rot of pear (*Pyrus*). Occasional. Essex, Worcs, Cambs (Groom & Hall, *Plant Path.* **6**, 1957, 95). The imperfect stage was also found growing saprophytically or as a wound parasite on twigs and branches in Essex and Kent.

Nummularia discreta (Schw.) Tul., *Sel. Fung. Carp.* ii, 1863, 45; Sacc., *Syll.* i, 398; Miller, *TBMS*, **17**, 1932, 132; Bisby & Mason, ibid. **24**, 1940, 153.

*Branch blister of apple. Rare. Sandsend, Yorks, 1910 (*J. Bd Agric.* **18**, 1911, 314; Smith & Ramsbottom, *TBMS*, **4**, 1913, 172).

Ochropsora ariae (Fuckel) Ramsb., *TBMS*, **4**, 1914, 337; Wilson & Bisby, ibid. **37**, 1954, 66; syn. *O. sorbi* (Oud.) Diet., *Ber. dtsch. bot. Ges.* **13**, 1895, 401; Sacc., *Syll.* xxi, 605 and vii, 592; Grove, *Brit. Rust Fungi*, 329.

*Uredo- and teleuto-stages on *Sorbus aucuparia* not seen in Britain, but the species was reported at Cardiff in 1944 on one leaf of the same crab-apple tree as in previous years.

The aecidial stage on *Anemone nemorosa* is also uncommon.

Ochropsora sorbi (Oud.) Diet. = *O. ariae*.

Oedocephalum lineatum Bakshi = stat. conid. of *Fomes annosus*.

Oedocephalum pallidum (Berk. & Br.) Cost. ex Thaxter, *Bot. Gaz.* **16**, 1891, 18.

*In mushroom (*Agaricus*) beds, invading the compost, Alcester, War, 1954. The fungus disappeared after damping the beds. *Oedocephalum* sp. was recorded on casing soil of mushroom beds at Welwyn, Herts, in 1946 and 1949.

Oidium begoniae Puttemans, *Bull. Soc. Bot. Belg.* **48**, 1912, 238; Sacc., *Syll.* xxv, 648.

*Begonia mildew. Widely distributed throughout Britain and at times severe; has become much more prevalent since 1950.

The first British record was a slight attack under glass at Cambridge in Feb. 1938 (*Report*, viii, 80) and it was not seen again until found in Yorks in 1950. The first Scottish record was at Cockburnspath, Berwick, in 1952.

Oidium chrysanthemi Rabenh., *Hedwigia*, **1**, 1862, 19; Sacc., *Syll.* iv, 43.

*Chrysanthemum mildew. Very common in all districts under glass; frequent out-of-doors. See W. G. Smith (*Gdnrs' Chron.* **22**, 1884, 685). Rhodes *et al.* (*AAB*, **45**, 1957, 221–5) obtained promising results in controlling it in small-plot trials using the antibiotic griseofulvin.

OIDIUM

Oidium ericinum Erikss., *Bidrag Känned. växt. sjukd.* 1885; Blumer, *Die Erysiphaceen Mitteleuropas,* 1933, 412.

*Erica mildew. Rare. The only British record is of a specimen received at Cheshunt (*Rep. exp. Res. Sta. Cheshunt for 1937,* 42).

Oidium erumpens Cooke & Mass., *Grevillea,* **16,** 1887, 49; Sacc., *Syll.* x, 520; Cooke, *Fung. Pests Cult. Pl.,* 60; Wakefield & Bisby, *TBMS,* **25,** 1941, 76.

*Mildew of *Rivea.* On cultivated plants of *R. hypocrateriformis,* Kew Gardens, 1887.

Oidium euonymi-japonicae (Arcang.) Sacc., *Syll.* xviii, 506 apud Salmon in *Ann. mycol., Berl.,* **3,** 1905, 5.

*Euonymus mildew. Common and widely distributed, especially in southern and western coastal areas of England. Salmon (*J. R. hort. Soc.* **29,** 1904–5, 434–42) described the disease and carried out cultural experiments with the fungus (*Ann. mycol., Berl.,* **3,** 1905, 1–15).

Oidium hortensiae Jørstad = stat. conid. of *Microsphaera polonica.*

Oidium lini Skoric, *Ann. pro Exp. Foresticus, Zagreb,* **1,** 1926, 108.

*Flax mildew. Sporadic and usually mild. Seen under glass at Cambridge, 1927 (Salmon & Ware, *Gdnrs' Chron.* **82,** 1927, 34–5), and then not until 1941. War, Wilts, Pembs, Kent, Norfolk.

The mildew also occurs in Northern Ireland (Colhoun & Muskett, *Gdnrs' Chron.* **110,** 1941, 30). There is still some doubt about the identity of the perfect stage. Two mildews have been recorded on flax (see *Rev. appl. Mycol.* **7,** 1928, 783); one of these is *Erysiphe polygoni* DC. and the other may be *E. cichoracearum* DC. (see *Report,* viii, 78).

Oidium oxalidis McAlp., *Proc. Roy. Soc. Victoria,* **7,** 1894, 219; Sacc., *Syll.* xiv, 1041.

*Oxalis mildew. On *Oxalis rosea* and *O. corniculata,* Forth.

Oidium quercinum Thüm. = stat. conid. of *Microsphaera alphitoides.*

Oidium tuckeri Berk., see *Uncinula necator*.

Oidium valerianellae Fuckel, *Symb. myc.* 1869, 358; Sacc., *Syll.* iv, 41.

*On corn salad (*Valerianella olitoria*), near Bodmin, Corn, 1945 (W. C. Moore, *TBMS*, **29**, 1946, 250).

Oidium sp.

*Potato (*Solanum*). Local. Until 1945 the only British record was of a slight attack under glass at Cambridge in 1932 (*Report*, vii, 31). In 1945 it was again found under glass at Cambridge and was prevalent in trial plots and field crops within a ten-mile radius of Cambridge (Thomas, *Nature, Lond.*, **158**, 1946, 417). Subsequently seen occasionally in Cambs, Beds, Bucks, Hants and Kent in outdoor crops during the hot summers of 1947 and 1955.

The fungus is usually regarded abroad as belonging to *Erysiphe cichoracearum* DC. Brooks (*Plant Diseases*, 1953, 132) lists it as *Oidium solani* Auct.

*Mint (*Mentha*). Rare. Berks, April 1937.
*Sage (*Salvia*). Worcs, 1955–6.
*Himalayan blackberry (*Rubus*). Ledbury, Herefs, 1953. New host record.
Anemone*. Uncommon. Slight in Devon and Corn since 1935, with a rather severe attack in 1937 (Gregory, *TBMS*, **32, 1950, 241–2).
Antirrhinum*. Not uncommon on *A. majus* and easily overlooked. First recorded in Bucks and Devon in 1928. Occurs chiefly under glass; occasional in the open. Salop, Cambs, Herts, Glam (W. C. Moore, *TBMS*, **31, 1947, 86), Sussex.
*Cineraria (*Senecio*). Frequent. Herts, Staffs, I. of Ely, War, Yorks, Jersey; St Andrews, Scotland (Macdonald, *Gdnrs' Chron.* **105**, 1939, 111; *Trans. bot. Soc. Edinb.* **32**, 1939, 558–9). (See also *Gdnrs' Chron.* **12**, 1852, 404.)
*Carnation (*Dianthus*). Common in many English districts; listed in Forth. Described by Mercer (*J. R. hort. Soc.* **41**, 1915, 227–9) and White (*Rep. exp. Res. Sta. Cheshunt for 1927*, 43–4).

Colutea arborescens. Harpenden, Herts, Aug. 1949. New British record.

Cornflower (Centaurea). Frequent in Cambs and Hunts.

Cyclamen. Affecting petals of *C. persicum* under glass, Bath, Som, 1955. Foliage not attacked. New host record.

Dahlia. Renfrew, 1955.

Gloxinia (Sinningia). On young plants of *S. grandiflora gigantea,* Badsey, Worcs, 1955. New host record.

Rock rose (Helianthemum). On *H. pilosum purpureum,* Edinburgh, Nov. 1947 (W. C. Moore & F. Joan Moore, *TBMS,* **32,** 1950, 274).

Laburnum. On *L. vulgare,* Harpenden, Herts, 1939, and subsequently most years on the same trees; Yorks, 1951; Dee.

Lilac (Syringa). Not uncommon in autumn in southern England for some years after 1948, but now apparently disappeared. First seen Oct. 1948 at Cheam, Surrey, and in the next two years found in Herts, Oxon, Cambs, Beds, Berks, Glam, Devon, Mon, I.W. (F. Joan Moore, *Plant Path.* **1,** 1952, 55). This mildew spread in epidemic fashion after 1939 in Europe (Blumer, *Phytopath. Z.* **17,** 1951, 477–88), but has also disappeared there. It may be the same as that called in America *Microsphaera alni* (Wallr.) Wint.

Nepeta. Causing extensive damage and leaf drop, Sunderland, Dur, Feb. 1956.

Nicotiana. On garden plant of *N. alata* var. *grandiflora,* Harpenden, Herts, Nov. 1939.

*On *Papaver somniferum* grown for oil, Mersea, Essex, 1951.

*On *Primula.* On *P. malacoides,* Flints, 1956.

*On *Schizanthus* sp. under glass, Harpenden, Herts, 1934; War, 1945; Forth, Devon, 1949. (See W. C. Moore, *TBMS,* **29,** 1946, 250.)

Stachys. Rare. On *S. olympica (lanata),* Beckenham, Kent, 1947; Ches, 1951.

Viola. Rather uncommon on viola and pansy. Known since 1916. Carns, 1922; Surrey, 1938 (on *V. tricolor*); Ches, Corn, Berks, Bucks, Herts, Northumb, Forth, Clyde, Dee.

Olpidium agrostidis Samps., *TBMS*, **17**, 1932, 192.

*In roots of *Agrostis stolonifera*. Yorks (Sampson, loc. cit. 182–94).

Olpidium brassicae (Woron.) Dangeard, *Ann. Sci. nat.* 7 Ser. **4**, 1886, 327; Sacc., *Syll.* vii, 312.

*In roots of brassicae. Probably fairly common on cabbage and cauliflower.

*In tomato (*Lycopersicon*) roots, Lancs, 1925; Ches and Glam, 1942, and perhaps common, but it causes little or no damage.

Sampson (*TBMS*, **23**, 1939, 199–205) studied this species and concluded that *Asterocystis radicis* de Wildeman and *Olpidium radicicolum* de Wildeman are synonymous with it. Bartlett (ibid. **13**, 1928, 221–38) claimed that it produces the hybridization nodules so familiar at times on swedes, but his experience could not be confirmed.

Olpidium majus Cook & Collins, *Ann. mycol., Berl.*, **33**, 1935, 72.
*In cucumber (*Cucumis*) roots, Glam, 1933. *Olpidium* sp. was also recorded in the roots of unthrifty plants in Sussex, 1949.

Olpidium radicicolum Wildeman = *O. brassicae*.

Oospora fimicola (Cost. & Matr.) Cub. & Megl. = *Scopulariopsis fimicola*.

Oospora lactis (Fres.) Sacc., *Syll.* i, 15.
*Rubbery rot of potato (*Solanum*). Associated with a dirty pink rot of the tubers in clamps and field, East Midlands, Essex and Glam, 1948–9. Parasitism uncertain. Waterlogging and high temperature are predisposing factors.

Oospora pustulans Owen & Wakef., *Kew Bull.* 1919, 297; Wakefield & Bisby, *TBMS*, **25**, 1941, 58.
*Skin spot of potato. Found mostly on Scotch and Irish seed potatoes, and is seasonal in occurrence. Sometimes causes losses of 10–60 % among seed potatoes, because they fail to sprout, and

is frequently of considerable economic importance. At Rothamsted it has recently been found to cause a brown cortical rot of the
roots and stolons.

This disease was evidently known in Scotland over a century
ago (Johnston, *The Potato Disease in Scotland*, 1846, 185). It
was named skin spot by Pethybridge (*J. Dep. Agric. Ire.* **15**, 1915,
524) and was carefully studied by Owen (*Kew Bull.* 1919, 289–
301) who gave earlier references. Shapovalov (*J. agric. Res.* **23**,
1923, 285–94) thought it was an early stage in the development of
powdery scab (*Spongospora subterranea*), but Millard and Burr
(*Kew Bull.* 1923, 273–87; *Gdnrs' Chron.* **73**, 1923, 355) disproved
this. Greeves and Muskett (*AAB*, **26**, 1939, 481–96) in Northern
Ireland, and Foister (ibid. **30**, 1943, 186–7) in Scotland, have
dealt with its control by tuber disinfection with organo-mercurial
compounds. (See also *Scot. J. Agric.* **15**, 1932, 191–6 and
Report, vii, 27.) Boyd (*AAB*, **45**, 1957, 284–92) discussed varietal
susceptibility, control, and other field aspects of the disease.
Kerr's Pink is particularly susceptible and Allen (ibid. **45**, 1957,
293–8) studied the development of skin spot in this variety.

Ives (*Plant Path.* **4**, 1955, 17–21) described an unusually deep
form of skin spot in tubers treated with a dust containing iso-
propyl phenyl carbamate (IPC) which appeared to hinder cork
formation.

Oospora scabies Thaxt. = *Streptomyces scabies*.

Ophiobolus graminis (Sacc.) Sacc., *Reliquiae Mycologicae Libertianae*, ii, no. 143; Sacc., *Syll.* ii, 349; Bisby & Mason, *TBMS*,
24, 1940, 194; as *Linocarpon cariceti* (Berk. & Br.) Petrak in
Sydowia, **6**, 1952, 387. Arx and Olivier, *TBMS*, **35**, 1952, 29–
33, like others, maintained that the fungus does not belong in
the genus *Ophiobolus* and they made it the type species of a new
genus *Gaeumannomyces* as *G. graminis* (Sacc.) Arx & Olivier.

*Take-all of cereals. Widely distributed in Britain on wheat,
barley and certain grasses, including *Dactylis*, *Agrostis*, *Poa* and
Lolium. Rye is fairly resistant and oats almost immune. The
disease occurs mainly on alkaline soils overlying chalk or lime-

stone, such as on the Yorkshire and Lincolnshire Wolds, the Chiltern Hills, and in East Anglia, Wilts and Hants. Rare in Wales (first record on wheat 1950). Occasional in lawn turf. It is not often serious unless rotations are too short. It came into special prominence during the 1939–45 War when frequent cropping with cereals on the same land was inevitable, and there was a devastating epidemic in 1948 following a sequence of conditions specially favourable to the development of the fungus.

Ophiobolus graminis may have been present in England a century ago (M.J.B., *Gdnrs' Chron.* **27**, 1867, 933), but the disease it causes was not clearly described here until 1913 (*J. Bd Agric.* **19**, 1913, 1020), and it is only in comparatively recent years that it has been thoroughly investigated. The relation of soil conditions to take-all was portrayed in a series of ten articles published in the *Annals of Applied Biology*, vols. **23–35** (1936–48) by S. D. Garrett, who also summarized existing knowledge about the disease in 1942 (*Tech. Commun. Bur. Soil Sci., Harpenden*, no. 41, 40 pp.) and 1946 (*Rep. Rothamsted Exp. Sta. for 1946*, 60–4). See also Samuel (*J. Minist. Agric.* **44**, 1937, 231–41) for a general account of the disease; Storey (*AAB*, **33**, 1947, 546–50) for its occurrence in Yorks; and W. C. Moore (*Agriculture, Lond.*, **55**, 1948, 383–5) for an account of the unusually severe epidemic of 1948.

Publications dealing with special aspects of the subject include: take-all and manuring (Glynne, *AAB*, **22**, 1935, 225–35; Garrett, *Agriculture, Lond.*, **53**, 1946, 223–5; Salt, *J. agric. Sci.* **48**, 1957, 326–35); weed hosts (Walker, *AAB*, **32**, 1945, 177–8); effect of temporary leys and rotations (Dillon-Weston, ibid. **25**, 1938, 209–10; Garrett, *Agriculture, Lond.*, **47**, 1940, 134–5; Buddin & Garrett, ibid. **51**, 1944, 108–10; Wehrle & Ogilvie, *Plant Path.* **4**, 1955, 111–13); physiologic races (Padwick, *AAB*, **23**, 1936, 45–56); control under the Chamberlain system of intensive barley growing (Garrett & Buddin, *Agriculture, Lond.*, **54**, 1947, 425–6; Garrett, ibid. **56**, 1950, 514–16); the nature of resistance in oats (Turner, *J. exp. Bot.* **7**, 1956, 80–92; *AAB*, **44**, 1956, 200); and the effect of amino-acids on the growth of the fungus (Turner, *J. gen. Microbiol.* **16**, 1957, 531–3). S. G. Jones (*Ann. Bot., Lond.*, **40**,

OPHIOBOLUS

1926, 607–29) studied the development of the perithecium of *Ophiobolus graminis*. (See also *O. graminis* var. *avenae*.)

Ophiobolus graminis (Sacc.) Sacc. var. **avenae** E. M. Turner, *TBMS*, **24**, 1930, 279.

*Take-all of oats (*Avena*). Largely restricted to those areas where oats is the chief cereal crop, viz. Wales, north and north-west England, and Scotland, where it is worst in the west and north. Elsewhere only occasional, Surrey, Som, Dorset, Devon, I.W.

The fungus is capable of attacking wheat and barley and is probably responsible for most of the take-all of wheat in Wales and Scotland. Davies (*TBMS*, **33**, 1950, 352–3), for instance, found it common in mid-Wales on wheat, barley and oats. He made extensive ascospore measurements, and confirmed the differences between it and the parent species *O. graminis*, as described by Turner (ibid. **24**, 1930, 269–81). In Scotland its occurrence on oats was dealt with by Garrett and Dennis (ibid. **26**, 1943, 146) and on wheat by Dennis (*AAB*, **31**, 1944, 100–1).

*Patch disease of turf. Occurs in many parts of Britain, especially in the west, in sports turf containing a high proportion of *Agrostis tenuis*, *A. stolonifera* and *Poa annua* (Drew Smith, *J. Sports Turf Res. Inst.* **8**, 1952, 140–3; **9**, 1956, 180–202). Also on *Festuca rubra* (ibid. **8**, 1953, 259–60).

Ophiobolus herpotrichus (Fr.) Sacc., *Reliquiae Mycologicae Libertianae*, ii, n. 144; *Syll.* ii, 352; Bisby & Mason, *TBMS*, **24**, 1940, 194.

*On wheat stubble, Rothamsted, Herts, 1935. Parasitism uncertain. It also occurs on wild grasses (Glynne, *TBMS*, **20**, 1936, 122). Webster and Hudson (ibid. **40**, 1957, 509–11) obtained a pycnidial stage in culture and occurring naturally on *Agropyron repens*.

Ophiobolus sativus Ito & Kuribay., *Trans. Sapporo nat. Hist. Soc.* **10**, 1929, 138; syn. *Cochliobolus sativus* (Ito & Kuribay.) Dastur, *Indian J. agric. Sci.* **12**, 1942, 733; stat. conid. *Helminthosporium sativum* Pamm., King & Bakke, *Bull. Ia agric. Exp.*

Sta. no. 116, 1910, 178; Wakefield & Bisby, *TBMS*, **25**, 1941, 92.

The perfect stage has not been seen in Britain.

*Foot rot of cereals. Infrequent, and causes no appreciable damage. Has been reported causing a leaf stripe in Scotland. On wheat, a few records only, Cambridge, Wicken Fen: on barley, Cambridge, 1923 (Smith, *Proc. Camb. phil. Soc. (Biol. Sci.*) **1**, 1924, 132) (see also Russell, *TBMS*, **16**, 1932, 253–69); Scotland, 1948; Dee, 1955: on rye, single record at Cambridge, Oct. 1944. *On grasses. Causing leaf blight of *Lolium perenne* in sports field (Drew Smith, *J. Sports Turf Res. Inst.* **8**, 1953, 259–60); on leaves of *Festuca rubra*, Gravesend, Kent, Oct. 1955.

Ophiostoma narcissi Limber, *Phytopathology*, **40**, 1950, 493.

*In rotting narcissus bulbs, Lincs, 1953 (Baker, *Plant Path.* **3**, 1954, 29) and subsequently elsewhere. Probably non-parasitic or at most a weak parasite.

Ophiostoma ulmi (Buism.) Nannf. = *Ceratostomella ulmi.*

Ovularia berberidis Cooke, *Grevillea*, **13**, 1885, 98; Sacc., *Syll.* iv, 144; Cooke, *Fung. Pests Cult. Pl.* 1906, 186; Wakefield & Bisby, *TBMS*, **25**, 1941, 76.

*On *Berberis asiatica*, Kew Gardens, London.

Ovularia clematidis Chittend., see *Erysiphe polygoni.*

Ovularia interstitialis (Berk. & Br.) Massee, see *Ramularia primulae.*

Ovularia nymphaearum Allesch., in Rabenh., *Krypt. Fl.* **7**, 1901, 510; Sacc., *Syll.* xiv, 1004 (as *Gloeosporium nymphaearum* Allesch.); Grove, *Coelomycetes*, ii, 217.

*Leaf spot of *Nymphaea.* Kew Gardens; Mon, 1932; Devon, 1937 (*Report*, viii, 91); Forth; Hants, 1953 (on *N. gonnere*).

Grove (loc. cit.) states 'many specimens (?all) so-named are only *Ovularia nymphaearum* Bres. & All. = *Ramularia nymphaeae* Bres. (Sacc., *Syll.* xi, 601) = *Ovulariella nymphaearum* Kab. & Bub. *Exs.* no. 585'.

Smith and Ramsbottom (*TBMS*, **5**, 1915, 166) named the fungus *Ramularia nymphaearum* (Allesch.) Ramsb., *J. R. hort. Soc.* **40**, 1914, cxv, and recorded it from Wicken Fen, Cambs, 1913.
The species would repay further study.

Ovularia primulana Karst., see *Ramularia primulae*.

Ovulinia azaleae Weiss, *Phytopathology*, **30**, 1940, 243.
*Petal blight of *Rhododendron*. Established in Scotland along the Solway coast. Also at Dunkeld, Perthshire. Usually on hybrid rhododendrons flowering in early June. The first British record was near Kilmarnock, 1950 (Paton, *Plant Path.* **3**, 1954, 50; *Gdnrs' Chron.* **139**, 1956, 233). Dennis (*Mycol. Pap.* **62**, 1956, 159) listed the species as *Sclerotinia azaleae* (Weiss) n.c., but in a footnote said that apothecia received as this species from Scotland did not agree with Weiss's diagnosis.

Pachybasium hamatum (Bon.) Sacc. var. **candidum** Sacc., *Fungi Algerienses, Tahitenses et Gallici*, 1885, 6; Sacc., *Syll.* iv, 150 and xx, 255; Wakefield & Bisby, *TBMS*, **25**, 1941, 77.
*On strawberry (*Fragaria*), associated with root rot, but doubtfully parasitic (Berkeley & Lauder-Thompsen, *J. Pomol.* **12**, 1934, 222, as *Pachybasium candidum* Sacc.).

Paneolus sub-balteatus Berk. & Br., *Ann. Mag. nat. Hist.* 3 Ser. **7**, 1861, 373, no. 923; Sacc., *Syll.* v, 1124; Rea, *Brit. Basid.* 372.
*Invading mushroom (*Agaricus*) beds, Surrey, 1948; Herefs, 1953.

Papulaspora byssina Hotson, *Bot. Gaz.* **69**, 1917, 270; Sacc., *Syll.* xxv, 850; Wakefield & Bisby, *TBMS*, **25**, 1941, 102.
*Brown plaster mould of mushroom (*Agaricus*) beds. Common and comparatively harmless unless the compost is wet and tight. Early records were given by Ware (*Gdnrs' Chron.* **96**, 1934, 463–5). (See also Wood, ibid. **97**, 1935, 161–2.)

Passalora graminis (Fuckel) v. Höhn., *Zbl. Bakt.* II, **60**, 1923, 6; syn. *Scolecotrichum graminis* Fuckel, *Symb. myc.* 1869, 107; Sacc., *Syll.* iv, 348; Wakefield & Bisby, *TBMS*, **25**, 1941, 87.

*On grasses. Produces chocolate spotting of leaves of *Alopecurus, Cynosurus, Deschampsia, Poa* and *Glyceria*. Widely distributed. Was formerly said to be local in east and south-east England on wheat (Carruthers, *J. R. agric. Soc.* **9**, 1898, 751; **3**, 1892, 794).

Paxillus giganteus (Sow.) Fr., *Hymenomycetes Europaei*, 1874, 401; Sacc., *Syll.* v, 983; Rea, *Brit. Basid.* 1922, 549.

*Fairy rings associated with this fungus have been found causing considerable damage to young plantations of Scots pine (*Pinus sylvestris*), Thetford Chase, Norfolk. The same species or *P. extenuatus* Fr. was concerned at Heggiesmuir, Fife (Peace, *Forestry*, **10**, 1936, 74–8).

Pellicularia filamentosa (Pat.) Rogers = *Corticium solani*.

Pellicularia praticola (Kotila) Flentje = *Corticium praticola*.

Penicillium corymbiferum Westl., *Ark. Bot.* **11**, 1911, 92; Sacc., *Syll.* xxv, 667; Wakefield & Bisby, *TBMS*, **25**, 1941, 78.

*On *Galtonia*. Rotting stored imported bulbs of *G. candicans* (Ghamrawy, *TBMS*, **18**, 1933, 249–52).
*On hyacinth. Rotting bulbs of *Hyacinthus orientalis*, Forth, Tay; on bases of pedicels of pot hyacinths showing much floret drop, 1941.
*On *Iris*. A species of *Penicillium*, probably this, commonly rots iris bulbs in storage (W. C. Moore, *Diseases of Bulbs*, 1939, 135).
*On snowdrop (*Galanthus*). Causing a bulb rot, Forth.

Penicillium cyclopium Westl., *Ark. Bot.* **11**, 1911, 90; Sacc., *Syll.* xxv, 665; Wakefield & Bisby, *TBMS*, **25**, 1941, 78.

*On hyacinth. Causing bulb rot (W. C. Moore, *Diseases of Bulbs*, 1939, 18); Devon, 1947.
*On lily. Found in imported bulbs (Ghamrawy, *TBMS*, **18**, 1933, 249).

PENICILLIUM

*On *Scilla*. Rotting imported bulbs of *S. campanulata* var. *albida* (Macfarlane, *Trans. bot. Soc. Edinb.* 32, 1939, 542–7) and imported bulbs of *S. nutans* and *S. campanulata* (Singh, *TBMS*, 25, 1941, 194–9). Singh found the same species on *Scilla* bulbs collected from woods in England.

Penicillium expansum Link em. Thom, *Bull. U.S. Bur. Anim. Ind.* no. 118, 1910, 27; Sacc., *Syll.* iv, 78 (as *P. glaucum* Link p.p.).
*Blue mould of apple (*Malus*). Common on stored apples and market supplies. (See Kidd and Beaumont, *TBMS*, 10, 1924, 98.)

Penicillium gladioli McCull. & Thom, *Science*, 67, 1928 (24 Feb.), 216; Wakefield & Bisby, *TBMS*, 25, 1941, 78. It antedates by four days the name *P. gladioli* Machacek, *Rep. Quebec Soc. Prot. Pl.* 19, 1926–7, 77 (distributed 28 Feb. 1928).
*Storage rot of *Gladiolus*. Known since about 1925 and not uncommon, though usually found in imported corms. Occasional on Tigridia bulbs (Kent, 1928) and on corms of *Crocus* and *Tritonia* (*Montbretia*). Described by W. C. Moore (*Diseases of Bulbs*, 1939, 119).

Penicillium hirsutum Dierckx, *Ann. Soc. sci. Brux.* 25, 1901, 89; Sacc., *Syll.* xvi, 1031; Biourge, *La Cellule*, 33, fasc. I, 1923, 157; Wakefield & Bisby, *TBMS*, 25, 1941, 78.
*On hyacinth. Rotting bulbs (W. C. Moore, *Diseases of Bulbs*, 1939, 18).
*On *Scilla*. Rotting stored bulbs of *S. patula excelsior*, Bucks, 1941 (W. C. Moore, *TBMS*, 26, 1943, 21).

Penicillium thomii Maire, *Bull. Soc. Hist. nat. Afr. N.* 8, 1917, 192; Sacc., *Syll.* xxv, 683 (as *Citromyces thomii* (Maire) Sacc.).
*Leaf blotch of *Cypripedium callosum*. St Albans, Herts, 1941 (W. C. Moore, *TBMS*, 25, 1941, 206–8).

Penicillium sp.
*On potato (*Solanum*). Attacking sprouts, Pembs, 1944.

*On cucumber (*Cucumis*). Associated with rot and canker around pruning snags, Ches, 1950.

*On tomato (*Lycopersicon*) fruits. Causing a rot in market supplies 1945.

*On hyacinth. Frequently associated with bulb rotting.

*On *Iris*. A sclerotium-forming species of *Penicillium* has been found associated with rotting of iris bulbs in Scotland (1931) and of bulbs imported into England on several occasions (W. C. Moore, *Diseases of Bulbs*, 1939, 136).

*On *Narcissus*. Causing a rot of the bulbs. Occasional. The species is near *Penicillium chrysogenum* Thom and was named *P. narcissi* A. Smith by Brooks (*Plant Diseases*, 1928, 119) but this is a nomen nudum.

*On *Ornithogalum*. Affecting the corms, Devon, 1957.

*On tulip. Sclerotium-forming species rotting the bulbs, Worcs, 1931. *Penicillium* is frequently associated with the condition in tulips known as chalking or chalkiness.

Peridermium coruscans Fr., *Summ. Veg. Scand.* 510; Sacc., *Syll.* vii, 835.

*Rust of silver fir (*Abies*). The only evidence of the presence of this fungus in Britain appears to be a statement by Massee (*Dis. Cult. Pl. Trees*, 322) that he had 'seen it on *Abies pinsapo* in England'.

Peridermium pini (Pers.) Lév., *Mém. Soc. Sc. Lille*, **4**, 1826, 212.

*Resin top of Scots pine (*Pinus sylvestris*). Aecidia on stems. Not common except locally in north and east Scotland. For occurrence in Ireland see Pethybridge (*J. Dep. Agric. Ire.* **11**, 1911, 500–2).

Peronoplasmopara humuli Miyabe & Takah. = *Pseudoperonospora humuli*.

Peronospora antirrhini Schroet., *Hedwigia*, **13**, 1874, 183; Sacc., *Syll.* vii, 255.

*Downy mildew of *Antirrhinum*. Widely distributed throughout Britain; common in coastal areas, especially in southern England.

Usually has a crippling effect on young seedlings; not damaging to older plants and easily overlooked.

This fungus was not recognized in any country on the cultivated antirrhinum until it was found at Carlow, Ireland, in May 1936 (Murphy, *Int. Bull. Pl. Prot.* 11, 1937, 176M). The following year it was found near Brighton (Green, *Gdnrs' Chron.* 102, 1937, 27; *J. R. hort. Soc.* 63, 1938, 159–65) and in 1940 at Aberystwyth. From 1947 the disease began to be noticed at many places, and by 1952 fifty attacks had been reported from twenty-five counties in England and Wales (W. C. Moore & F. Joan Moore, *Plant Path.* 1, 1952, 135–6). The mildew was first seen in Scotland in 1948 (Wallace, *Gdnrs' Chron.* 124, 1948, 21) and is now known to be widely distributed there.

McKay (*Gdnrs' Chron.* 126, 1949, 28) demonstrated the prolific production of oospores in the cortical tissues of the stems and roots.

Peronospora arborescens (Berk.) de Bary, *Ann. Sci. nat.* 4 Ser. 20, 1863, 119; Sacc., *Syll.* vii, 251; syn. *Botrytis arborescens* Berk., *J. hort. Soc.* 1, 1846, 31.

*Downy mildew of *Meconopsis* and *Papaver*. Widely distributed in Britain and not uncommon. The fungus has been reported on *Meconopsis napaulensis*, *M. betonicifolia* and var. *baileyi*, which are particularly susceptible, *M. cambrica*, *Papaver somniferum*, *P. rhoeas* and *P. alpinum*. It has been described on *Meconopsis* by Cotton (*Gdnrs' Chron.* 85, 1929, 143–4) and by Alcock (*New Flora and Silva*, 5, 1933, 279–82). W. G. Smith (*Gdnrs' Chron.* 26, 1886, 140) recorded it on wild poppy (*Papaver dubium*) at Dunstable, Beds.

Peronospora arthuri Farlow, *Bot. Gaz.* 8, 1883, 315; Sacc., *Syll.* vii, 248.

*Downy mildew of *Clarkia*. Perth, 1954.

Peronospora corollae Tranzsch., *Hedwigia*, 34, 1895, 214; Sacc., *Syll.* xiv, 459.

*Downy mildew of *Campanula*. On *C. cochlearifolia* (*pulsilla*) and var. *alba*, Cambridge (Stearn, *Gdng ill.* 51, 1929, 565).

233

Peronospora coronillae Gäum., see *P. cytisi.*

Peronospora cytisi Magn., *Hedwigia,* **31,** 1892, 419; Sacc., *Syll.* xi, 243. There is also *P. cytisi* Rostr., *Z. PflKrankh.* **2,** 1892, 1, and Gäumann, *Beitr., Monogr. Gattung Peronospora,* 1923, 191, lists both as synonyms of *P. coronillae* Gäum.

*Downy mildew of laburnum. This fungus was reported on a seedling at Worplesdon, Surrey, in Sept. 1953, but no specimen exists and the record must remain a doubtful one.

Peronospora dipsaci Tul., *C.R. Acad. Sci., Paris,* **38,** 1854, 1103; Sacc., *Syll.* vii, 258.

*Downy mildew of teazle (*Dipsacus*). On leaves of *D. sylvestris* in garden, Dundry, Som, 1941; in a wood, Herts, 1946.

Peronospora destructor (Berk.) Casp., in Berk., *Outl.* 1860, 349; Sacc., *Syll.* vii, 257 (as *P. schleideni* Unger); syn. *P. schleideniana* W. G. Smith, *Gdnrs' Chron.* **21,** 1884, 418.

*Downy mildew of onion (*Allium*). Becomes prominent in the open in July and August, or earlier in frames. Occurs in most parts of Britain and is the most serious onion disease in south-east England. Severe also in the onion-growing areas of Lincs and East Anglia, but rarely damaging in Beds or Worcs. Special surveys showed that in the 1930's about 25 % of the onion acreage in England and Wales was infected annually.

The disease occurs on shallots (*A. ascalonicum*), but less commonly, except in mid-Wales and western England.

The parasite was first named *Botrytis destructor* by Berkeley (*Ann. Mag. nat. Hist.* **6,** 1841, 436), who later (*Gdnrs' Chron.* **22,** 1862, 689) described the disease caused by it. W. G. Smith (ibid. **21,** 1884, 418) found the oospores, and Murphy and McKay (*Nature, Lond.,* **108,** 1921, 304; *Sci. Proc. R. Dublin Soc.* **18,** 1926, 237–61; *J. Dep. Agric. Ire.* **26,** 1926, 115–23) proved the presence of perennial mycelium in infected bulbs. Later they made further observations on the disease (ibid. **31,** 1932, 60–76), and McKay (*Nature, Lond.,* **135,** 1935, 306; **139,** 1937, 758) succeeded in germinating the oospores. McKay (*J. R. hort. Soc.* **64,** 1939,

272–85) continued his observations and (*Sci. Proc. R. Dublin Soc.* **27**, 1957, 295–307) studied the longevity of the oospores, while Hickman, Marsh and Wilkinson (*AAB*, **30**, 1943, 179–83) tried unsuccessfully to find an effective method of control using oil-soluble copper sprays.

Peronospora effusa (Grev. ex Desm.) Rabenh., *Herb. Myc.* 1880; Sacc., *Syll.* vii, 256.

*Downy mildew of spinach (*Spinacia*). Widely distributed both under glass and in the open, especially in south and east England. (See W. G. Smith, *Gdnrs' Chron.* **23**, 1885, 480.)

Peronospora ficariae Tul., *C.R. Acad. Sci., Paris*, **38**, 1854, 1103; Sacc., *Syll.* vii, 251.

*Downy mildew of *Anemone*. Widely distributed and in some seasons very damaging in Devon and Corn; also prevalent in the Cheddar area of Som; Suffolk and Sussex, 1955; Oxon, Mon and Cards, 1957. Also occurs in Jersey.

Until about 1950 this disease was rarely seen. First recorded on *A. coronaria*, Penzance, Corn, 1935. Also at Newton Abbot, Devon, 1936 and Corn, 1938 (*Report*, viii, 79). Gregory (*TBMS*, **32**, 1950, 242–3) briefly described it. By 1953 it was widespread in the south-west and two years later prevalent around Cheddar.

Peronospora galligena Blumer, *Mitt. naturf. Ges. Bern, 1937*, 17, 1938.

*Downy mildew of *Alyssum*. On *A. saxatile*. First recognized at Slough, Bucks, in Nov. 1946 (W. C. Moore, *TBMS*, **32**, 1949, 95–7) and now known to be widely distributed in Britain. Glam, Glos, Hants, Berks, Oxon, Worcs, Yorks, Lancs, Inverurie.

Peronospora gei Sydow, in Gäumann, *Beiträge zu einer Monographie der Gattung Peronospora Corda*, 1923, 291.

*Downy mildew of *Geum*. On cultivated varieties, Daglingworth, Glos, 1945 (Dennis & Wakefield, *TBMS*, **29**, 1946, 153); Evesham, Worcs, 1951.

Peronospora grisea Unger, *Bot. Ztg*, **5**, 1847, 315; Sacc., *Syll.* vii, 255.

*Downy mildew of *Veronica*. Not uncommon in south and south-west England and the Isles of Scilly. It was recorded on *Hebe* (*Veronica*) *hulkeana* by Ivy Massee (*Gdnrs' Chron.* **55**, 1914, 335).

Peronospora hyoscyami de Bary, *Ann. Sci. nat.* 4 Ser. **20**, 1863, 123; Sacc., *Syll.* vii, 261.

*Downy mildew of henbane (*Hyoscyamus*). Not uncommon. Described by W. G. Smith (*Gdnrs' Chron.* **23**, 1885, 176).

Peronospora leptoclada Sacc., *Mich.* ii, 530; *Syll.* vii, 250.

*Downy mildew of rock rose (*Helianthemum*). On *H. nummularium* (*vulgare*), Bloxham, Oxon, Oct. 1947 (W. C. Moore & F. Joan Moore, *TBMS*, **32**, 1950, 274).

Peronospora myosotidis de Bary, *Ann. Sci. nat.* 4 Ser. **20**, 1863, 112; Sacc., *Syll.* vii, 245.

*Downy mildew of *Myosotis*. Occasional. Clyde, Dee, Herefs.

Peronospora oerteliana Kühn, *Hedwigia*, **23**, 1884, 173; Sacc., *Syll.* ix, 342.

*Downy mildew of *Primula*. Widely distributed but not common. On *P. wanda*, Staffs, 1928; on *P. pulverulenta* and hybrid of *P. juliae*, Southampton, 1936; Scotland. Early records were of *Peronospora candida* Fuckel (e.g. Smith, *Gdnrs' Chron.* **25**, 1886, 564).

Peronospora parasitica (Fr.) Tul., *C.R. Acad. Sci., Paris*, **38**, 1854, 1103; Sacc., *Syll.* vii, 249.

*Downy mildew of crucifers. Fairly common in Britain on turnip and swede; often troublesome in cauliflower and other cruciferous seedlings raised under glass; sometimes occurs on the sprouts of Brussels sprouts; and frequent on stock (*Matthiola*), wallflower (*Cheiranthus*) and watercress (*Nasturtium*).

Has occasionally been recorded on radish (Som, Sussex, Essex), seakale (Glam), marrow-stem kale (Mon), rape (Essex, Devon), horse-radish (Ches), and aubretia (Inverurie, 1952).

Peronospora pulveracea Fuckel, *Symb. myc.*, 67; Sacc., *Syll.* vii, 261.

*Downy mildew of Christmas rose (*Helleborus*). Occasional. Bucks, 1936; Norfolk, 1953. (See also W. G. Smith, *Gdnrs' Chron.* 4, 1888, 16–17.)

Peronospora schachtii Fuckel, *Symb. myc.* 71; Sacc., *Syll.* vii, 262.

*Downy mildew of sugar beet (*Beta*). A serious disease in the midlands and East Anglia where seed and root crops are grown together. Sporadic elsewhere in England, chiefly in years of severe epidemics. Rare in Scotland; Berwick, 1950; east Lothian, 1957. Occasionally seen on mangold, garden beet and seakale beet. The wild beet (*B. maritima*) is very susceptible.

Early records were from Ireland (Carruthers, *J. R. agric. Soc.* 3 Ser. **1**, 1890, 836), I. of Ely and Spalding, Lincs, 1921 (*Report*, iv, 39; Stirrup, *Bull. Minist. Agric., Lond.*, no. 93, 1935, 49–52), and southern England (Salmon & Ware, *J. Minist. Agric.* **32**, 1925, 833–8). Later, when it became an important economic disease, Hale, Watson and Hull (*AAB*, **33**, 1946, 13–28) gave an account of the field symptoms. Cornford (*Plant Path.* **3**, 1954, 82–3) estimated that up to 37 % loss in yield could follow from July infections. A general account of the disease was given by Hull (*Bull. Minist. Agric., Lond.*, no. 142, 1950, 26–9) and by McKay in his Research Bulletin, *Field Studies on the Downy Mildew of Sugar Beet*, 1957.

Peronospora schleideniana W. G. Smith = *P. destructor*.

Peronospora sparsa Berk., *Gdnrs' Chron.* **22**, 1862, 308; Sacc., *Syll.* vii, 263.

*Downy mildew of rose. Not often reported and probably more frequent than the records indicate. On seedling briars and adjacent roses, Oxon, 1939; causing defoliation and lesions on young shoots budded on briar stocks, I. of Ely, 1952; Corn, 1954, 1956; Yorks, 1956; Leics, under glass, 1957.

Notes on the disease were published by Cooke (*Gdnrs' Chron.* **8**, 1877, 7 July Suppl. vii–viii) and by Green (*J. R. hort. Soc.* **59**, 1934, 470).

Peronospora trifoliorum de Bary, *Ann. Sci. nat.* 4 Ser. **20**, 1863, 117; Sacc., *Syll.* vii, 252.

*Downy mildew of clover (*Trifolium*). This species is not uncommon in England and Wales on clover, trefoil (*Medicago lupulina*) and lucerne (*M. sativa*), but it does not assume economic proportions. Occasional in Scotland on white, red and alsike clovers. The fungus was briefly described by W. G. Smith (*Gdnrs' Chron.* **22**, 1884, 84).

Peronospora viciae (Berk.) Casp., *Monatsber. Kgl. Preuss Akad. Wiss. Berl.* 1855, 308; Sacc., *Syll.* vii, 245.

*On leaves of field and broad bean (*Vicia faba*). Uncommon. Hants, Herefs, 1922 (*Report*, v, 34); Lincs, 1939; Yorks, 1945; on the pods, Bristol (Fraymouth, *TBMS*, **39**, 1956, 94).

*On pea (*Pisum*). Widely distributed throughout Britain and very common in eastern England.

*On vetches (*Vicia* spp.). Sometimes severe. Cambs, Herts, Hants, Som. Occurs in Scotland on *V. sepium*, *V. cracca* and *V. sativa*.

*On sweet pea (*Lathyrus*). Not common. Worcs, Lancs, Tweed, Forth, Moray.

Peronospora violae de Bary, *Ann. Sci. nat.* 4 Ser. **20**, 1863, 125; Sacc., *Syll.* vii, 251.

*Downy mildew of *Viola*. Uncommon. Reported new to Britain by Cooke (*Gdnrs' Chron.* **5**, 1876, 118); on *Viola* sp., Yorks (*Report*, vii, 95); on *Viola* sp., Cambridge Botanic Garden, April 1949; Som, 1952. It has been listed on *V. tricolor* and *V. arvensis* from Clyde, Dee and Orkney.

Pestalotia decolorata Speg., *Fung. Arg.* i, 190; Sacc., *Syll.* iii, 784.

*Leaf spot of myrtle (*Myrtus*). On *M. communis* in southern England, 1950 (Hewlett, *J. R. hort. Soc.* **77**, 1952, 413–18).

Pestalotia gracilis Kleb., *Myc. Zbl.* **4**, 1912, 10; Sacc., *Syll.* xxv, 609; as *Pestalotiopsis gracilis* (Kleb.) Steyaert, *Bull. Jard. bot. Brux.* **19**, 1949, 310.

*On moribund saxifrage plants, Wheatley, Oxon, 1953; on *Saxifraga oppositifolia*, Maidenhead, Berks, 1953.

Pestalotia guepini Desm., *Ann. Sci. nat.* **13**, 1840, 182; Sacc., *Syll.* iii, 794; Grove, *Coelomycetes*, ii, 347; Cooke, *Fung. Pests Cult. Pl.* 164; as *Pestalotiopsis guepini* (Desm.) Steyaert, *Bull. Jard. bot. Brux.* **19**, 1949, 312.

*Leaf spot of *Camellia*. On *C. japonica* under glass in England and Scotland; on *C. reticulata*, Devon, 1951.
*Leaf spot of *Rhododendron*. Surrey, 1953.

Pestalotia hartigii Tub., *Beiträge zur Kenntnis der Baumkrankheiten*, 1888, 40; Sacc., *Syll.* x, 490; Grove, *Coelomycetes*, ii, 349.

*Strangling disease of conifers. Fischer (*J. econ. Biol.* **4**, 1909, 72-7) studied the biology of the fungus but could not demonstrate its pathogenicity. Reported killing young *Cupressus allumii*, Surrey, 1927 (*Report*, vi, 64). Frequent on dying seedlings of Sitka and Norway spruce (*Picea*), but there is no proof of pathogenicity.

Pestalotia macrotricha Kleb., *Myc. Zbl.* **4**, 1914, 6; Sacc., *Syll.* xxv, 601; Grove, *Coelomycetes*, ii, 350.

*Graft disease of *Rhododendron*. Fungus often associated with a stem disease of hybrid rhododendrons, but Howarth and Chippindale (*Gdnrs' Chron.* **86**, 1929, 471; *Mem. Manch. lit. phil. Soc. 1930–1*, **75**, 1931, 95–103) concluded the primary cause is imperfect grafting. A species of *Pestalotia* has also been found associated with disease in rhododendrons in Corn, 1935, and Hants, 1936.

Pestalotia sp.

*Associated with a crown disease of rose, Northumb, 1934.

Petriella asymmetrica Curzi, *Boll. Staz. Pat. veg. Roma*, **10**, 1930, 392.

*On tomato (*Lycopersicon*), associated with brown root rot, especially in the early part of the season (Ebben, *Rep. exp. Res.*

Sta. Cheshunt for 1949, 28–32; Williams & Ebben, ibid. *for 1950*, 23–6) but apparently not parasitic (Ebben & Williams, *AAB*, **44**, 1956, 425–36).

Pezicula corticola Nannf., Inop. Disc., *Nova Acta Soc. Sci. upsal.* 4 Ser. **8**, 1932, 94; stat. conid. *Myxosporium corticola* Edgert., *Ann. mycol., Berl.*, **6**, 1908, 51; Sacc., *Syll.* xxii, 1195; Grove, *Coelomycetes*, ii, 253; syn. *Cryptosporiopsis corticola* (Edgert.) Nannf., *Nova Acta Soc. Sci. upsal.* 4 Ser. **8**, 1932, 94.

The perfect stage has not been found in Britain.

*Surface canker of apple (*Malus*)*. Infrequent, or negligible and overlooked. Widely distributed, Sussex, Som, Herefs, Worcs, Ches, Berwick, Perth. First recorded in Som by Wiltshire (*Rep. Long Ashton Res. Sta. for 1920*, 81) and studied by Gilchrist (*TBMS*, **8**, 1923, 230–43) and Briton-Jones (*J. Pomol.* **4**, 1925, 162–83).

Peziza vesiculosa Bull. = *Aleuria vesiculosa*.

Pezizella lythri (Desm.) Shear & Dodge, see *Hainesia lythri*.

Phacidiella coniferarum Hahn, see *Phomopsis pseudotsugae*.

Phacidiella discolor (Mouton & Sacc.) Potebn., *Z. PflKrankh.* **22**, 1912, 147; Sacc., *Syll.* xxiv, 1261; stat. conid. *Phacidiopycnis malorum* Potebn., loc. cit. 144; Sacc., *Syll.* xxv, 230; syn. *Fuckelia conspicua* Marchal, *Bull. Soc. Bot. Belg.* **54**, 1921, 124; Ramsbottom & Browne, *TBMS*, **34**, 1951, 106 (as *Phacidium discolor* Mout. & Sacc.).

*Bark canker of apple (*Malus*) and pear (*Pyrus*)*. Not uncommon. Glos, Som, Herefs, Worcs, War, Cambs, Yorks, Ches, Dee. A weak parasite.

First reported in Herefs in 1926 as *Fuckelia conspicua* (*Report*, vi, 42) and seen shortly afterwards in Worcs (Nattrass, *Rep. Long Ashton Res. Sta. for 1927*, 99–100). A fungus provisionally identified as *Cytosporella fructorum* El. & Em. Marchal, to which a die-back of apple trees in Cambs was attributed by Southee and

Brooks (*TBMS*, **11**, 1926, 213–19), was later shown by Brooks (ibid. **13**, 1928, 75–81) to be the pycnidial stage of *Phacidiella discolor*.

Phacidiopycnis malorum Potebn. = stat. conid. of *Phacidiella discolor*.

Phacidiopycnis pseudotsugae, see *Phomopsis pseudotsugae*.

Phacidium discolor Mout. & Sacc., see *Phacidiella discolor*.

Phaeobulgaria inquinans (Fr.) Nannf., Inop. Disc., *Nova Acta Soc. Sci. upsal.* 4 Ser. **8**, 1932, 311; Sacc., *Syll.* viii, 636 (as *Bulgaria inquinans* (Pers.) Fr.); Ramsbottom & Browne, *TBMS*, **34**, 1951, 72; Dennis, *Mycol. Pap.* **62**, 1956, 167.
*On beech (*Fagus*). Usually saprophytic on the bark of felled oak and beech, but sometimes acts as a parasite. An attack on pollard beech at Burnham Beeches, Bucks, was described by Tabor and Barratt (*AAB*, **4**, 1917, 20–7) and attributed to *Bulgaria polymorpha* Wettst., a name frequently used erroneously for *Phaeobulgaria inquinans*, the biology of which was studied by Biffen (*Ann. Bot., Lond.*, **15**, 1901, 119).

Phaeocryptopus gäumannii (Rohde) Petrak, *Ann. mycol., Berl.*, **36**, 1938, 22.
*Swiss needle cast of Douglas fir (*Pseudotsuga*). Abundant in the west of England, Wales and parts of Scotland, but apparently absent from eastern England. Seldom damaging. Occurrence in Eire described by Liese (*Quart. J. For.* **33**, 1939, 247–52).

Phialea mucosa E. G. Gray = *Gloeotinia temulenta*.

Phialea temulenta Prill. & Delacr. = *Gloeotinia temulenta*.

Phleospora aceris (Lib.) Sacc., *Syll.* iii, 577; Grove, *Coelomycetes*, i, 432; Cooke, *Fung. Pests Cult. Pl.* 1906, 200; syn. *Septoria aceris* Berk. & Br., *Ann. Mag. nat. Hist.* 1850, 379.
*On leaves of maple (*Acer campestre*) and sycamore (*A. pseudoplatanus*). Common throughout Britain. (See also *Phyllosticta platanoidis*.)

Phleospora mori (Lév.) Sacc. = *Septogloeum mori.*

Phleospora oxyacanthae (Kunze & Schm.) Wallr., *Flora Crypto-gamica Germaniae*, 1833, 177; Sacc., *Syll.* iii, 578; Grove, *Coelomycetes*, i, 434.

*On leaves of hawthorn (*Crataegus*). Widely distributed and sometimes harmful.

Phleospora pseudoplatani Bubák, *Pilz Montenegro*, 1903, 16; Sacc., *Syll.* xviii, 489; Grove, *Coelomycetes*, i, 432; syn. *Septoria pseudoplatani* Rob. & Desm., *Ann. Sci. nat.* **8**, 1847, 21; Sacc., *Syll.* iii, 478.

*On leaves of *Acer*. Common. Also occurs on the fruits and cotyledons. (See also *Phyllosticta platanoidis*.)

Phleospora robiniae (Lib.) Höhnel, *Ann. mycol., Berl.*, **3**, 1905, 334; Sacc., *Syll.* xxii, 1235; Grove, *Coelomycetes*, i, 436.

*On leaflets of *Robinia pseudoacacia*, Kew.

Phleospora ulmi Wallr. = *Septogloeum ulmi.*

Phlyctaena linicola Speg. = stat. conid. of *Sphaerella linorum.*

Pholiota squarrosa (Müll. ex Fr.) Sacc., *Syll.* v, 749; Rea, *Brit. Basid.* 1922, 117.

*On apple (*Malus*), Som, 1951.
*On cherry (*Prunus*). Perhaps a weak parasite on old trees, Kent (Salmon & Ware, *J. S.-E. agric. Coll. Wye*, **40**, 1937, 18–26). In *Report*, iv, 68, *P. spectabilis* (Fr.) Sacc., *Syll.* v, 751, was said to be fairly common at the base of old trees in Worcs and Herefs.
*On Norway spruce (*Picea abies*), producing a soft brown rot (Peace, *Quart. J. For.* **32**, 1938, 81–104).

Phoma alternariacearum Brooks & Searle, *TBMS*, **7**, 1921, 193; Grove, *Coelomycetes*, i, 105.

*On tomato (*Lycopersicon*), producing a rot of immature and ripe fruits. Grove (loc. cit.) considered it to be merely an early stage of *Ascochyta lycopersici*.

Phoma apiicola Kleb., *Z. PflKrankh.* **20**, 1910, 22; Sacc., *Syll.* xxii, 880; Grove, *Coelomycetes*, i, 66.

*Root rot of celery (*Apium*). Fairly common in English celery-growing areas but of little economic importance. Occasional in east Solway. First recognized in 1924 (*Report*, v, 47). See Stirrup and Ewan, *Bull. Minist. Agric., Lond.*, no. 25, 1931, 34 pp.

Phoma araucariae Trav., see *Phomopsis araucariae*.

Phoma berberidicola Vestergr., *Öfv. Svensk. Vet. Akad. Förh.* 1897, 38; Sacc., *Syll.* xiv, 866; Grove, *Coelomycetes*, i, 67.

*On branches of *Berberis vulgaris* in hedges, Porlock, Som; Fowey, Corn.

Phoma betae Frank = stat. conid. of *Pleospora betae*.

Phoma caryophylli Cooke = *Phomopsis caryophylli*.

Phoma cinerascens Sacc. = *Phomopsis cinerascens*.

Phoma destructiva Plowr., *Gdnrs' Chron.* **16**, 1881, 621; Sacc., *Syll.* x, 175.

*Fruit rot of tomato (*Lycopersicon*). Formerly reported frequently from many districts, especially in southern England, as the cause of much late loss in outdoor fruits. Listed also in Scotland but not often reported there. It may be doubted if the fungus commonly identified in Britain as *Phoma destructiva* is specifically distinct from *Diplodina lycopersici* (q.v.). Dennis (*TBMS*, **29**, 1946, 11–42) considered the true *Phoma destructiva* a distinct species. Grove (*Coelomycetes*, i, 314) lists it as a synonym of *Ascochyta lycopersici* Brun., but this requires confirmation. There is little doubt that, as Hickman (*J. Pomol.* **22**, 1946, 69–75) showed, there are at least two distinct forms of tomato fruit rot.

Phoma devastatrix Berk. & Br., *Ann Mag. nat. Hist.* **3**, 1859, 359; Sacc., *Syll.* iii, 132; Cooke, *Fung. Pests Cult. Pl.* 1906, 53; Grove, *Coelomycetes*, i, 90.

*On *Lobelia*, Shrublands, Suffolk and elsewhere. Can be destructive.

Phoma eupyrena Sacc., see *P. foveata* and *P. tuberosa*.

Phoma foveata Foister, *Trans. bot. Soc. Edinb.* **33**, 1940, 65.

*Gangrene of potato (*Solanum*). This disease was first recognized in Scotland (Alcock & Foister, *Scot. J. Agric.* **19**, 1936, 252–7), where it is a source of annual loss among stored tubers, and especially boxed tubers. It has been found in potatoes from Eire and Northern Ireland, and has become of increasing importance and severity in recent years in England and Wales, usually on seed tubers derived from Scotland. It seems to be particularly common in potatoes from the north and east of Scotland.

The disease was described by Foister (*Scot. J. Agric.* **23**, 1940, 63–7), who also dealt with its distribution and prevalence (*Plant Path.* **1**, 1952, 85–6) and, with others, showed that it is soil borne (Foister *et al.*, *Nature, Lond.*, **155**, 1945, 793). See also Rosser and Jones (*Plant Path.* **5**, 1956, 148–9). The fungus was studied by Dennis (*TBMS*, **26**, 1946, 11–42); it has been seen rarely on living potato stems (Perthshire, 1952), but is much less frequent on dead potato haulm than *Phoma eupyrena* Sacc., *Mich.* i, 526; *Syll.* iii, 127, and *P. tuberosa* (q.v.).

P. tuberosa not uncommonly causes lesions of the gangrene type, but *P. eupyrena*, which can frequently be isolated from gangrene lesions, is regarded as a camp follower rather than a primary parasite.

Phoma herbarum Westend., *Exs.* no. 965; Sacc., *Syll.* iii, 133; Grove, *Coelomycetes*, i, 60.

*Phoma wilt of hop (*Humulus*). Rare. Sussex, 1926; Wye, 1935 (Salmon, *J. Inst. Brew.* **41**, 1935, 235).

Phoma lavandulae Gabotto, *N. Giorn. bot. Ital.* **12**, 1905, 69; Sacc., *Syll.* xviii, 258; Grove, *Coelomycetes*, i, 89.

*Lavender shab. Widely distributed but only locally troublesome in the lavender areas at Hitchin, Herts and Mitcham, Surrey. The variety Dwarf French and *Lavandula dentata* appear to be immune. Studied by Brierley (*Kew Bull.* 1916, 113–31) and by Metcalfe (*J. Minist. Agric.* **36**, 1929, 640–5; *J. R. hort. Soc.* **55**, 1930, 271–5; *TBMS*, **16**, 1931, 149–76; *Gdnrs' Chron.* **85**, 1929, 296; **89**, 1931, 221–2).

Phoma lingam (Fr.) Desm., *Ann. Sci. nat.* **11**, 1849, 281; Sacc., *Syll.* iii, 119; Grove, *Coelomycetes*, i, 69; syn. *P. napobrassicae* Rostr., *Tidsskr, Landøkon.* **2**, 1892, 330; Sacc., *Syll.* xi, 488; *P. oleracea* Sacc., *Mich.* i, 913; *Syll.* iii, 135; Grove, *Coelomycetes*, i, 63.

*Dry rot and canker of turnip and swede (*Brassica*). Occurs in all districts but damaging mainly in Scotland, northern England, Wales and Norfolk. Responsible for substantial losses in some years.

Potter (*J. Bd Agric.* **6**, 1900, 448–56) described this disease as new to England when he found it at Corbridge in the winter of 1896–7. Later, Carruthers (*J. R. agric. Soc.* **69**, 1908, 308–20) gave a careful description of it. Murphy and Hughes (*Nature, Lond.*, **122**, 1928, 13; *J. Dep. Agric. Dublin*, **29**, 1929, 29–40) and later Hughes alone (*Sci. Proc. R. Dublin Soc.* **20**, 1933, 495–530) studied it in Eire, and Whitehead and Jones (*Welsh J. Agric.* **5**, 1929, 159–75; **6**, 1930, 289–95) in Wales, while after a special investigation carried out in England, Buddin (*Bull. Minist. Agric., Lond.*, no. 74, 1934, 47 pp.) published a comprehensive account of the disease, with particular reference to its occurrence in the swede seed crop. Since then Dennis (*Scot. J. Agric.* **22**, 1939, 226–30; *AAB*, **26**, 1939, 627–30; *Scot. Fmr*, 25 Jan. 1941, 332) has dealt with it and emphasized the importance of seed-borne infection.

*Canker of *Brassicae*. Widely distributed in Britain and for a long time the cause of serious loss in Yorks and northern England. Most common on broccoli, but also on Brussels sprouts, cabbage, cauliflower, kale, and occasionally marrow-stem kale (Millard, *Agriculture, Lond.*, **52**, 1945, 39–42). The introduction of a service for the hot-water treatment of commercial samples of broccoli seed, described by Bant, Beaumont

and Storey (*N.A.A.S. Quart. Rev.* no. 9, 1950, 45–6), soon led to a marked decrease in the intensity of the disease in the north. Canker became prominent on broccoli in Kent about 1951.
*On stock (*Matthiola*). Occasional. Yorks, Glos, Surrey, Clyde.
*On wallflower (*Cheiranthus*). Occasional.

Phoma linicola March. & Verplancke, see *Ascochyta linicola*.

Phoma mali Schulz & Sacc., *Micromycetes Slavonici novi*, 1884, no. 42; Schulz, *Illustrationes Fungorum Slavonicorum*, no. 783; Sacc., *Syll.* iii, 75.
*Phoma rot of apple (*Malus*). Colhoun (*AAB*, **25**, 1938, 93–4) reported this species as of considerable importance in bringing about loss in stored apples in Northern Ireland, especially late in the season. It does not appear to have been reported specifically from Britain, but in 1941 *Phoma* sp. was recorded causing a stalk end rot of apple fruits in Kent.

Phoma mororum Sacc., *Mich.* i, 525; *Syll.* iii, 95; Grove, *Coelomycetes*, i, 94.
*On mulberry (*Morus*). On twigs of *M. alba*, Kew Gardens.

Phoma napobrassicae Rostr. = *P. lingam*.

Phoma oleracea Sacc. = *P. lingam*.

Phoma phlogis Roum., *Rev. mycol.* **6**, 160; Sacc., *Syll.* iii, 129.
*Stem rot of *Phlox*. Forth.

Phoma phormii (Cooke) Sacc. = *Coniothyrium phormii*.

Phoma rhodorae Cooke, see *Phyllosticta rhododendri*.

Phoma rostrupii Sacc., *Syll.* xi, 490; Grove, *Coelomycetes*, i, 80.
*This species causes greyish brown cankers on the roots and stems of carrot (*Daucus carota*) in Denmark, and it has been described as British, apparently erroneously, on the basis of Massee's reference to it in *Dis. Cult. Pl. Trees*, 408.

PHOMA

Phoma solanicola Prill. & Delacr. fide Köhler = *P. tuberosa.*

Phoma syringae (Fr.) Sacc., *Syll.* iii, 82.
*On lilac (*Syringa*), killing the shoots, Sussex, 1922.

Phoma tuberosa Melh., Rosenb. & Schultz, *J. agric. Res.* 7, 1916, 251; Sacc., *Syll.* xxv, 115; Grove, *Coelomycetes,* i, 107; syn. *P. solanicola* Prill. & Delacr. fide Köhler, *Angew. Bot.* 10, 1928, 113; Grove, *Coelomycetes,* i, 106.

*Phoma rot of potato (*Solanum*). Further investigation of this fungus is required. At the end of summer *Phoma* is commonly present on potato stems, and a closely related fungus causes tuber lesions in storage. The stem fungus has usually been called *P. solanicola* and the tuber fungus *P. tuberosa,* but I can see no specific difference between them, and Dennis (*TBMS,* 29, 1946, 11–42) grouped both species with other strains of *Phoma.* The tuber rot has been recognized since 1923 (*Report,* v, 28), and reported as such it is uncommon. Modern thought, however, places *P. tuberosa* along with *P. foveata* (q.v.) as a cause—if a minor one—of potato gangrene, at least around Edinburgh. It is also commonly present on dead potato haulm in late summer.

A species of *Phoma,* commonly identified as *P. eupyrena* Sacc. (*Mich.* i, 526; *Syll.* iii, 127), is sometimes associated with the tuber disease known as skin necrosis (*Report,* viii, 24), which appears to be related in some way to gangrene.

Phoma sp.
*On eggplant (*Solanum melongena*), associated with a rot of the fruits, Berks, 1945. Possibly a form of *Diplodina lycopersici.*
*On stems of mint (*Mentha*). Som, Glos.
*Infrequent as the cause of one form of parsnip canker. Mentioned by Cotton (*Kew Bull.* 1918, 8) and by Ogilvie and Mulligan (*Rep. Long Ashton Res. Sta. for 1933,* 115) who regarded it as an *Ascochyta.*
*On stored vegetable marrows (*Cucurbita*). Leics, 1941.
*Foot rot of flax (*Linum*), see *Ascochyta linicola* and *Colletotrichum linicola.*

247

*Foot rot of scabious. On *Scabiosa caucasia*, Wilts, 1949.

*Associated with a foot rot of statice (*Limonium*). Berks, Oxon. The fungus did not agree with previous species of *Phoma* recorded on this host.

*Leaf and stem rot of *Viola*. Edinburgh; Perth, 1951.

Phomopsis araucariae Grove, *Coelomycetes*, i, 181 and 456. Grove (loc. cit. 73) regarded *Phoma araucariae* Trav., *Malpighia*, **14**, 1900, 469; Sacc., *Syll.* xvi, 876, as an immature stage of this.

*On *Araucaria*. On fallen scale-leaves of *A. araucana* (monkey puzzle), Foxcote, nr. Ilmington, War; west Scotland and Perthshire, 1945. Doubtfully parasitic.

Phomopsis asteriscus (Berk.) Grove, *Kew Bull.* 1917, 53; *Coelomycetes*, i, 234; Sacc., *Syll.* iii, 126 (as *Phoma asteriscus* Berk.).

*On seed of parsnip (*Pastinaca*), Scotland, 1948.

Phomopsis aucubicola Grove, *Kew Bull.* 1917, 67; Sacc., *Syll.* xxv, 123; Grove, *Coelomycetes*, i, 170.

*On branches of *Aucuba japonica*, Botanic Gardens, Edgbaston, War.

Phomopsis caryophylli (Cooke) Grove, *Kew Bull.* 1917, 54; *Coelomycetes*, i, 183; syn. *Phoma caryophylli* Cooke, *Grevillea*, **13**, 1885, 94; Sacc., *Syll.* x, 176.

*On *Dianthus*. On calyces, peduncles and stems of cultivated pinks and carnations (*D. caryophyllus*), Shrewsbury, Twycross, King's Cliffe, Perranzabulae Church, Fordingbridge (Hants).

On dead flowering stems of sweet william (*D. barbatus*), Salop, 1936.

Phomopsis cinerascens (Sacc.) Trav., *Flora italica Cryptogama*, ii, 1906, 278; Grove, *J. Bot., Lond.*, 1917, 55; *Coelomycetes*, i, 186; syn. *Phoma cinerascens* Sacc., *Mich.* i, 521; *Syll.* iii, 96; *Libertella ulcerata* Mass., *Gdnr's Mag.* 1898, 475; *J. Bd Agric.* **17**, 1910, 47-9.

500

PHOMOPSIS

*Fig canker. Frequent in southern England, occasional elsewhere. Was serious in the Worthing area of Sussex in 1920-5. Described by Salmon and Wormald (*AAB*, 3, 1916, 1-12).

Phomopsis conorum (Sacc.) Died., see *P. pseudotsugae*.

Phomopsis controversa (Sacc.) Trav., see *Diaporthe eres*.

Phomopsis gardeniae Budd. & Wakef., *Gdnrs' Chron.* 103, 1938, 45. The name antedates *P. gardeniae* Hans. & Barrett, *Mycologia*, 30, 1938, 15, by a week or two.
*Stem canker of *Gardenia*. Uncommon. Middx, Surrey (Buddin & Wakefield, *Gdnrs' Chron.* 101, 1937, 226; 103, 1938, 45); Sussex, 1951; Glam, 1956.
Cooke (*Gdnrs' Chron.* 15, 1894, 605; *J. R. hort. Soc.* 28, 1904, 327) may have described the same disease.

Phomopsis occulta (Sacc.) Trav., *Flora italica Cryptogama*, ii, 1906, 221; Sacc., *Syll.* iii, 150 (as *Phoma occulta* Sacc.); Grove, *Coelomycetes*, i, 179. It was listed by Wehmeyer, *TBMS*, 17, 1933, 251, as an imperfect stage of *Diaporthe eres* Nits. (q.v.)
*On spruce (*Picea*). Common on fallen cones, dead leaves and stems, but occasionally parasitic on seedlings of Norway and Sitka spruce.

Phomopsis perniciosa Grove = stat. conid. of *Diaporthe perniciosa*.

Phomopsis pseudotsugae Wilson, *Trans. bot. Soc. Edinb.* 28, 1920, 47; Sacc., *Syll.* xxv, 122; Grove, *Coelomycetes*, i, 180.
Hahn (*Mycologia*, 49, 1957, 226-39) described the perfect stage in North America as *Phacidiella coniferarum* Hahn (ibid. p. 227) and the imperfect stage as *Phacidiopycnis pseudotsugae* (Wilson) Hahn (ibid. p. 230).
*Phomopsis disease of conifers. Central and south Scotland. Causing shoot die-back and stem canker of Douglas fir (*Pseudotsuga taxifolia*). Occasionally reported on *Larix decidua, L.*

leptolepis, Tsuga heterophylla, Abies grandis and other conifers, up to about 25 years old. Often follows initial frost injury.

See Wilson (*Trans. R. Scot. arb. Soc.* **34**, 1920, 145–9; **35**, 1921, 73–4; *Bull. For. Comm., Lond.*, **6**, 1925, 34 pp.). Wilson (*Gdnrs' Chron.* **88**, 1930, 412–13) also described the disease on *Cedrus atlantica* in Hants and Scotland. Wilson and Hahn (*TBMS*, **13**, 1928, 261–78) discussed the relationship of this fungus with *Phoma pitya* Sacc. and *P. abietiana* Hart., and (*Phytopathology*, **19**, 1929, 979–92) reviewed the history and distribution of the fungus in Europe.

The species may be confused superficially with *Phomopsis conorum* (Sacc.) Died., *Ann. mycol., Berl.*, **9**, 1911, 22; syn. *Phoma conorum* Sacc., *Mich.* ii, 615; *Syll.* xxii, 903, which has been found occurring as a saprophyte on Douglas fir in Britain (Hahn, *TBMS*, **13**, 1928, 278–86). Hahn (ibid. **15**, 1930, 32–93) also made a careful study of eight species of *Phomopsis* occurring on conifers.

Phomopsis scobina v. Höhn. = stat. conid. of *Diaporthe eres*.

Phomopsis tuberivora Güssow & Foster, *Canad. J. Res.* **6**, 1932, 253.

*Stem end hard rot of potato (*Solanum*). Occasional in late autumn or in stored tubers after hot dry summers. Lincs, Bucks, 1947; Lincs, Essex, 1949; Cambs, Norfolk, Leics, Corn, Devon, 1955. The disease may be primarily a drought effect with the fungus secondary or at most weakly pathogenic (Lester, *Plant Path.* **5**, 1956, 114).

Phomopsis sp.

*Associated with leaf spot of lucerne (*Medicago*). Mon, 1943.
*On tomato fruits, War, 1952. Closely resembling but not identical with *P. vexans*.
*Associated with a foot rot of statice (*Limonium*). Berks, 1934–7 (*Report*, viii, 89).

Phragmidium disciflorum (Tode) James = *P. mucronatum*.

Phragmidium mucronatum (Pers.) Schlecht., *Flora berolinensis*, ii, 1824, 156; Sacc., *Syll.* vii, 746 (as *P. subcorticium* (Schrank) Wint.?); Wilson & Bisby, *TBMS*, **37**, 1954, 66; syn. *P. disciflorum* (Tode) James, *Contr. U.S. nat. Herb.* **3**, 1895, 276; Grove, *Brit. Rust Fungi*, 293.

*Rose rust. Common and widely distributed throughout Britain on cultivated and wild roses. Sometimes causes defoliation but most of the older, very susceptible varieties are no longer grown.

An early account of rose rust is that by W. G. Smith (*Gdnrs' Chron.* **26**, 1886, 76). Investigations begun at the Cheshunt Experimental Station in 1932, particularly on the relative susceptibility of different rose stocks, on the occurrence of different strains of the rust, and on its control with oil in copper emulsions, were described by Bewley (*Rose Annu.* 1933, 110–12; *Sci. Hort.* **6**, 1938, 97–101), by Williams (*AAB*, **25**, 1938, 730–41) and in the *Reps. exp. Res. Sta. Cheshunt for 1933–6*. Muskett and Taylor (*Rose Annu.* 1933, 135–9; *J. Minist. Agric. N. Ire.* **4**, 1933, 62–6) tested a number of sprays against the rust in Northern Ireland, where it is often destructive. Notes on the parasitism of the rust by *Tuberculina persicina* (Ditm.) Sacc. were published by Deacon (*Rose Annu.* 1939, 136).

Phragmidium potentillae (Pers.) Karst., *Myc. fenn.* iv, 1878, 49; Sacc., *Syll.* vii, 743; Grove, *Brit. Rust Fungi*, 291; Wilson & Bisby, *TBMS*, **37**, 1954, 67.

*On *Potentilla*. Occurs on various cultivated species as well as on wild ones, but not commonly.

Phragmidium rubi-idaei (DC.) Karst., *Myc. fenn.* iv, 1878, 52; Sacc., *Syll.* vii, 748; Grove, *Brit. Rust Fungi*, 298; Wilson & Bisby, *TBMS*, **37**, 1954, 67.

*Raspberry rust. Widely distributed in Scotland and southern England, and not uncommon, but seldom serious.

Phragmidium violaceum (Schultz) Wint. *Pilze*, 231; Sacc., *Syll.* vii, 744; Grove, *Brit. Rust Fungi*, 295; Wilson & Bisby, *TBMS*, **37**, 1954, 67.

*Blackberry rust. Frequent locally on cultivated blackberry and on *Rubus laciniatus*. Kent, Som, Northumb.
*Loganberry rust. Occasional. Kent, Devon.

Phycomyces nitens (Agardh) Kunze, *Mykologische*, Hefte ii, 1823, 113; Sacc., *Syll.* vii, 205.
*Overgrowing seedlings of cress (*Lepidium*) in a dish, Sheffield, 1949.

Phyllachora graminis (Fr.) Fuckel, *Symb. myc.*, 216; Sacc., *Syll.* ii, 602; Bisby & Mason, *TBMS*, **24**, 1940, 206.
*Black leaf spot of grasses. Widely distributed in Britain, and common, but harmless. Occurs chiefly on cocksfoot (*Dactylis*) and *Agrostis*, especially in autumn and winter; occasional on *Agropyron* (Forth) and *Bromis inermis* (Forth).

Phyllachora pastinacae Rostr., *Plantepatologie*, 1902, 511; Bisby & Mason, *TBMS*, **24**, 1940, 206; stat. conid. *Cylindrosporium pastinacae* Lind, *Danish Fungi*, 1913, 493; Sacc., *Syll.* xi, 583 (as *C. pimpinellae* Massal. var. *pastinacae* Sacc.).
*On leaves of parsnip (*Pastinaca*). In conidial stage only, Mickleton, Glos, 1917 (Cotton, *Kew Bull.* 1918, 17).

Phyllachora sylvatica Sacc. & Speg., *Mich.* i, 410; Sacc., *Syll.* ii, 603.
*Black leaf spot of *Festuca*. Locally abundant on *F. rubra* and *F. ovina* in England in recent years. Also Solway, Moray.

Phyllachora ulmi (Duv.) Fuckel = *Systremma ulmi*.

Phyllactinia corylea (Pers.) Karst., *Acta Soc. Fauna Flora fenn.* **2**, 1885, 92; Sacc., *Syll.* i, 5 (as *P. suffulta* (Reb.) Sacc.); Bisby & Mason, *TBMS*, **24**, 1940, 135.
*Mildew of ash (*Fraxinus*).
*Mildew of *Corylus*. Not uncommon on cob nuts, but rarely damaging.

Phyllosticta aceris Sacc., *Mich.* i, 147; *Syll.* iii, 14; Grove, *Coelomycetes*, i, 3.

*Leaf spot of maple (*Acer*). Kent, Sussex, Surrey, War, Worcs, etc. (See also *Phyllosticta platanoidis.*)

Phyllosticta ajacis Thüm., *Fl. mycol. Litor.* no. 329; Sacc., *Syll.* iii, 38.

*Leaf spot of *Delphinium*. Clyde. First seen 1935.

Phyllosticta angulata Wenzl, *Phytopath. Z.* 9, 1936, 349–56.

*On apple (*Malus*). Was originally described as the cause of angular leaf spot, but is saprophytic on spots caused by the frog-hopper *Cercopis sanguinea* Geoff. (See W. C. Moore, *TBMS*, 22, 1939, 264; 24, 1940, 345, and Wormald, *Rep. E. Malling Res. Sta. for 1939*, 63–6.)

Phyllosticta antirrhini Syd., *Hedwigia*, 38, 1899, 134; Sacc., *Syll.* xvi, 839; Grove, *Coelomycetes*, i, 6.

*Stem rot of *Antirrhinum*. Not uncommon in southern England; north Wales, Solway, Forth. Described by Buddin and Wakefield (*Gdnrs' Chron.* 76, 1924, 150–2).

Phyllosticta apii Halst., *Rep. N. J. agric. Exp. Sta.* 1891, 253; Sacc., *Syll.* xi, 478; Grove, *Coelomycetes*, i, 6.

*Leaf spot of celery (*Apium*). Sussex, 1909 (*J. Bd Agric.* 16, 1010); Lincs, 1910 (ibid. 17, 301). Doubtfully distinct from *Phoma apiicola* (q.v.).

Phyllosticta aquifolina Grove, *Coelomycetes*, i, 23 and 455 (1935).

*On leaves of holly (*Ilex aquifolium*), Kew Gardens.

Phyllosticta argyrea Speg., *Fung. Arg.* ii, 121; Sacc., *Syll.* iii, 29; Grove, *Coelomycetes*, i, 15.

*On leaves of *Elaeagnus*. On *E. pungens variegata*, Hayling Island, Hants; Kent; this or *P. elaeagni* (Sacc.) Allesch. on *E. aurea* and other species, Harpenden, Herts, 1926.

Phyllosticta atrozonata Voss, see *Coniothyrium hellebori*.

Phyllosticta begoniae Brun., *Sphéropsidées recoltées jusqu'à ce jour dans la Charente-Inférieure*, 1889, 10; Sacc., *Syll.* xiv, 851.
*Leaf spot of *Begonia*. Forth.

Phyllosticta betulina Sacc., *Mich.* i, 154; *Syll.* iii, 32; Grove, *Coelomycetes*, i, 8.
*On leaves of *Betula*. Not uncommon in England.

Phyllosticta bolleana Sacc., *Syll.* iii, 15; Grove, *Coelomycetes*, i, 16.
*Leaf spot of *Euonymus*. On *E. japonicus*, Richmond and Wisley, Surrey, 1916 (Smith & Ramsbottom, *TBMS*, 6, 1920, 366); Polperro, Corn.
Grove (loc. cit.) considered it may be identical with *P. euonymi* Sacc., recorded on *E. japonicus* and *E. europaeus* in Kew Gardens and Ayrshire.

Phyllosticta brassicicola McAlp. = stat. conid. of *Mycosphaerella brassicicola*.

Phyllosticta buxina Sacc., *Mich.* i, 137; *Syll.* iii, 24; Grove, *Coelomycetes*, i, 9.
*On leaves of box (*Buxus*), Box Hill, Surrey.

Phyllosticta camelliae Westend., in Kickx, *Fl. Cr. Fl.* i, 1867, 416; Sacc., *Syll.* iii, 25; Grove, *Coelomycetes*, i, 10.
*Leaf spot of *Camellia*. On *C. japonica*, near Birmingham, War, 1885; Cound, Salop; east Scotland.

Phyllosticta camelliaecola Brun. var. **meranensis** Bubák, *Öst. bot. Z.* **55**, 1905, 80; Sacc., *Syll.* xviii, 224; Moore, *TBMS*, 31, 1947, 88.
*On leaves of *Camellia*, St Albans, Herts, 1939. Doubtfully parasitic and associated with spots believed to be caused by insects.

Phyllosticta carpathica Allesch., *Hedwigia*, **36**, 1897, 157; Sacc., *Syll.* xiv, 854; Grove, *Coelomycetes*, i, 11.

*Leaf spot of *Campanula*. On *C. persicifolia*, Ayrshire. (See also *Ascochyta bohemica*.)

Phyllosticta cucurbitacearum Sacc., *Mich.* i, 145; *Syll.* iii, 52; Grove, *Coelomycetes*, i, 14.

*Leaf spot of cucumber (*Cucumis*). Rare. Lea Valley, Essex. Recorded in 1926 and not seen again until 1935.

Phyllosticta cytisi Desm., see *Ascochyta kabatiana*.

Phyllosticta dahliicola Brun. = *Ascochyta dahliicola*.

Phyllosticta destructiva Desm., *Ann. Sci. nat.* **8**, 1847, 29; Sacc., *Syll.* iii, 40; Grove, *Coelomycetes*, i, 28.

*On leaves of *Malva*. On *M. sylvestris*, Kew Gardens; Dartford, Kent; Lyndhurst, Hants.

Phyllosticta dianthi Westend., see *Ascochyta dianthi*.

Phyllosticta digitalis Bell, in Westend. *Exs.* no. 1053; Sacc., *Syll.* iii, 47; Grove, *Coelomycetes*, i, 15.

*Leaf spot of *Digitalis*. On *D. purpurea*, Shere, Surrey; Lyndhurst, Hants; Dartford, Kent (Spilsbury, *TBMS*, **36**, 1953, 344–5).

Phyllosticta draconis Berk., *Cryptogamic Plants Collected in Portugal by Dr F. Welwitsch*, 1853, 5; Sacc., *Syll.* iii, 60; Grove, *Coelomycetes*, i, 54; Cooke, *Fung. Pests Cult. Pl.* 169.

*On leaves of *Dracaena*. On *D. cooperi* and *Cordyline* (*Dracaena*) *terminalis* in conservatories.

Phyllosticta euonymi Sacc., see *P. bolleana*.

Phyllosticta forsythiae Sacc., *Syll.* iii, 27; Grove, *Coelomycetes*, i, 18.

*Leaf spot of *Forsythia*. Rarely recorded. Seamile, Ayrshire, 1907, on *F. suspensa*; Berks and Hants, 1937; Fife, 1957.

255

Phyllosticta fraxinicola Curr., *Synopsis of Simple Sphaeriaceae*, 1859, no. 388; Sacc., *Syll.* iii, 21; Grove, *Coelomycetes*, i, 18.
*On leaves of ash (*Fraxinus*). Common.

Phyllosticta fuchsiicola Speg., *Fungi Chilenses*, 1910, 138; Sacc., *Syll.* xxii, 839; Grove, *Coelomycetes*, i, 18.
*Leaf and stem spot of *Fuchsia*. On stems of *F. coccinea*, Ayrshire (Smith & Ramsbottom, *TBMS*, **6**, 1918, 48); Orkney, 1951.

Phyllosticta garryae Cooke & Harkn., *Grevillea*, **9**, 1881, 84; Sacc., *Syll.* iii, 24; Grove, *Coelomycetes*, i, 19.
*On leaves of *Garrya elliptica*, Kew Gardens.

Phyllosticta gei Bres., see *Gnomonia fructicola*.

Phyllosticta grandimaculans Bubák & Krieg., see *Gnomonia fructicola*.

Phyllosticta grossulariae Sacc., *Mich.* i, 136; *Syll.* iii, 17; Grove, *Coelomycetes*, i, 37.
*Leaf spot of gooseberry (*Ribes*). Kent, Forth, Clyde, Shetland, Dee.

Phyllosticta hedericola Dur. & Mont., *Sylloge generum cryptogamarum*, 1856, 279; Sacc., *Syll.* iii, 20; Grove, *Coelomycetes*, i, 20.
*Leaf blotch of ivy (*Hedera*). Very common throughout Britain on *H. helix*. Fungus studied by Dennis (*TBMS*, **29**, 1946, 28).

Phyllosticta helleborella Sacc. var. **nigra** Cooke, see *Coniothyrium hellebori*.

Phyllosticta hepaticae Brun., *Act. Soc. linn. Bordeaux*, **44**, 1890, 33; Sacc., *Syll.* xi, 477; Grove, *Coelomycetes*, i, 5.
*On leaves of *Hepatica americana* (*Anemone hepatica*), Ayrshire.

Phyllosticta heucherae Brun., *Act. Soc. linn. Bordeaux*, **44**, 1890, 57; Sacc., *Syll.* xiv, 853, forma **sanguineae** Grove, *J. Bot., Lond.*, **60**, 1922, 16; *Coelomycetes*, i, 22.

*On *Heuchera*. On *H. sanguinea*, Ayrshire.

Phyllosticta hoyae Allesch. in Rabenh., *Krypt. Fl.* **1** (6), 1901, 48; Sacc., *Syll.* iii, 104 (as *Phoma bolleana* Thüm.); *Grevillea*, **14**, 1885, 39; Grove, *Coelomycetes*, i, 22. Not *Phyllosticta bolleana* Sacc., *Syll.* iii, 15.

*On leaves of *Hoya*. On *H. carnosa*, Neatishead.

Phyllosticta impatientis Fautr., *Rev. myc.* **20**, 109; Sacc., *Syll.* iii, 62 (as *Depazea impatientis* Kirchn.); Grove, *Coelomycetes*, i, 24.

*On leaves of *Impatiens*. On *I. parviflora*, Kew Gardens; on *I. noli-tangere*, Ayrshire.

Phyllosticta lauri Westend., *Exs.* no. 650; Sacc., *Syll.* iii, 17; Grove, *Coelomycetes*, i, 25.

*On leaves of *Laurus*. On *L. nobilis*, Swanscombe, Highgate, Scarborough.

Phyllosticta ligustri Sacc. = stat. conid. of *Mycosphaerella ligustri*.

Phyllosticta limbalis Pers., *Champignons comestibles*, 1818, 148; Sacc., *Syll.* iii, 24; Grove, *Coelomycetes*, i, 10.

*White spot of box (*Buxus*). Occasional on leaves of *B. sempervirens*, Surrey, Hants.

Phyllosticta lonicerae Westend., *Bull. Acad. Brux.* 1851, 399; Sacc., *Syll.* iii, 18 (as *P. vulgaris* Desm.); Cooke, *Fung. Pests Cult. Pl.* 48; Grove, *Coelomycetes*, i, 25.

*Leaf spot of honeysuckle (*Lonicera*). Common, including garden forms.

Phyllosticta maculiformis Sacc., *Mich.* ii, 538; *Syll.* iii, 35; Grove *Coelomycetes*, i, 11.

*On leaves of oak (*Quercus*). Usually on fading or dead leaves, but has been seen on living leaves in Argyll.

Phyllosticta magnoliae Sacc., *Mich.* i, 139; *Syll.* iii, 25; Grove, *Coelomycetes*, i, 27.

*On leaves of *Magnolia*. On *M. grandiflora*, Kew Gardens; Worcs.

Phyllosticta mahoniae Sacc. & Speg., *Mich.* i, 153; Sacc. *Syll.* iii, 25; Grove, *Coelomycetes*, i, 27. Grove considered *P. mahoniana* (Sacc.) Allesch., in Rabenh., *Krypt. Fl.* i (6), 1901, 57, to be the same.

*Leaf spot of *Mahonia*. Widely distributed and occasional on *M. aquifolium* and *M. japonica*.

Phyllosticta mahoniana (Sacc.) Allesch., see *P. mahoniae*.

Phyllosticta mali Prill. & Delacr., *Bull. Soc. mycol. Fr.* **6**, 1890, 181; Sacc., *Syll.* x, 109; Grove, *Coelomycetes*, i, 41.

*Leaf spot of apple (*Malus*). Not uncommon in England. Also recorded from Ross-shire (Dennis & Wakefield, *TBMS*, **29**, 1946, 155–8). Parasitism uncertain.

Leaf spotting of apple, attributed to this or other species of *Phyllosticta*, has been reported from time to time in different parts of the country (e.g. Salmon, *Gdnrs' Chron.* **42**, 1907, 305–6; *J. S.-E. agric. Coll. Wye*, **17**, 1908, 316–19). It is not easy to distinguish this form of spot from the non-parasitic Cox spot or from spotting caused by *Botrytis cinerea*. The *Phyllosticta* may often be merely a secondary saprophyte.

Phyllosticta narcissicola von Keissl., see *Ramularia vallisumbrosae*.

Phyllosticta nuptialis Thüm., *Contr. ad floram mycologiam lusitanicam*, no. 585; Sacc., *Syll.* iii, 9; Grove, *Coelomycetes*, i, 30.

*Leaf spot of myrtle (*Myrtus*). On *M. communis*, Torquay (Smith & Ramsbottom, *TBMS*, **5**, 1917, 424); Foxcote, near Ilmington, War.

Phyllosticta paulowniae Sacc., *Mich.* i, 143; *Syll.* iii, 27; Grove, *Coelomycetes*, i, 31.

*On leaves of *Paulownia*. On *P. tomentosa*, Kew Gardens.

Phyllosticta paviae Desm., *Ann. Sci. nat.* **8**, 1847, 32; Sacc., *Syll.* iii, 4; Grove, *Coelomycetes*, i, 4. *Guignardia aesculi* (q.v.) is probably the perfect stage.

*Leaf blotch of *Aesculus*. On *A. parviflora*, Kew Gardens.

Phyllosticta pentestemonis Cooke, *Grevillea*, **14**, 1886, 90; Sacc., *Syll.* x, 130; Grove, *Coelomycetes*, i, 32.

*On leaves of *Penstemon*. On *P. grandiflorus*, Kew Gardens.

Phyllosticta phillyreae Sacc., *Mich.* i, 531; *Syll.* iii, 23; Grove, *Coelomycetes*, i, 32; Cooke, *Fung. Pests Cult. Pl.* 1906, 184.

*On leaves of *Phillyrea*, Kew Gardens.

Phyllostica pirina Sacc. = *Coniothyrium pirinum*.

Phyllosticta platanoidis Sacc., *Mich.* i, 360; *Syll.* iii, 13; Grove, *Coelomycetes*, i, 3.

*On sycamore (*Acer*), Surrey; Oxon; Gower, Glam; Ayrshire, etc.

There is confusion about the leaf-spotting genera *Phyllosticta*, *Phleospora* and *Leptothyrium* on *Acer* and research is necessary to determine how many genera and species are separable.

Phyllosticta populina Sacc., *Mich.* i, 155; *Syll.* iii, 33; Grove, *Coelomycetes*, i, 33.

*On leaves of poplar (*Populus*). On *P. nigra* and *P. serotina*, Holloway, Ches, Northumb., Dur. Sometimes prevalent.

Phyllosticta primulicola Desm., *Ann. Sci. nat.* **8**, 1847, 30; Sacc., *Syll.* iii, 56; Grove, *Coelomycetes*, i, 34.

*Leaf spot of *Primula*. Not uncommon on *P. vulgaris* and garden hybrids; on polyanthus, Bristol, 1951 (Robertson, *Plant Path.* **1**, 1952, 102).

Phyllosticta rhododendri Westend., *Bull. Acad. Brux.* 1851, 399; Sacc., *Syll.* iii, 23; Grove, *Coelomycetes*, i, 36.

*Leaf spot of *Rhododendron*. Kew Gardens, Pembs, Ches, Salop, Forth, Tay, Jedburgh.

Several other species of *Phyllosticta*, as well as *Phoma rhodorae* Cooke (*Grevillea*, **14**, 3; Sacc., *Syll.* x, 148), have been recorded on *Rhododendron*, including two or three in Britain, but their relationships with this species and with one another are uncertain.

Phyllosticta richardiae Brooks, *AAB*, **19**, 1932, 16.

*Leaf blotch of arum lily (*Zantedeschia*). Widely distributed and fairly common but rarely harmful unless conditions are very humid as in the Isles of Scilly and Corn; Jersey, 1956–7. Described by Brooks (loc. cit.) and, in Scotland, by Macdonald (*Gdnrs' Chron.* **100**, 1936, 321; *Trans. bot. Soc. Edinb.* **32**, 1939, 556–9).

Phyllosticta rosarum Passer. = *Sphaceloma rosarum*.

Phyllosticta sanguinea (Desm.) Sacc., *Syll.* iii, 6; Cooke, *Fung. Pests Cult. Pl.* 185; Grove, *Coelomycetes*, i, 38.

*On leaves of *Cotoneaster*. On *C. frigida*, Kew Gardens.

Phyllosticta subnervisequa Allesch., in Rabenh., *Krypt. Fl.* i (6), 1901, 42; Sacc., *Syll.* iii, 113 (as *Phoma subnervisequa* Desm.); Grove, *Coelomycetes*, i, 17.

*On *Euonymus*. Causing damage to *Euonymus* hedges, Isles of Scilly, 1923 (*Report*, v, 88); on *E. latifolius*, Polperro, Corn.

Phyllosticta syringae Westend., *Bull. Acad. Roy. Belg.* **18**, 1851, 399; Sacc., *Syll.* iii, 22; Grove, *Coelomycetes*, i, 48.

*Leaf spot of lilac (*Syringa*). Kew Gardens, Kent, Clyde, Dee, etc.

Phyllosticta tiliae Sacc. & Speg., *Mich.* i, 158; Sacc., *Syll.* iii, 27; Grove, *Coelomycetes*, i, 49.

*On leaves of *Tilia*. On *T. europaea* and *T. cordata*, Hants, Ches, Forres, Aberdeen.

Phyllosticta tinea Sacc., *Mich.* i, 135; *Syll.* iii, 16; Grove, *Coelomycetes*, i, 51.

*On leaves of *Viburnum*. On *V. tinus*, Kew Gardens; Swanscombe; Corn.

PHYLLOSTICTA

Phyllosticta violae Desm., *Ann. Sci. nat.* **8**, 1847, 29; Sacc., *Syll.* iii, 38; Grove, *Coelomycetes*, i, 52.

*Leaf spot of violet and viola. Occurs regularly in the south and west; occasional elsewhere in Britain.

Since 1904 leaf spot of violet in Britain has commonly been attributed to *Ascochyta violae* Sacc. & Speg., *Mich.* i, 163; Sacc., *Syll.* iii, 397, but the description given to the English fungus (see, for example, Massee, *Dis. Cult. Pl. Trees*, 431) does not correspond to the original description of *A. violae*, which has longer, fusoid spores. The fungus now usually found causing violet leaf spot is different from *A. violae*. It produces a varying percentage of two-celled spores and may therefore be rightly named an *Ascochyta*, but in general it agrees with the description of *Phyllosticta violae* Desm.

Phyllosticta sp.

*Leaf spot of mint (*Mentha*). Long Ashton, Bristol, 1936.
*On leaves of black currant (*Ribes*), Shetland, 1952.
*On *Anemone*. Frequently associated with one of the forms of winter browning. Has been shown to be pathogenic.
*On leaves of *Buddleia farreri*, Isles of Scilly, 1945.
*Leaf spot of *Chrysanthemum*. Glam, Forth.
*On *Cyclamen*, associated with leaf spotting, Devon, 1952.
*Causing leaf browning of *Erica*. Devon, 1950.
*Leaf spot of lupin. Aberdeen, 1953.
*Leaf spot of sumach (*Rhus*). Tweed.

Phyllostictina hysterella (Sacc.) Petr. = stat. conid. of *Physalospora gregaria* var. *foliorum*.

Physalospora gregaria Sacc. var. **foliorum** Sacc., *Mich.* i, 11; *Syll.* i, 435; Bisby & Mason, *TBMS*, **24**, 1940, 146; stat. conid. *Phyllostictina hysterella* (Sacc.) Petr. in Petrak & Sydow, *Die Gattungen der Pyrenomyzeten, Sphaeropsideen und Melanconieen Beih.* **42**, 1927, 1.

*Leaf scorch of yew (*Taxus*). Forth, Clyde (Callen, *TBMS*, **22**, 1938, 100–2); on golden yew (*T. baccata* var. *stricta*), Insch, Scotland, 1955.

Physalospora miyabeana Fukushi, *Ann. phytopath. Soc. Japan*, **1**, 1921, 1; Bisby & Mason, *TBMS*, **24**, 1940, 146.

*Black canker of willow (*Salix*). Widely distributed and fairly common in southern England, particularly on osiers in Som; occasional elsewhere in Britain.

There has been some confusion between this fungus and *Fusicladium saliciperdum*, and it has not yet been fully cleared up. Alcock (*Trans. R. Scot. arb. Soc.* **38**, 1924, 128–30; *TBMS*, **11**, 1926, 161–7) described willow scab in Lanarkshire as caused by *F. saliciperdum*, which she said was often followed successively on the rods by two less vigorous parasites *Cryptomyces maximus* and *Scleroderris fuliginosa*, and ultimately by saprophytes. Nattrass (ibid. **13**, 1928, 286–304; see also Nattrass and Hutchinson, *J. Minist. Agric.* **36**, 1929, 363–9) pointed out that the leaf symptoms of scab were identical with those of black canker occurring in Som, which he proved to be caused by *Physalospora miyabeana*, with the *Fusicladium* often secondary. Previous to this, canker had been attributed by Johnson (*Proc. Roy. Dublin Soc.* **10**, 1904, 153–6) to *Physalospora (Botryosphaeria) gregaria* Sacc., a fungus looked upon as a wound parasite or a secondary organism to *Fusicladium saliciperdum*. Dennis (*TBMS*, **16**, 1931, 76–84) confirmed Nattrass's findings and emphasized that in the west of England *F. saliciperdum* did not injure willow leaves or rods, but Brooks and Walker (*New Phytol.* **34**, 1935, 64–7) maintained that in the neighbourhood of Cambridge *F. saliciperdum* is the sole cause of a disease otherwise indistinguishable from black canker. There the matter still rests.

Physalospora mutila (Fr.) Stevens, see *P. obtusa.*

Physalospora obtusa (Schw.) Cooke, *Grevillea*, **20**, 1892, 86; Sacc., *Syll.* xi, 292; Bisby & Mason, *TBMS*, **24**, 1940, 147; stat. conid. *Sphaeropsis malorum* Peck, *Rep. N.Y. St. Mus.* **34**, 1881, 36; Sacc., *Syll.* xi, 294. See also Grove, *Coelomycetes*, ii, 53–4.

*Black rot and leaf spot of apple (*Malus*) and pear (*Pyrus*). The leaf-spot stage is not uncommon on apple throughout Britain, but is rarely serious; occasional on pear; also reported on *Prunus incisa*, Tay, Scotland. The fruit-rot stage is uncommon. The fungus is usually found in the imperfect state, and the perfect stage has only rarely been seen.

There has been some confusion about this fungus. *Sphaeropsis malorum* Berk., *Outl.* 1860, 316, was reported in England in 1836 (Berkeley, *English Flora*, **5**, 257). Salmon (*Gdnrs' Chron.* **42**, 1907, 305–6) attributed an apple leaf spot to *Sphaeropsis* sp., but when later (ibid. **47**, 1910, 258–9; *J. S.-E. agric. Coll. Wye*, **19**, 1910, 358–61) he found a pear canker caused by the same fungus he identified the disease as 'New York canker' and the fungus as *Sphaeropsis malorum* Peck. The same species was reported later as the cause of a leaf spot (*J. Bd Agric.* **20**, 1913, 513–15) and fruit rot (Kidd & Beaumont, *TBMS*, **10**, 1924, 106). Stevens (*Mycologia*, **28**, 1936, 330–6) reported finding two species of *Physalospora* in England. One was *P. obtusa*, the perfect stage of *Sphaeropsis malorum* Peck, and he found this on apple at Saltash (Corn) and on hawthorn at St Ives (Hunts) and Plymouth; the other, which he named *Physalospora mutila* (Fr.) Stevens n.c. (p. 333), was the perfect stage of *Sphaeropsis malorum* Berk., a synonym of *Diplodia mutila* (Fr.) Mont., *Ann. Sci. nat.* II, **1**, 1834, 302, and was found on apple and ash (*Fraxinus*) at Saltash. The two species apparently differed only in the size of the ascospores.

Crosse and Bennett (*Rep. E. Malling Res. Sta. for 1950*, 137–8) described severe attacks of *P. obtusa* on fruit in two apple orchards in Kent.

Physarum cinereum (Batsch) Pers., *Syn. Fung.* 1801, 170; Sacc., *Syll.* vii, 344.

*On strawberry (*Fragaria*). A non-parasitic Myxomycete found fruiting on strawberry leaves, East Malling, Kent, 1935 (Wormald, *Gdnrs' Chron.* **98**, 339); West Malling, Kent, 1945.

Physoderma graminis (Büsgen) de Wildeman, *Bull. Soc. Roy. Belg.* **35**, 1896, 59.

*On couch grass (*Agropyron*). Infecting tillers of *A. repens*. Uncommon. Herefs, Bucks, Cards, Yorks.

Phytophthora cactorum (Leb. & Cohn) Schröt., *Kr. Fl. Schles.* iii, 1886, 236; Sacc., *Syll.* vii, 238; Waterhouse & Blackwell, *Mycol. Pap.* **57**, 1954, 1.

The life history of this species was dealt with by Blackwell (*TBMS*, **26**, 1943, 71–89), who also described methods of germinating its oospores (ibid. **26**, 1943, 93–103). Stamps (ibid. **36**, 1953, 248–53) studied variation in a strain of this species originally isolated from onion.

*Phytophthora fruit rot of apple. First seen in Surrey and Sussex, 1915 (Wormald, *AAB*, **6**, 1919, 89–100). Not uncommon. See also *P. syringae*. The same fungus causes a collar rot of apple trees. South Benfleet, Essex, 1952 (H. C. Smith, *Plant. Path.* **2**, 1953, 85–6); Kent, Sussex, Essex, Worcs (Sewell, *Rep. E. Malling Res. Sta. for* 1956, 168–9).

*Fruit rot of pear (*Pyrus*). Not common. Herts, 1912; Devon, 1935.

*Leathery rot of strawberry fruit. Local in south-west England since 1930. The fungus also causes wilting of the plants in the same area. Rare in Scotland; Ross, 1950; Perth, 1952. Along with *P. megasperma* Dreshsl. causing crown rot, Botley, Hants, 1951.

*On hop (*Humulus*), invading the roots and rootstocks. Not infrequent in south-east England (Keyworth, *Gdnrs' Chron.* **113**, 1943, 238). Occasional in Worcs. Waterhouse (*TBMS*, **40**, 1957, 352) considers the species concerned is *P. citricola*, q.v.

*Foot rot of *Clarkia*. Solway, Tay, Edinburgh.

*On *Erica*. With *P. cinnamomi*, occasional as a cause of wilt in cultivated plants (*Report*, viii, 83).

*Collar rot of flowering currant (*Ribes*). Staffs, 1954.

*Shanking of *Gladiolus*. In early flowering varieties, Som, 1929 (W. C. Moore, *Diseases of Bulbs*, 1939, 124).

*Root rot of *Godetia*. Tay.

*Foot rot of lily. Clyde.

*Root rot of *Meconopsis wallichi*. Tay.

*Crown rot of *Primula obconica*. Tay. An unidentified species also reported on *P. malacoides*, Clyde.

*Root rot of *Viola*. Forth, Middx. Similar disease in Kent, east Scotland and Dee, but species of *Phytophthora* not identified.
*Also recorded on larch (*Larix*) and by Collinge (*Rep. econ. Biol.* **2**, 1912, 41), presumably as *Phytophthora omnivora*, on *Acer pseudoplatanus*, *Fraxinus* (ash), *Picea* sp. and *Pinus sylvestris*.

Phytophthora cactorum var. **applanatum** Chester = *P. citricola*.

Phytophthora cambivora (Petri) Buism., *Root Rots caused by Phycomycetes*, Haarlem, 1927, 13; Waterhouse & Blackwell, *Mycol. Pap.* **57**, 1954, 1.
*On cineraria (*Senecio*). With *P. cinnamomi* (q.v.) causing a root rot, Ayrshire.
*On beech (*Fagus*). Associated with *P. syringae* in Phytophthora root rot. Notably in Som, 1930–2.
*Ink disease of chestnut (*Castanea*). First recognized in 1930 and by 1932 (Day, *Forestry*, **6**, 182; *Rep. Imp. Forest. Inst.* **10**, 1934, 20) had been found in virulent form in Hants and Herefs. The disease is common in chestnut coppice, especially on the Hastings beds in south-east England. Day (*Forestry*, **12**, 1938, 101–16; **13**, 1939, 46–58) studied symptoms, cause and control, and the relation of the disease to soil conditions.

Phytophthora cinnamomi Rands, *Meded. Inst. PlZiekt. Buitenz.* **54**, 1922, 5; Waterhouse & Blackwell, *Mycol. Pap.* **57**, 1954, 1.
*Root rot of cineraria (*Senecio*). Ayrshire, with *P. cambivora* also associated (Munro, *TBMS*, **28**, 1945, 115–26); Kelso, 1948 (*Phytophthora* sp.). See also *P. cryptogea*.
*Wilt of *Erica*. On cultivated plants, London, 1934 (Oyler & Bewley, *Reps. exp. Res. Sta. Cheshunt for 1934, 1935 and 1937*; *AAB*, **24**, 1937, 1–16); Leics, Norfolk, Sussex; on *Calluna vulgaris*, Ayr, 1950. The disease may also be caused by *Phytophthora cactorum* (q.v.).
*Ink disease of chestnut (*Castanea*). Along with *P. cambivora*.

Phytophthora citricola Sawada, *Rep. Gov. Res. Inst. Formosa*, **27**, 1927, 22; Ito & Tokunaga, *Trans. Sapp. nat. Hist. Soc.* **14**, 1935, 14; Waterhouse, *TBMS*, **40**, 1957, 349; syn. *P. cactorum*

265

PARASITES

(Leb. & Cohn) Schröt. var. *applanatum* Chester, *J. Arnold Arbor.* **13**, 1922, 232; Waterhouse & Blackwell, *Mycol. Pap.* **57**, 1954, 1.

*Root rot and die-back of raspberry (*Rubus*). Forth, Clyde, Tay, East Lothian. Described by Waterston (*Trans. bot. Soc. Edinb.* **32**, 1937, 251–9).

*On hop (*Humulus*), see under *Phytophthora cactorum*.

*On *Antirrhinum* (Waterhouse, *TBMS*, **40**, 1957, 352).

*Foot rot of *Primula*. On *P. japonica*, Tweed.

Phytophthora cryptogea Pethybr. & Laff., *Sci. Proc. R. Dublin Soc.* **15**, 1919, 498; Sacc., *Syll.* xxiv, 1333; Waterhouse & Blackwell, *Mycol. Pap.* **57**, 1954, 1.

This species causes foot rot, wilt or damping-off in a wide range of hosts, notably tomato and many ornamental plants.

*Seedling rot of rape (*Brassica*). Lancs, 1953.

*On cauliflower (*Brassica*), Northumb, 1929 (*Report*, vii, 52).

*Damping-off of celery seedlings (*Apium*) (*Report*, iv, 50).

*Foot rot and damping-off of tomato (*Lycopersicon*). Widespread and destructive when seedlings are raised in unsterilized soil. The species is not always distinguished from *Phytophthora parasitica* (q.v.) which is perhaps almost an equally common cause.

The disease was described by Spinks (*Rep. Long Ashton Res. Sta. for 1917*, 25–7) and by Bewley (*J. Minist. Agric.* **27**, 1921, 670–3). Brown (*Gdnrs' Chron.* **109**, 1941, 55) described an effective method of preventing the trouble by soil treatment with a formalin-sawdust powder. Cheshunt compound is also commonly used, but if applied indiscriminately it can damage the roots (Glasscock & Dermott, *J. hort. Sci.* **25**, 1950, 151–4).

*Foot rot of *Antirrhinum*. Occasional. (See Bewley, *AAB*, **7**, 1920, 159.)

*Root rot of carnation (*Dianthus*). West Lothian, Edinburgh, Corn, Bucks.

*Foot rot of *Celosia*. Bucks, 1953.

*Foot rot of China aster (*Callistephus*). Common and widely distributed in Britain; develops from June onwards. Studied by Robinson (*AAB*, **2**, 1915, 125–37).

*Foot rot of *Chrysanthemum* (see Oyler, *Rep. exp. Res. Sta. Cheshunt for 1936*, 59). Sheffield, 1950. *Phytophthora* sp. has also been reported on rooted cuttings on occasions.

*Foot rot of cineraria (*Senecio*). Not uncommon.

*On *Dahlia*, Carlisle, 1931.

*Wilt of gloxinia (*Sinningia*). Forth.

*On lupin, Bewley (*AAB*, **7**, 1920, 159).

*Foot rot of African marigold (*Tagetes*). Devon, Hants, Kent.

*Root rot of *Nemesia*. East Lothian, 1954; Glam, 1951 (as *Phytophthora* sp.).

*Foot rot of *Petunia*. Frequent. Was recorded as natural host by Pethybridge and Lafferty (*Sci. Proc. R. Dublin Soc.* **15**, 1919, 487–503). Beds, Glos, Middx, Sussex, Kent, Lancs, Forth.

*Wilt of *Schizanthus*. Forth; Carlisle, Cumb, 1956.

*Foot rot of stock (*Matthiola*). Frequent. Sussex, Middx, Glos, Northumb, Forth, Perthshire.

*Foot rot of sweet sultan (*Centaurea moschata*). Peebles, 1952; Sussex, 1956 (*Phytophthora* sp.)

*Shanking of tulip. Along with *P. erythroseptica*. Widely distributed in England and Wales under glass and not infrequently causes heavy losses. Not uncommon in Scotland. The disease has not yet been recognized out-of-doors.

Shanking was fully described by Buddin (*AAB*, **25**, 1938, 705–29) and recorded from Scotland by Foister (*Gdnrs' Chron.* **87**, 1930, 171–2). Ashby (*TBMS*, **14**, 1929, 254–60) published cultural notes on the fungus.

*Foot rot of *Zinnia*. Southern England, Forth.

Phytophthora cryptogea Pethybr. & Laff. var. **richardiae** (Buism.) Ashby = *P. richardiae*.

Phytophthora erythroseptica Pethybr., *Sci. Proc. R. Dublin Soc.* **13**, 1913, 547; Sacc., *Syll.* xxiv, 37; Waterhouse & Blackwell, *Mycol. Pap.* **57**, 1954, 1.

Murphy (*Ann. Bot., Lond.*, **32**, 1918, 115–53) described the morphology and cytology of the sexual organs of this fungus, and McKay (*Nature, Lond.*, **139**, 1937, 802) induced the ready production of zoosporangia with the aid of potassium permanganate, while Gregg (*Nature, Lond.*, **180**, 1957, 150) obtained ready germination of the oospores after they had passed through the digestive tract of the garden snail, *Helix aspersa*. McKay, Loughnane and Kavanagh (*TBMS*, **40**, 1957, 407–8) found and illustrated the haustoria of the fungus.

*Pink rot of potato tubers. First recognized in Scotland in 1919 and in England in 1921 (Cotton, *J. Minist. Agric.* **28**, 1922, 1126–30). Now widely distributed in both countries, but seldom serious. In some seasons causes wilting of the plants in eastern England.

P. megasperma has also been found causing pink rot in Northern Ireland by Cairns and Muskett (*Nature, Lond.*, **131**, 1933, 277; *AAB*, **20**, 1933, 381–403), who studied the effect of environment on the disease (ibid. **26**, 1939, 470–80). Symptoms similar to those of pink rot may be caused as a result of damage by sulphuric acid (Boyd, *Plant Path.* **6**, 1957, 114).

*Shanking of tulip. Along with *P. cryptogea*, q.v.

Phytophthora erythroseptica var. **atropae** Alcock, *Pharm. J.* **116**, 1926, 232; Waterhouse & Blackwell, *Mycol. Pap.* **57**, 1954, 1.

*Root rot and wilt of belladonna (*Atropa belladona*). Forth. An unidentified species of *Phytophthora* associated with similar symptoms was recorded near Bristol in 1917 and in Herts in 1942.

Phytophthora fragariae Hickman, *J. Pomol.* **18**, 1940, 103; Waterhouse & Blackwell, *Mycol. Pap.* **57**, 1954, 1.

*Red core of strawberry (*Fragaria*). Widely distributed in Britain and locally severe. Occurs chiefly in Lanarkshire, the Tamar Valley, Sussex, Hants and Kent, with minor outbreaks elsewhere.

This disease began to attract attention in Lanarkshire about 1921, and soon the so-called 'Lanarkshire Disease' became more than a menace to the strawberry industry there. Early publications about it included those by Wardlaw (*Scot. J. Agric.* **10**, 1927, 156–

65; **11**, 1928, 65–71; *AAB*, **14**, 1927, 197–201; *Ann. Bot., Lond.,* **41**, 1927, 817–18; *The Lanarkshire Strawberry Industry,* Glasgow, R. MacLehose and Co. 1928, 53 pp.) and O'Brien and M'Naughton (*Res. Bull. W. Scotl. Coll. Agric. Dep. Pl. Husb.* **1**, 1928, 32 pp.; *Scot. J. Agric.* **11**, 1928, 286–97). Probably several diseases were being confused, and though the trouble was assigned to various causes, the chief of these was at first completely overlooked. It was Alcock (*Gdnrs' Chron.* **86**, 1929, 14–15) who first became convinced that red core root rot, as it was now called, was caused by a species of *Phytophthora*, and later Scottish literature swung over to this view (Alcock, Howells & Foister, *Scot. J. Agric.* **13**, 1930, 242–51; Howells, ibid. **17**, 1934, 287–93; Alcock & Howells, *Sci. Hort.* **4**, 1936, 52–8). Meanwhile the disease was recognized in England (Hants) in 1931, and was attributed primarily to impeded drainage (Neville & Kay, *Rep. Botley Exp. Fruit Sta. on Strawberries 1923–1937*, 1938, 64–6). A special investigation was begun in 1937, and Hickman (*J. Pomol.* **18**, 1940, 89–118; *TBMS*, **23**, 1939, 210–11) proved conclusively that red core was caused by a fungus which he named *Phytophthora fragariae*.

BIOLOGY. Hickman and English (*TBMS*, **34**, 1951, 223–36) investigated the influence of soil moisture and pH on the development of the disease and reported (ibid. **34**, 1951, 356–9) on varietal susceptibility and the occurrence of physiologic races. Later, Hickman and Goode (*Nature, Lond.*, **172**, 1953, 211–12) devised a new method of testing for pathogenicity. Goode (*TBMS*, **38**, 1955, 172; **39**, 1956, 367–77) followed zoospore infection of the roots in susceptible and resistant varieties, and Gregg (ibid. **38**, 1955, 169) studied oospore production in pure cultures.

BREEDING FOR RESISTANCE. Reid (*Scot. J. Agric.* **23**, 1941, 264–72; **27**, 1948, 218–23; *Fruit Gr.* **97**, 1944, 9–10, 29–30) succeeded in raising varieties which proved highly resistant to red core, at least for a time. He also dealt generally with the problems of breeding strawberry varieties for disease resistance in a number of articles (*Agriculture, Lond.*, **55**, 1949, 476–82; *Plant*

Dis. Reptr, **36**, 1952, 395–405; *Rep. 13th Int. Hort. Congr. London 1952*, ii, 1953, 739–50).

LEGISLATION. Under the Red Core Disease of Strawberry Plants Order, 1957, red core is a notifiable disease, and infected plants, or plants grown in infected land, may not be sold; nor may they be planted except in land already infected.

Phytophthora infestans (Mont.) de Bary, *J. R. agric. Soc.* **12**, 1876, 239; Sacc., *Syll.* vii, 237; Waterhouse & Blackwell, *Mycol. Pap.* **57**, 1954, 1.

*Potato blight. First recorded in Britain, 1845 (Salter, *Gdnrs' Chron.* **5**, 1845, 561), and since then one of the comparatively few, constantly recurring diseases of high economic importance in the British Isles. Very dependent on weather conditions. Epidemics causing considerable loss by premature defoliation of unsprayed crops occur almost every year in south-west coastal regions and parts of Ireland, but only 4 or 5 years out of 10 in southern England, and rarely in the north of the country. Attack on the tubers varies less from year to year, but is usually at a low level except on relatively few crops here and there.

There have been innumerable scientific papers on potato blight published in the British Isles and only some can be mentioned. The following references cover most of the more significant advances made from time to time.

PRE-1900 LITERATURE. The first report of the appearance of potato blight took the form of a letter from Dr Bell Salter in the Isle of Wight, published in the *Gdnrs' Chron.* for 16 Aug. 1845, p. 561, and within a surprisingly short time Berkeley (*J. hort. Soc.* **1**, 1846, 9–34) published his now famous and accurate observations on the potato 'murrain'. Carruthers (*J. R. agric. Soc.* **9**, 1873, 248–53) summarized existing knowledge about blight just before the controversy started about W. G. Smith's erroneous discovery of the alleged resting spores of the fungus (*Gdnrs' Chron.* for 1875–6).

GENERAL ACCOUNTS. Helpful general accounts of blight include those by Pethybridge (*Rep. Int. Potato Conf.* 1921, 112–26),

Small (*Potato Blight and its Prevention in Jersey*, Bigwood Ltd. 1938, 46 pp.), Large (*Agriculture, Lond.*, **48**, 1941, 22–8) and Dillon-Weston and Taylor (ibid. **51**, 1944, 111–16). McKay (*J. Dep. Agric. Ire.* **53**, 1956–7 (1958), 5–10) published a retrospect of 50 years' outbreaks in Ireland from 1907 to 1956.

BIOLOGY. Pethybridge and Murphy (*Sci. Proc. R. Dublin Soc.* **13**, 1913, 566–88) obtained oospores in pure cultures of the fungus, and Murphy (ibid. **18**, 1927, 407–12) found them occurring naturally on the tubers. Murphy (ibid. **16**, 1922, 442–66) also studied the vitality of the conidia, and with McKay (*J. Dep. Lds Agric. Dubl.* **24**, 1924, 103–16) showed that the fungus could live an independent life in the soil for only a very short period.

Dickinson and Keay (*Nature, Lond.*, **162**, 1948, 32) and Keay (*Plant Path.* **2**, 1953, 103) devised methods of growing the fungus readily on artificial media. The haustoria of this and related species were studied by Blackwell (*TBMS*, **36**, 1953, 138–58). Müller (*Nature, Lond.*, **166**, 1950, 231–2) produced tumours on potato tubers by inoculation with *Phytophthora infestans*, and also (ibid. **166**, 1950, 392–5) followed the reaction of various flowering plants to inoculation. Grainger (*Phytopathology*, **46**, 1956, 445–56) studied the relation between host composition and attack by the fungus.

SOURCE OF THE DISEASE. Massee (*Kew Bull.* 1906, 110–12) postulated that blight made its annual reappearance from the presence of dormant mycelium in the planted tubers which grew up into the stems, but Pethybridge (*Sci. Proc. R. Dublin Soc.* **13**, 1911, 12–27) refuted this, though more recently Keay's experiments (*Plant Path.* **2**, 1953, 68–71; **3**, 1954, 88–9) have rather supported Massee. Brooks (*New Phytol.* **18**, 1919, 187–200) studied the problem in the field. Later, ample evidence became available to prove that blight begins each year mainly from infected tubers (Murphy & McKay, *J. Dep. Agric. Ire.* **25**, 1925, 10–21; *Sci. Proc. R. Dublin Soc.* **18**, 1927, 413–22; Salmon & Ware, *AAB*, **13**, 1926, 289–300; Alcock & McIntosh, ibid. **14**, 1927, 440–1). Hirst (*Plant Path.* **4**, 1955, 44–50) attempted to trace the course of blight epidemics from the time of planting infected tubers, while Clayson and Robertson (ibid. 5, 1956, 30–1) demonstrated that the fungus can survive in stem lesions for upwards of forty days during hot, dry summer weather, and (ibid. 6, 1957, 123–7) tried to assess the relative importance of foci arising from infected tubers, from wind-blown spores and from long-lived stem infections in determining an epidemic.

EFFECT OF AGE AND WEATHER. At one time, there was a general belief that the potato plant did not become susceptible to blight until it reached a certain age, but work by Pethybridge (*Rep. Int. Potato Conf.* 1921, 112–26) and Beaumont (*AAB*, 21, 1934, 23–47) disproved this. On the other hand, there is a close correlation with weather conditions, as demonstrated by Wiltshire (*Quart. J. R. met. Soc.* 57, 1931, 304–16), by Napper (*J. Pomol.* 11, 1933, 177–84), and by Beaumont (*Reps. Dep. Pl. Path. Seale-Hayne agric. Coll.* for 1930–37) in a series of experiments extending over many years, and later summarized (*TBMS*, 31, 1947, 45–53). Richardson and Doling (*Nature, Lond.*, 180, 1957, 866) showed that blight developed more readily on leaf-roll infected than on healthy plants, due presumably to differences in microclimate.

FORECASTING. Beaumont showed that blight usually appears 7–21 days after periods of at least 48 hours during which the temperature is not less than 50° F. and the relative humidity is not below 75 %. This temperature-humidity rule was tested by Grainger (*TBMS*, 33, 1950, 82–91) for Scotland during 1944–8, and since 1950 it has formed the basis for countrywide collaborative work between the Ministry of Agriculture, Fisheries and Food and the Meteorological Office in a successful attempt to develop a system of regional forecasting of blight based on macroclimate data. The results of this work were summarized by Large for 1950–2 (*Plant Path.* 2, 1953, 1–15) and for 1953–5 (ibid. 5, 1956, 39–52). Warnings based on observations in screens at synoptic weather stations provide a workable basis for forecasting, and no great advantage is gained by measuring the weather in the crops themselves (Hirst, ibid. 5, 1956, 135–40).

Smith (ibid. **5**, 1956, 83–7) applied a different humidity criterion (viz. at least 11 hours with humidity 90 % or more) to the same data but obtained only a little more accuracy. Grainger (*Weather*, **10**, 1955, 213–22) described a thermohydrographic instrument suitable for use in forecasting.

LOSSES CAUSED. Surveys to estimate the precise amount of damage caused by Blight formed an integral part of the collaborative work and the results are given in the papers by Large cited above. The surveys were based on the use of a simple key (*TBMS*, **31**, 1947, 140–1) for damage assessment and the preparation of curves showing the progress of the disease in potato crops (Large, *Plant Path.* **1**, 1952, 109–17), from which accurate estimates of losses can be made when they are considered in conjunction with other data, such as the rate of bulking-up of the crop (Doncaster & Gregory, *Rep. agric. Res. Coun.* no. 7, 1948, 1–8). The results obtained by these simplified methods were shown to be comparable with those obtained by careful yield trials in Yorks (Beaumont, Bant & Storey, *Plant Path.* **2**, 1953, 56–60), around the Wash (Large *et al.*, ibid. **3**, 1954, 40–8), and in the west midlands (Rosser, ibid. **6**, 1957, 77–84).

RESISTANCE AND PHYSIOLOGIC RACES. Breeding for resistance was carried out by Salaman (*Deuxième Cong. Path. Comp. Paris*, **2**, 1931, 436–7; *Rothamsted Conference*, xvi, *Problems of Potato Growing*, 1934, 9–17) and more recently Black has devoted a great deal of attention to this subject, which is complicated by the existence of physiologic races of the fungus. Black's chief papers are published in *Proc. roy. Soc. Edinb.* Sect B, **62**, 1945 171–81; **63**, 1949, 290–301; **64**, 1950, 216–28; **64**, 1952, 312–52; *AAB*, **32**, 1945, 279–80; **34**, 1947, 631–3; and *J. Minist. Agric.* **54**, 1947, 198–200. With Müller (*Z. PflZücht.* **31**, 1952, 305–18) he summarized developments in breeding potatoes for resistance to blight and virus diseases during the past century.

Other papers concerned mainly with physiologic races include those by O'Connor (*Gdnrs' Chron.* **93**, 1933, 104–5) and Salaman (*Agric. Progr.* **11**, 1934, 77–86) in England; by Black (*Trans. Roy. Soc. Edinb.* **61**, 1943, 137–47) and Black and Haigh (*Scot. J.*

Agric. **27**, 1947, 49–50) in Scotland; and by Doling (*Nature, Lond.*, **177**, 1956, 230; *AAB*, **45**, 1957, 299–303) in Northern Ireland.

Black (*Proc. roy. Soc. Edinb.* **65**, 1952, 36–51) proposed a genetical basis for the classification of physiologic races, which was accepted internationally (*Euphytica*, **2**, 1953, 173–9); and Howard (*AAB*, **40**, 1953, 584–93) discussed the use of differential hosts for determining the races. Black (*Rep. Scot. Pl. Breed. Sta.* 1957, 43–9) studied the races from many countries and showed that race 4 is now predominant in commercial varieties.

Müller and Behr (*Nature, Lond.*, **163**, 1949, 498) discussed the mechanism of resistance, and the nature of field resistance has been studied by Müller and Haigh (ibid. **171**, 1953, 781–3), by Müller (*J. nat. Inst. agric. Bot.* **6**, 1953, 346–60) and by Deshmukh and Howard (*Nature, Lond.*, **177**, 1956, 794). Müller, Cullen and Kostrowicka (*J. nat. Inst. agric. Bot.* **7**, 1955, 341–54) described laboratory and greenhouse methods of testing the 'true resistance' of varieties and seedlings to Blight.

Boyd and Henderson (*Plant Path.* **2**, 1953, 113–16) showed that resistance of the tubers increases with maturity: Müller and Munro (*AAB*, **38**, 1951, 765–73) dealt with the reaction to Blight of virus infected plants.

SPRAYING AND DUSTING. The earliest spraying trials in England were described by Whitehead (*J. R. agric. Soc.* **2**, 1891, 828–35; **3**, 1892, 761–71) and Voelcker (ibid. **3**, 1892, 771–83) and those conducted during and shortly after the 1914–18 War were summarized by Brooks (*J. Bd Agric.* Suppl. **18**, 1919, 60–8), Petherbridge (*J. Bd Agric.* **25**, 1919, 1166–72; **27**, 1920, 282–6), and Pennington and Robinson (*Bull. Univ. Coll. Reading*, **30**, 1921, 8 pp.). Spraying was compared with dusting by Muskett and Cairns (*J. Minist. Agric. N. Ire.* **2**, 1929, 54–62; **3**, 1931, 117–23) in Northern Ireland, and by Murphy and McKay (*J. Dep. Agric. Ire.* **32**, 1933, 30–48) in the Irish Republic; and Thompson (*J. Minist. Agric.* **45**, 1938, 418–19; *Kirton agric. J.* **3**, 1939, 7–28) endeavoured to lay down appropriate standards for the copper content of dusts.

The modern approach to the subject began during the 1939–45 War, when accurate data about the economic value of spraying in south-west England were provided by Beaumont and Large (*Agriculture, Lond.*, **48**, 1942, 235–40; **51**, 1944, 71–5) and by Large (*AAB*, **32**, 1945, 319–29). Large and Beer (ibid. **33**, 1946, 406–13) also tried fungicides of low copper content. Since then Beaumont, Bant and Storey (*Plant Path.* **2**, 1953, 56–60) have shown that protective spraying is scarcely an economic proposition in Yorkshire. In the area of the Wash, Large *et al.* (ibid. **3**, 1954, 40–8) demonstrated that on the average over a long series of years (10) it is scarcely economic to spray King Edward and uneconomic to dust them. Three of the ten years were 'blight' years when it paid handsomely to spray, but three other years were virtually 'no blight' years, when yields were substantially depressed on account of damage by the spraying machines.

During the 1939–45 War simplified methods of spraying small areas with a watering-can (Marsh & Martin, *Rep. Long Ashton Res. Sta. for 1940*, 60–75; *for 1941*, 79–82) and of dusting (Hickman, ibid. *for 1940*, 76–9) were devised for allotment holders and gardeners. The practice of applying the spray from above became widespread some years later when light low-volume spraying machines—at first introduced for the application of growth regulator weedkillers to cereal crops—were adopted for use in potato spraying.

Until about 1940 Bordeaux Mixture or simple proprietary copper fungicides were used in the trials made, and in practice copper oxychloride and cuprous oxide sprays had already replaced the more troublesome Bordeaux Mixture. Later, Large (*Nature, Lond.*, **151**, 1943, 80) tried metallic copper, Montgomery and Shaw (*Rep. E. Malling Res. Sta. for 1942*, 68–70) phenyl mercury chloride, and Kingston (*J. Sci. Fd Agric.* **7**, Suppl. 1956, S 16–S 19) a mixture of phenyl mercury chloride and copper oxychloride. Kearns and Morgan (*Rep. Long Ashton Res. Sta. for 1954*, 169–70) obtained promising results with small-volume air-flow spraying. Brenchley (*Agriculture, Lond.*, **64**, 1958, 475–9) compared the value of ground spraying and dusting with spraying from the air.

18-2

SPRAY DEPOSITS. Large, Beer and Patterson (*AAB*, 33, 1946, 54–63) looked into the question of spray retention on the leaves, and later Large and Taylor (*Plant Path.* 2, 1953, 93–8) investigated the distribution of copper spray deposits in low-volume spraying. Hebblethwaite (ibid. 5, 1956, 26–8) also described a colorometric method for measuring dosage distribution of copper compounds using nigrosine.

TUBER DIPS. Efforts to prevent blight developing in seed tubers during transit or storage by dipping them in fungicidal solutions immediately after lifting have been described by Small (*Nature, Lond.*, 130, 1932, 367; *J. Minist. Agric.* 43, 1937, 1162–8; *AAB*, 22, 1935, 16–22, 469–78), by Greeves (*AAB*, 24, 1937, 26–32), and on a commercial scale by Wakely and Mellor (*Nature, Lond.*, 150, 1942, 769).

HAULM KILLING. It was Murphy and McKay (*J. Dep. Lands Agric. Dublin*, 24, 1924, 103–16) who really laid the foundation of future haulm-killing experiments when they demonstrated that blight in the clamp was a result of infection at digging time. Early trials with sulphuric acid were made by MacDowell (*Scot. J. Agric.* 18, 1935, 243–9) and Bates and Martin (*J. Minist. Agric.* 42, 1935, 231–5). Findlay and Sykes (ibid. 43, 1936, 457–9; 44, 1937, 546–51) tested sulphuric acid against copper sulphate, and Small (*Rapp. États Jersey*, 1938, 21–35, 1939) various compounds. The various alternatives were set out by Samuel (*Agriculture, Lond.*, 50, 1943, 214–16; 51, 1944, 277–9). Preliminary results with tar acid (T.A.C.) compounds (ibid. 52, 1945, 215–17) were followed by a general review of the whole problem, with the results of comparative tests, by Wilson, Boyd, Mitchell and Greaves (*AAB*, 34, 1947, 1–33). Main and Grainger (*Scot. J. Agric.* 27, 1947, 14–17) used sodium chlorate. Later, when the supply of sulphur was threatened, co-operative experiments were carried out by the Ministry of Agriculture to find satisfactory substitutes for sulphuric acid. The results were summarized by Large (*Plant Path.* 1, 1952, 2–9, 56–9; 3, 1954, 90–8). In general, however, the value of haulm destruction lies in directions other than control of tuber blight; it clears the ground and facilitates

PHYTOPHTHORA

harvesting, and may be used to restrict tuber size for seed purposes and to avoid late infection by viruses. (See also Rosser, ibid. 6, 1957, 77–84.)

*Blight of tomato (*Lycopersicon*). Not common or serious under glass. In the open it occurs annually in August–September wherever the crop is grown in Britain, and epidemics develop under the same conditions as with potato blight, especially in southern and south-western districts where the bulk of the outdoor crop is grown.

Outbreaks of this disease have occasionally been reported for many years. Allard (*Plant Dis. Reptr*, 31, 1947, 231) referred to destructive attacks in England in 1875 and 1878, and for other early references see W. G. Smith (*Gdnrs' Chron.* 16, 1881, 346) and *J. Bd Agric.* 14, 1907, 481. The relation between tomato and potato blight, and the possibility of physiologic races on the two hosts have been discussed from time to time, for instance, by G. Smith (*J. S.-E. agric. Coll. Wye*, 22, 1913, 494–6), Wiltshire (*Rep. Long Ashton Res. Sta. for 1915*, 93–2), Salmon and Wormald (*Gdnrs' Chron.* 69, 1921, 311–12) and Small (*AAB*, 25, 1938, 271–6). See also a summary of this literature in *Report*, vii, 53.

It was not until the 1939–45 War, however, that tomato blight was given serious attention. During that period outdoor spraying trials, mainly with copper compounds, were carried out by Read (*Rep. exp. Res. Sta. Cheshunt for 1942*, 43–7), Wain and Wilkinson (*Rep. Long Ashton Res. Sta. for 1942*, 56–8), Large (*Fmr & Stockbreeder*, 57, 1943, 1279), Green and his colleagues (*J. R. hort. Soc.* 68, 1943, 179–83; 70, 1945, 211–14), and Croxall (*Rep. Long Ashton Res. Sta. for 1943*, 95–9; *for 1944*, 161–6). Kearns (ibid. *for 1941*, 70) suggested a method of spraying commercial outdoor crops with spray pumps on the headlands.

Phytophthora infestans has also been observed occurring naturally in the British Isles on a number of other hosts, viz.

*On *Anthocercis*, Dublin (Berkeley, *Gdnrs' Chron.* 8, 1848, 557).
*On *Datura*. On *D. stramonium*, Harpenden, Herts, 1956 (Hirst & W. C. Moore, *Plant Path.* 6, 1957, 76).

277

*On *Petunia*, Bristol (Berkeley, *Gdnrs' Chron.* **16**, 1856, 644); Harpenden, Herts, 1956 (Hirst & W. C. Moore, *Plant Path.* **6**, 1957, 76); Jersey, 1954 (*Phytophthora* sp.).

*On *Lycium*. On *L. halimifolium*, Bodmin, Corn (W. C. Moore, *TBMS*, **28**, 1945, 130).

*On *Solanum crispum*. Newton Abbot, Devon, 1936, 1938; Dundry, Som, 1939 (*Report*, viii, 12).

*On *Solanum demissum*. See Lindley (*J. hort. Soc.* **3**, 1848, 65).

*On *Solanum dulcamara*. See Smith (*J. S.-E. agric. Coll. Wye*, **22**, 1913, 494), who also artificially infected *S. aviculare*; Glam, 1944.

*On *Solanum edinense*. See Salaman (*J. Genet.* **1**, 1910, 7).

*On *Solanum laciniatum*, Canterbury, Kent (Masters, *Gdnrs' Chron.* **6**, 1846, 661); Cambridge (Henslow, ibid, **8**, 1848, 573).

*On *Solanum maglia*. See Sutton (*J. R. hort. Soc.* **18**, 1896, 420).

Phytophthora megasperma Drechsl., *J. Wash. Acad. Sci.* **21**, 1931, 525; Waterhouse & Blackwell, *Mycol. Pap.* **57**, 1954, 1.

*Pink rot of potato, see *P. erythroseptica*.

*On sugar beet. Rotting the roots, east midlands, 1936; Lincs, 1954.

*On brassicae. Causing root rot in savoy cabbage and cauliflower, Som, 1949; in marrow-stem kale, Glos, 1950–53 (Robertson & Ogilvie, *Plant Path.* **2**, 1953, 15); in Brussels sprouts, Essex, 1953; in broccoli, Hants, 1955.

*Crown rot of strawberry (*Fragaria*). Along with *P. cactorum*, Botley, Hants, 1951.

*On *Campanula*. A strain or variety of this species associated with root rot of Canterbury bell, Hartlepool, Dur, 1935.

*On *Narcissus*. Causing a bulb rot (*Rep. exp. Res. Sta. Cheshunt for 1939*, 12).

Phytophthora parasitica Dastur, *Mem. Dep. Agric. India, Bot.* **5**, 1913, 226; Waterhouse & Blackwell, *Mycol. Pap.* **57**, 1954, 1.

*Damping-off of melon (*Cucumis*). Tay. Also in England as *Phytophthora* sp. (*Report*, v, 82).

*Foot rot and damping-off of tomato (*Lycopersicon*). Along with *P. cryptogea* a very common cause in unsterilized soil.

Frequent also as the cause of buckeye rot of the fruit, and occasionally a stem rot (Williams & Sheard, *Gdnrs' Chron.* **114**, 1943, 96–7).

*Stem rot of *Begonia*. Forth.

*Wilt of China aster (*Callistephus*). Along with *Phytophthora cryptogea*. Tweed.

*Foot rot of cineraria (*Senecio*). Forth.

*On gloxinia (*Sinningia*), rotting the stems and corms, Newcastle, 1935; Middx, 1955–6.

*On *Lilium*. Associated with foot rot of *L. longiflorum* var. *takesima* under glass. Middx, 1935.

*Damping-off of *Meconopsis*. Forth.

*Root rot of *Nemesia*. Forth.

*Root rot of pansy (*Viola*). I.W. 1955.

*Damping-off of *Phlox*. Forth; Aberdeen, 1951 (as *Phytophthora* sp.).

*Foot rot of *Primula*. Moray.

*Foot rot of *Solanum capsicastrum*. Herts, Isle of Man, 1938. Described by Orchard (*Rep. exp. Res. Sta. Cheshunt for 1934*, 54–5) and by Oyler (ibid. *for 1956*, 58–9).

*Stem rot of *Verbena* (Oyler, *Rep. exp. Res. Sta. Cheshunt for 1936*, 58–9).

Phytophthora porri Foister, *Trans. bot. Soc. Edinb.* **30**, 1931, 277; Waterhouse & Blackwell, *Mycol. Pap.* **57**, 1954, 1.

*White tip of leek (*Allium*). Common locally, especially in northern England and the Bristol and Evesham areas. Occasional elsewhere. Surrey, mid- and south Wales. Now rare in Scotland, where formerly common in the south.

First found in the Midlothians near Edinburgh in 1928 (Foister, *Gdnrs' Chron.* **85**, 1929, 106) and fully described three years later (Foister, *Trans. bot. Soc. Edinb.* **30**, 1931, 257–81). (See also Ogilvie and Mulligan, *Gdnrs' Chron.* **89**, 1931, 360.)

*Crown rot of *Campanula persicifolia*, Evesham (Legge, *TBMS*, **34**, 1951, 293–303).

Phytophthora primulae Tomlinson, *TBMS*, **35**, 1952, 233; Waterhouse & Blackwell, *Mycol. Pap.* **57**, 1954, 1.

*Brown core of *Primula*. Widely distributed in the southern half of England on *P. polyantha* Mill. (polyanthus) and occasional on other species. Also recorded from Northumb, Yorks, Lancs, War and Inverness-shire. It was first seen at Lewes, Sussex, in May 1949. The disease was investigated by Tomlinson (*TBMS*, **35**, 1952, 221–35).

Phytophthora richardiae Buism., *Tijdschr. PlZiekt*. **33**, 1927, 17; *Root Rots caused by Phycomycetes*, Thesis, Haarlem, 1927, 13; Waterhouse & Blackwell, *Mycol. Pap*. **57**, 1954, 1; syn. *P. cryptogea* Pethybr. & Laff. var. *richardiae* (Buism.) Ashby, *TBMS*, **14**, 1929, 254.

*Root rot of arum lily (*Zantedeschia*). Widely distributed and not uncommon in England and Wales. First reported in 1927 from Berks and Kent (Salmon & Ware, *Gdnrs' Chron*. **81**, 1927, 234–5).

Phytophthora syringae (Kleb.) Kleb., *Krankheiten des Flieders*, Berlin, 1909, 75; Sacc., *Syll*. xxi, 860; Waterhouse & Blackwell, *Mycol. Pap*. **57**, 1954, 1.

*Phytophthora fruit rot of apple (*Malus*) and pear (*Pyrus*). Not uncommon. Described by Ogilvie (*Rep. Long Ashton Res. Sta. for 1930*, 147–50) in the west of England, and by Lafferty and Pethybridge (*Sci. Proc. R. Dublin Soc*. **17**, 1922, 29–43) in Eire. Colhoun (*AAB*, **25**, 1938, 90) described the disease in stored apples. This disease may also be caused by *Phytophthora cactorum* (q.v.).

*Wilt of lilac (*Syringa*). Forth, Dee, Clyde. It was recorded from Aberdeen by Berkeley (*Gdnrs' Chron*. **16**, 1881, 665) as *Ovularia syringae* n.sp. (See also W. G. Smith, ibid. **20**, 1883, 439.)

*With *P. cambivora* causes Phytophthora root disease of beech (*Fagus*).

Phytophthora verrucosa Alcock & Foister, *Trans. bot. Soc. Edinb*. **33**, 1940, 65; Waterhouse & Blackwell, *Mycol. Pap*. **57**, 1954, 1.

*Toe rot of tomato (*Lycopersicon*). Frequent in Scotland since 1934. Occasional in England. Northumb, Lancs, Ches, War,

Norfolk, Lincs, Glos, Herts. The disease was studied by Howells (*Scot. J. Agric.* **19**, 1936, 47–50).
*Root rot of *Meconopsis*. Forth.
*On seedlings of *Primula*, Forth.

Phytophthora sp.
*Associated with a wet rot of mangold roots (*Beta*). Som, 1950.
*Damaging garden peas (*Pisum*). Leeds, 1947.
*Root rot of broad bean (*Vicia faba*) sown between rows of an affected anemone crop, Corn, 1953.
*Root rot of trefoil (*Medicago*). On a single plant, Glam, 1955.
*Damping-off of ryegrass (*Lolium*). Devon, 1944.
*Root rot of *Asparagus*. Corn, 1953.
*Damping-off of cucumber seedlings (*Cucumis*). Common in the south. Observations by Williams (*Rep. exp. Res. Sta. Cheshunt for 1925*, 72–5).
*Shanking of onion and shallot (*Allium*). Widely distributed in the midlands and south of England. (See Hickman, *Gdnrs' Chron.* **114**, 1943, 40; *Rep. Long Ashton Res. Sta. for 1943*, 100–7.)
*Damping-off of *Anemone*. Dawlish, Devon, 1950; Corn, 1953.
*Basal rot of cactus. Newcastle, 1937; Clyde, 1957.
*Wilt of *Calceolaria*. Forth.
*Rotting stem bases of *Eschscholtzia*. Som, 1936.
*Root rot of *Gerbera*. Bideford, Devon, 1950.
*On *Iris*. A leaf disease of bulbous iris, caused by a species of *Phytophthora* not unlike *P. cyperi-rotundati* Sawada, has been destructive in wet seasons in the Isles of Scilly since 1928 (Gibson & Gregory, *TBMS*, **24**, 1940, 251–4); also occurs in west Corn. Particularly prevalent on var. Wedgwood and *Iris tingitana*.
*On Japanese larch (*Larix leptolepis*), Llantrissant Forest, Wales (*Rep. imp. For. Inst. Oxford for 1944–5*, 8–10).
*Damping-off of marigold (*Calendula*). Worcs, 1954.
*Black leg of *Pelargonium*. Occasional.
*Killing hedges of privet (*Ligustrum*). Som, 1936.
*Foot rot of *Romneya coultheri*. Clyde.
*Root rot of rose. Forth.
*Foot rot of *Salpiglossis*. Harpenden, Herts, 1932.

Plasmodiophora brassicae Woron., *Pringsh. Jahrb.* **11**, 548; Sacc., *Syll.* vii, 464.

*Club root of crucifers. Widely distributed throughout Britain on soils deficient in lime, and therefore less serious in districts where the soil overlies chalk or limestone. On turnip and swede it attracts most attention in the north, west midlands and Wales.

Commonest and most severe on turnip, swede, cabbage, cauliflower and broccoli. Not uncommon on radish (*Raphanus*) and wallflower (*Cheiranthus*); rather troublesome on seakale (*Crambe*) (*Report*, iv, 57; Cayley, *Gdnrs' Chron.* **59**, 1916, 31); occasional on marrow-stem kale, kohl-rabi, rape (*Brassica napus*), stock (*Matthiola*), mustard (*Brassica nigra*), Hesperis and honesty (*Lunaria*). Webb (*Nature, Lond.*, **163**, 1949, 608) mentioned noncruciferous weeds, *Holcus lanatus*, and strawberry as possible hosts for the zoosporangia.

This is one of the oldest known diseases in Britain. Ellis (*The Modern Husbandman*, July 1742, 26) said he found it in 1736 in Norfolk and Suffolk, where it was called Anbury. Frequent references may be found to it in the *Gardeners' Chronicle* from vol. **1**, 1843 onwards, and among the earlier accounts of it are those by Curtis (*J. R. agric. Soc.* **4**, 1843, 121–32), Buckman (ibid. **15**, 1854, 125–35), Voelcker (ibid. **20**, 1859, 101–5) and Whitehead (*Report on Insects and Fungi injurious to crops, Bd Agric. Lond.* 1893, 49–54. Colhoun (*Phytopath. Pap.* **3**, 1958, 116 pp.) published a monograph on the disease.

BIOLOGY. Cook and Schwartz (*Phil. Trans.* B, **218**, 1930, 283–314) studied the life history, cytology and method of infection of *Plasmodiophora brassicae*; Samuel and Garrett (*AAB*, **32**, 1945, 96–101) devised a method of estimating its activity in the soil, and Warne (*Nature, Lond.*, **152**, 1943, 509; *J. R. hort. Soc.* **69**, 1944, 45–47) brought forward some evidence of its transmission with the seed.

PHYSIOLOGIC RACES. Walker (*Phytopathology*, **32**, 1942, 18) postulated the existence of physiologic races of the parasite when he found that two varieties of turnip, resistant in America, succumbed to the disease in England, and Macfarlane (*AAB*, **43**, 1955, 297–306) distinguished three such races in Britain.

SOIL CONDITIONS. The relation of the disease to soil conditions was given attention by Atkins (*Sci. Proc. R. Dublin Soc.* **16**, 1922, 427–34) and Potts (*TBMS*, **19**, 1935, 114–27). Colhoun (*Nature, Lond.*, **169**, 1952, 21–2) studied the effect of temperature and soil moisture on the disease, followed its development in limed soils when temperatures were favourable (*AAB*, **40**, 1953, 639–44), and subsequently discussed its epidemiology, and by investigating the various factors influencing infection was able to explain many anomalous and contradictory experiences with the disease (ibid. **40**, 1953, 262–82). He also (ibid. **45**, 1957, 559–65) devised a method of examining soil for the presence of the pathogen. Macfarlane (ibid. **39**, 1952, 239–56) has considered the factors affecting the survival of the parasite in the soil.

RESISTANT VARIETIES. Variety-resistant trials with swedes were carried out in Wales by Whitehead (*J. Minist. Agric.* **29**, 1922, 362–7; *Welsh J. Agric.* **1**, 1925, 176–84; **9**, 1933, 208–33) and by Davies, Griffiths and Evans (ibid. **4**, 1928, 295–303), while the resistance of the Bruce turnip was studied by Findlay (*Scot. J. Agric.* **14**, 1931, 173–83) and by Hendrick (*Trans. Highl. agric. Soc. Scot.* **44**, 1932, 52–63).

CONTROL. Many experiments have been carried out in Britain on the control of club root by liming (Whitehead, *Welsh J. Agric.* **12**, 1936, 183–92) and by chemical treatment of the soil. Notable among these are Preston's experiments on cabbage, cauliflower, etc., first with mercuric chloride (ibid. **4**, 1928, 280–95; *J. Minist. Agric.* **38**, 1931, 272–84; **41**, 1934, 329–35) and later with calomel, mercury-zinc amalgam and brassisan containing pentachloronitrobenzene (*AAB*, **28**, 1941, 351–9; *Gdnrs' Chron.* **115**, 1944, 128). Holmes Smith (ibid. **87**, 1930, 371; **90**, 1931, 35) and Baillie and Muskett (*J. Minist. Agric. N. Ire.* **4**, 1933, 44–6) used mercuric chloride, Shewell Cooper (*Gdnrs' Chron.* **91**, 1932, 387; **92**, 1932, 83) this and calcium cyanamide, and Smieton (*J. Pomol.* **17**, 1939, 195–217) chlorinated nitrobenzene, while Ogilvie, Mulligan and Brian (*Rep. Long Ashton Res. Sta. for 1934*, 178–9) and Green and Ashworth (*J. R. hort. Soc.* **68**, 1943, 111–15; **69**, 1944, 144–7) tested a variety of substances. Woodman, Brenchley and Hanley (*J. Soc. chem. Ind., Lond.*, **53**, 1934, 35 T) studied the effect of mercuric chloride treatment on seedling growth in uncontaminated soil. Colhoun (*AAB*, **41**, 1954, 290–304) devised a technique for evaluating soil fungicides used for control. Rosser (*Plant Path.* **6**, 1957, 42–4) found pentachloronitrobenzene and mercuric chloride gave good, and griseofulvin and cadmium chloride poor control.

Plasmopara nivea (Ung.) Schröt., *Kr. Fl. Schles.* 237; Sacc., *Syll.* vii, 240.

*Downy mildew of carrot (*Daucus*). Rare. Bere Alston, Devon, Nov. 1945.

*Downy mildew of parsley (*Petroselinum*). Rare. Reported by M. J. B[erkeley] (*Gdnrs' Chron.* **14**, 1880, 338); Surrey, 1903, 1946; Hants, 1904.

*Downy mildew of parsnip (*Pastinaca*). Not infrequent. Corn, Som, Glos, Worcs, Glam, Bucks, Kent. Reported by W. G. Smith (*Gdnrs' Chron.* **22**, 1884, 716) and by Cotton (*Kew Bull.* 1918, 8–21).

Plasmopara pygmaea (Ung.) Schröt., *Kr. Flor. Schles.* 239; Sacc., *Syll.* vii, 240.

*Downy mildew of *Anemone*. On *A. coronaria*. Uncommon. Corn, 1937 (*Report*, viii, 79); Penzance, Corn, 1945; Yorks, 1951; Northumb, 1955. On *Anemone* sp., Ayrshire, 1951.

Plasmopara viticola (Berk. & Curt. ex de Bary) Berl. & de Toni, in Sacc., *Syll.* vii, 239.

*Downy mildew of vine (*Vitis*). At one time rare, but has become more prevalent in recent years in south-east England. Derby, Kent, Surrey, Dorset. Occasional in Jersey. On *V. coignetiae*, Surrey, 1932 (*Report*, vii, 87); on *V. vinifera* var. *purpurea*, Yalding, Kent, 1947.

This mildew was first reported by Cooke (*Gdnrs' Chron.* **15**, 1894, 689–90) at two places more than 100 miles apart (one was

Derby). Subsequently there was no authentic record of it until 1926 when it was seen at Wye, Kent (Harrison & Ware, ibid. **80**, 1926, 448–9). It reappeared on the same vine in 1932 and four years later was found on another vine 100 yards away. W. C. Moore (*TBMS*, **32**, 1949, 97) summarized information about the first nine records up to 1948. Hyams (*Gdnrs' Chron.* **129**, 1951, 4–5) gave some details of the prevalence of the mildew in outdoor vineries at Canterbury, Kent and Oxted, Surrey, in 1950.

Plectodiscella veneta Burkh. = *Elsinoe veneta.*

Pleiochaeta setosa (Kirchn.) Hughes, *Mycol. Pap.* **36**, 1951, 39; syn. *Ceratophorum setosum* Kirchn., *Z. PflKrankh.* **2**, 1892, 324–7; Sacc., *Syll.* xi, 622; Wakefield & Bisby, *TBMS*, **25**, 1941, 88.
*Die-back of *Cytisus*. Causing death of cuttings of *C. scoparius* and *C. pallidus* (Green & Hewlett, *J. R. hort. Soc.* **74**, 1949, 310–12).
*Brown spot of *Laburnum*. Surrey, 1954, on seedlings of *L. anagyroides* (*vulgare*); Yorks, 1956.
*Leaf spot of lupin. Occasional. Wisley, Surrey, on *Lupinus cytisoides* (Green, *J. R. hort. Soc.* **58**, 1933, 144–5); Berks, 1936; Devon 1936; I. of Ely, 1950; Salop, 1950, 1955.

Pleospora betae Björling, *Bot. Not.* **2**, 1944, 215; stat. conid. *Phoma betae* Frank, *Z. Rübenzuckerind.* **42**, 1892, 903; Sacc., *Syll.* xi, 492; Grove, *Coelomycetes*, i, 68.
The perfect stage was not seen in Britain until 1957.
*Black leg of sugar beet and mangolds (*Beta*). The commonest cause of blackleg in seedlings. Also parasitic on the leaves, stems and roots of older plants, and often secondarily associated with heart rot (boron deficiency). Commonly present on beet seed samples. The apparent cause of blackened areas on petioles of seakale-beet, Harpenden, Herts, 1940.
Control by seed treatment was obtained by Woodward and Dillon Weston (*Gdnrs' Chron.* **85**, 1929, 229; *AAB*, **16**, 1929, 542–66) and by Hughes (*Sci. Proc. R. Dublin Soc.* **21**, 1935, 205–12) using organo-mercury compounds; and by Garner and Sanders

J. Minist. Agric. **38**, 1931, 8–9; **39**, 1933, 986–7) using sulphuric acid. Gates and Hull (*AAB*, **41**, 1954, 541–61) tested a wide range of seed disinfectants, including organo-mercury compounds, thiram (TMTD) and aldrin.

For many years all sugar-beet seed sown in Great Britain has been treated centrally with an organo-mercury dust before planting.

Pleospora calvescens (Fr.) Tul., *Sel. Fung. Carp.* ii, 1863, 266; Sacc., *Syll.* ii, 279 (as *Pyrenophora calvescens* (Fr.) Sacc.); stat. conid. *Helminthosporium papaveris* Hennig, *Z. PflKrankh.* **17**, 1907, 282; Sacc., *Syll.* xxv, 829.

The perfect stage has not been seen in Britain.

*Leaf blight of poppy (*Papaver*).* First British record at Woodley, Berks, Sept. 1950; conidial stage on *P. somniferum* (opium poppy) grown for oil seed. Also at Winchester, Hants, 1950; Wendover, Bucks, 1950.

Pleospora herbarum (Fr.) Rabenh., in Klotzsch, *Herbarium vivum mycologicum,* 2nd ed. 547; Sacc., *Syll.* ii, 247; Bisby & Mason, *TBMS,* **24**, 1940, 191; stat. conid. *Stemphylium botryosum* Wallr., *Flora Cryptogamica Germaniae,* 1833, ii, 300; Sacc., *Syll.* iv, 522; Wakefield & Bisby, *TBMS,* **25**, 1941, 100; syn. *Macrosporium parasiticum* Thüm., *Myc. univ.,* 667; Sacc., *Syll.* iv, 537.

For the taxonomy of this fungus see Wiltshire *TBMS,* **21**, 1938, 211–39.

*Leaf spot of mangold. Mon, 1943 (Hughes, *TBMS*, **28**, 1945, 91–3).

*Net blotch of field and broad bean (*Vicia*). Frequent in the south-west, occasional elsewhere. Wilts, Som, Glos, Dorset, Devon and Corn, Mon, eastern England. On tic beans, East Lothian, 1953. See Justham and Ogilvie (*Plant Path.* **2**, 1953, 140).

*Leaf spot of clover (*Trifolium*). Suffolk, 1954. The fungus was common on samples of red clover seed in south-west England from the 1948 crop; on red and white clover and on alsike, Dee; on red and crimson clover, Hebrides.

*Leaf spot of lucerne (*Medicago sativa*). Widely distributed in Britain in recent years, but not damaging. Found on many seed samples from the 1949 harvest in Scotland.

*Ring spot of sainfoin (*Onobrychis*). Glam (Hughes, *TBMS*, **28** 1945, 86–90), where common; Yorks, 1948; Cambs, 1948.

*On trefoil (*Medicago lupulina*). Detected at Edinburgh on many seed samples from the 1948 harvest.

*On *Lupinus angustifolius*, Inverness-shire, Hebrides, 1956.

*Leaf spot of endive (*Cichorium*). Penzance, Corn, 1933.

*Leaf spot of lettuce (*Lactuca*). Occasional in south-west England, Lancs. (See Ogilvie and Mulligan, *Gdnrs' Chron.* **89**, 1931, 35.)

*On onion (*Allium*), causing lesions on the stalks of the seed heads, Lincs, 1948.

*On tomato (*Lycopersicon*), associated with a foot rot. Jersey, 1953; Staffs, 1957, on imported fruit.

*Rotting stored apple fruits. See *Pleospora pomorum*. *Pleospora* sp. was reported rotting apple fruits at Long Ashton, Bristol, in 1938.

*On flax (*Linum*) straw, Tarves, Scotland, 1947.

*Associated with a leaf spot on *Arctostaphylos manzanita* (Briant & Martyn, *TBMS*, **14**, 1929, 221–5).

Pleospora pomorum Horne, *J. Bot., Lond.*, **58**, 1920, 239; Bisby & Mason, *TBMS*, **24**, 1940, 191.

*Fruit rot of apple (*Malus*). Recorded on stored fruit, but rarely. Kent (Horne & Horne, *Gdnrs' Chron.* **68**, 1920, 216–17; *AAB*, **7**, 1920, 183–201). Salmon and Ware (*J. S.-E. agric. Coll. Wye*, **37**. 1936, 15) collected the fungus on overwintered apple leaves. It may be synonymous with *P. herbarum* (q.v.). Kidd and Beaumont (*TBMS*, **10**, 1924, 100–1) recorded both species on stored apples, but found *P. herbarum* only once. Horne (*Rep. Imp. bot. Conf. Lond. 1924*, 1925, 363–72) studied the physiologic relations of *P. pomorum* and other fungi causing rots in stored apples.

Pleospora rehmiana Staritz, see *Ascochyta imperfecta*.

Pleospora trichostoma (Fr.) Ces. & de Not., see *Pyrenophora teres*.

Plicaria fulva Schneider, see *Botrytis gemella*.

Plowrightia ribesia (Fr.) Sacc., *Syll.* ii, 635; syn. *Dothidella ribesia*
(Pers. ex Fr.) Theiss. & Syd., *Ann. mycol., Berl.*, **13**, 1915, 309;
Bisby & Mason, *TBMS*, **24**, 1940, 206.

*Black pustule of currant (*Ribes*). Common on red currant; less
frequent on black currant and occasional on gooseberry (*J. Bd
Agric.* **14**, 1908, 680–1). A wound parasite, described by Massee
(*Gdnrs' Chron.* **27**, 1900, 290) and Hoggan (*TBMS*, **12**, 1927,
27–44).

Plowrightia virgultorum (Fr.) Sacc., *Syll.* ii, 636; Bisby & Mason,
TBMS, **24**, 1940, 206.

*Black knot of birch (*Betula*). Kent, Yorks, Scotland, and perhaps not uncommon (Massee, *Kew Bull.* 1914, 322–3).

Podosphaera leucotricha (Ell. & Everh.) Salm., *Monogr. Erysiph.*
1900, 40; Sacc., *Syll.* ix, 365 (as *Sphaerotheca leucotricha*
Ell. & Everh.); Bisby & Mason, *TBMS*, **24**, 1940, 135.

*Apple mildew. Very common in all districts, but less prevalent
where sulphur compounds are used as routine sprays.

The disease has been known on the shoots for many years (see,
for example, *Gdnrs' Chron.* **29**, 1869, 665). The perithecia were
first seen by Massee (ibid. **36**, 1904, 349) in Surrey, and soon
afterwards Salmon (ibid. **42**, 1907, 166) reported mildew on the
fruits. Woodward (*TBMS*, **12**, 1927, 173–204) discussed primary
infection from perennating mycelium, and Petherbridge and
Dillon-Weston (ibid. **14**, 1929, 109–11) secondary infection. The
effect of spraying on mildew, and its control with lime sulphur,
has been studied by Bedford and Pickering (*Rep. Woburn exp.
Fruit Fm*, **8**, 1908, 100–2), M. H. Moore (*J. Pomol.* **8**, 1930, 283–
90; **10**, 1932, 271–94; **12**, 1934, 57–79), and Marsh (*Rep. Long
Ashton Res. Sta. for 1939*, 42–51). Corner (*New Phytol.* **34**, 1935,
180–200) studied the behaviour of the fungus on resistant varieties.

*On *Malus* (*Pyrus*) *floribunda*, Long Ashton, Bristol, 1937.

*Mildew of medlar (*Mespilus*). Not uncommon, but insignificant. It was described by Cooke (*Gdnrs' Chron.* 7, 1877, 730) as *Oidium mespili* n.sp.

*Pear mildew. Widely distributed but not common.

*Mildew of quince (*Cydonia*). Not uncommon but usually slight.

Podosphaera oxyacanthae (DC.) de Bary, *Beiträge zur Morphologie und Physiologie der Pilze*, iii, 1870, 48; *Hedwigia*, **10**, 1871, 68; Sacc., *Syll.* i, 2; Bisby & Mason, *TBMS*, **24**, 1940, 135.

*Mildew of hawthorn (*Crataegus*). Common everywhere.

*Cherry mildew. Moray, Scotland, 1927.

Podosphaera oxyacanthae var. **tridactyla** (Wallr.) Salm., *Monogr. Erysiph.* 1900, 36; Sacc., *Syll.* i, 2 (as *P. tridactyla* (Wallr.) de Bary); Bisby & Mason, *TBMS*, **24**, 1940, 135.

*Plum mildew. Common on nursery stock and on suckers, especially in southern England; otherwise not often seen. Found rarely on the fruit. Recorded from Moray, Scotland, and on *Prunus padus* in Tay, Dee and Moray.

*Mildew of cherry laurel (*Prunus laurocerasus*). Frequent in the *Oidium* stage in south Wales (W. C. Moore & F. Joan Moore, *TBMS*, **32**, 1950, 274–5). Occasional elsewhere. Kew (Salmon, *J. R. hort. Soc.* **31**, 1906, 142–6). The perithecial stage was found in 1951 at Aberystwyth (F. Joan Moore, *Plant Path.* **1**, 1952, 54).

Polyopeus purpureus Horne, *J. Bot., Lond.*, **58**, 1920, 238; Kidd & Beaumont, *TBMS*, **10**, 1924, 105.

*Occasional as a cause of spotting of apple fruits, but doubtfully distinct from *Phoma*. See *Phoma mali*.

Polyporus adustus (Willd.) Fr., *Syst. Myc.* i, 363; Sacc., *Syll.* vi, 125; Rea, *Brit. Basid.* 587.

*On apple (*Malus*). A wound parasite in the Wisbech area of Cambs (Brooks, *TBMS*, **10**, 1925, 225–6).

*On beech (*Fagus*). A contributory cause of 'beech snap disease' on dry chalk escarpments and ravines. Selborne Hanger, Hants (Prior, *J. econ. Biol.* **8**, 1913, 249–63).

Polyporus betulinus Fr., *Syst. Myc.* i, 358; Sacc., *Syll.* vi, 139; Cooke, *Fung. Pests Cult. Pl.* 214; Rea, *Brit. Basid.* 584.

*Heart rot of birch (*Betula*). Frequent, and often killing the tree. Macdonald (*AAB*, **24**, 1937, 289–310) dealt with its biology.

Polyporus dryadeus Fr., *Syst. Myc.* i, 374; Sacc., *Syll.* vi, 136; Rea, *Brit. Basid.* 584.

*White butt rot of oak (*Quercus*). Widespread and fairly common, especially on heavy clay soil.

Polyporus giganteus Fr., *Syst. Myc.* i, 356; Sacc., *Syll.* vi, 99; Rea, *Brit. Basid.* 583.

*Root rot of beech (*Fagus*).

Polyporus hispidus Fr., *Syst. Myc.* i, 362; Sacc., *Syll.* vi, 129; Cooke, *Fung. Pests Cult. Pl.* 206; Rea, *Brit. Basid.* 584.

*On apple (*Malus*). Occasional on old trees. Doubtfully parasitic. Mon, Worcs, Devon, Kent.
*On plum (*Prunus*). Middx, Oxon.
*Heart rot of ash (*Fraxinus*). Common on *F. excelsior* and *F. nigra* (Nutman, *AAB*, **16**, 1929, 40–64).

Polyporus quercinus (Schrad.) Fr., *Epicr.* 441; Sacc., *Syll.* vi, 138; Rea, *Brit. Basid.* 584.

*On beech (*Fagus*). Rare. On a living tree, Woodstock, Oxon, 1949 (Cartwright, *TBMS*, **34**, 1951, 604–6).

Polyporus radiatus Fr., *Syst. Myc.* i, 369; Sacc., *Syll.* vi, 247 (as *Polystictus radiatus* (Sow.) Fr.); Rea, *Brit. Basid.* 586.

*Heart rot of alder (*Alnus*).

Polyporus schweinitzii Fr., *Syst. Myc.* i, 351; Sacc., *Syll.* vi, 76; Rea, *Brit. Basid.* 582.

*Butt rot of conifers. Fairly common on Scots pine, spruce (*Picea*) and larch (*Larix*). Also on Douglas fir (*Pseudotsuga*).

A rot of middle age and maturity, attacking chiefly trees 50 years or more old (Peace, *Quart. J. For.* **32**, 1938, 81–104).

Polyporus squamosus Fr., *Syst. Myc.* i, 343; Sacc., *Syll.* vi, 79; Rea, *Brit. Basid.* 579.

*Common as a wound parasite, causing white heart rot of broad-leaved trees, especially elm (*Ulmus*) and sycamore (*Acer*). Also reported on cherry. Buller (*J. econ. Biol.* **1**, 1906, 101–38) dealt with its biology, and Campbell and Munson (*AAB*, **23**, 1936, 453–64) studied the black lines produced by the fungus in elm wood.

Polyporus sulphureus Fr., *Syst. Myc.* i, 357; Sacc., *Syll.* vi, 104; Rea, *Brit. Basid.* 581.

*Heart rot of fruit trees. Occasional on apple, cherry, pear, plum and walnut.
*Heart rot of chestnut (*Castanea*), yew (*Taxus*) and willow (*Salix*). Common in cricket-bat willow.
*Brown cubical or heart rot of oak (*Quercus*). The most serious cause of decay of parkland oaks.

Polyspora lini Laff., *Sci. Proc. R. Dublin Soc.* **16**, 1921, 248–74; Grove, *Coelomycetes*, ii, 206; Wakefield & Bisby, *TBMS*, **25**, 1941, 53.

*Browning and stem-break of flax (*Linum*). First reported in England, 1941, and later fairly common in Pembs and the east midlands, though rarely serious. Declined with the reduction in flax acreages. It has been reported also on seed from Scotland.
  The disease was fully described by Lafferty (loc. cit.). Colhoun (*AAB*, **33**, 1946, 255–9) studied its effect on seed production and (ibid. **45**, 1957, 268–75) the effect on the disease of nutrient treatments. For observations on the disease by Pethybridge and others, and on its control by Muskett and his colleagues, see the literature cited under *Colletotrichum linicola*.

Polystictus velutinus Fr., *Syst. Myc.* i, 368; Sacc., *Syll.* vi, 258; Rea, *Brit. Basid.* 608.

*On privet (*Ligustrum*). Associated with patchy death of a privet hedge, north Scotland, 1950. The fungus is generally regarded as a saprophyte.

Polystigma rubrum (Pers. ex Fr.) Chev., *Comment. Mus. Hist.* 3, 1817, 330; Sacc., *Syll.* ii, 458; Cooke, *Fung. Pests Cult. Pl.* 1906, 127; Bisby & Mason, *TBMS*, **24**, 1940, 204.

*Leaf blotch of plum (*Prunus*). This species is common in Britain on blackthorn (*Prunus spinosa*) and bullace (*P. insititia*), but is very rare or non-existent on the cultivated plum. Blackman and Welsford (*Ann. Bot., Lond.*, **26**, 1912, 761–7) described the development of the perithecium.

Polythrincium trifolii Fr. = stat. conid. of *Cymadothea trifolii*.

Poria obliqua (Pers. ex Fr.) Karst., *Rev. Mycol., Paris*, 3, pt. 9, 1881, 19; Rea, *Brit. Basid.* 606.

*On birch (*Betula*), Aviemore, 1938 (Findlay, *TBMS*, **23**, 1939, 169–70); on *B. pubescens*, Perth, Braemar, Cumb (Batko, ibid. **33**, 1950, 105). Parasitism uncertain.

Pseudobalsamia microspora Diehl. & Lambert, *Mycologia*, **22**, 1930, 223.

*Mushroom bed truffle. The fungus invades mushroom beds from the casing soil. It was first seen at Cheshunt, Herts, in 1936 (Williams, *Gdnrs' Chron.* **100**, 147), is not uncommon in south-east England (Glasscock & Ware, *AAB*, **28**, 1941, 85–90), and appears to be on the increase. Yorks, Norfolk, Essex, Worcs. The first Scottish record was in Perthshire in 1953.

Atkins and La Touche (*Mushroom Diseases Leafl.* Yaxley, Peterboro, no. 2, 1948, 3 pp.) summarized existing information about the fungus, and Duncan (*Rep. Mushroom Res. Sta. Yaxley for 1949*, 42–4; *for 1950*, 31–6; *for 1951*, 44–50; *Mushroom Science*, **2**, 1953, 167–74) studied it.

Pseudodiscosia antirrhini (Wakef.) Buddin & Wakef. = *Heteropatella antirrhini*.

Pseudodiscosia dianthi Hösterm. & Laub. = *Heteropatella valtellinensis*.

Pseudoperonospora cubensis (Berk. & Curt.) Rostov., *Flora*, **92**, 1903, 422; Sacc., *Syll.* xvii, 520 (as *Plasmopara cubensis* (Berk. & Curt.) Humphrey).

*Downy mildew of cucurbits. Very rare. Massee (*Gdnrs' Chron.* **17**, 1895, 656) reported it from three localities in England in 1895. In one instance the entire contents of a large melon house were said to have been destroyed by it for three years in succession. No host was mentioned for the other two localities, but in *Diseases of Cultivated Plants*, 1915, 121, Massee stated that the mildew had been found on cucumber in Britain. There appears to be no other British record of this fungus, and there must be some doubt about the authenticity of the above records.

Pseudoperonospora humuli (Miyabe & Takah.) G. W. Wilson, *Mycologia*, **6**, 1914, 194; Sacc., *Syll.* xxiv, 64 (as *Peronoplasmopara humuli* Miyabe & Takah.).

*Downy mildew of hop (*Humulus*). Develops every year in the hop areas of south-east England and the west midlands, but is scarce in dry years, rarely becomes really destructive unless the rainfall in both July and August is above normal, and is usually controlled adequately by spraying. The disease begins in spring and early summer in the form of basal spikes from infected 'hills'. In Scotland listed from Dee. It is surprising that the disease was not seen in Eire until 1937 (Murphy, *Int. Bull. Pl. Prot.* **12**, 1938, 53), and then on a 'wild' hop in an experimental garden, Co. Cork.

Considerable attention was given to this disease for many years after its first appearance in England at Wye, Kent, in 1920 (Salmon & Wormald, *J. Minist. Agric.* **30**, 1923, 430; Salmon & Ware, *Gdnrs' Chron.* **76**, 1924, 265). The annual course of the disease in England between 1924 and 1938 was followed by Salmon and Ware (for 1924 in *AAB*, **12**, 1925, 121–51; for 1925–7 in *J. Minist. Agric.* vols. **33**–**4**; for 1928–31 in *J. Inst. Brew.* vols. **35**–**8**; and for 1932–8 in *J. S.-E. agric. Coll. Wye*, vols. **32**–**44**.

General accounts, including control, were published by Salmon and Ware (*J. S.-E. agric. Coll. Wye*, 1927, 28 pp.; 1931, 15 pp.), and special aspects of the disease, investigated at Wye or at the

East Malling Research Station, include: perennial mycelium (Salmon & Ware, *Nature, Lond.*, **116**, 1925, 134; Ware, *TBMS*, **11**, 1926, 91–107), artificial production of spikes (Ware, *Ann. Bot., Lond.*, **43**, 1929, 683–710), comparison with nettle mildew (Salmon & Ware, *AAB*, **15**, 1928, 352–70), effect of chemicals on zoospores (Goodwin, Salmon & Ware, *J. agric. Sci.* **19**, 1929, 185–200), copper content of sprayed hops (Harman, *J. Inst. Brew.* **38**, 1932, 197), incidence on new seedlings (Beard, *J. Pomol.* **15**, 1937, 205–25; *Rep. Dep. Hop Res. Wye for 1956*, 54–64), the effect of weather on the disease (Ware, *AAB*, **29**, 1942, 322), and the association of the fungus with rotting of the rootstock (Beard & Derbyshire, *Rep. Dep. Hop Res. Wye for 1956*, 65–72).

Pseudopeziza medicaginis (Lib.) Sacc., *Syll.* viii, 724; Ramsbottom & Browne, *TBMS*, **34**, 1951, 107.

*Leaf spot of lucerne and trefoil (*Medicago*). Very common and often damaging.

Pseudopeziza ribis Kleb., *Z. PflKrankh.* **16**, 1906, 76; Sacc., *Syll.* xxii, 743; Ramsbottom & Browne, *TBMS*, **34**, 1951, 107; stat. conid. *Gloeosporium ribis* (Lib.) Mont. & Desm. in Kickx, *Flor. Cr. Flanders*, ii, 95; Sacc., *Syll.* iii, 706; Grove, *Coelomycetes*, ii, 224.

*Leaf spot of currant and gooseberry (*Ribes*). Widely distributed in the southern half of England and Wales. Common. In wet seasons it may cause serious defoliation and sometimes spots the fruit. In Scotland widely distributed on red and black currant, local on gooseberry (Solway), occasional on Worcesterberry (Dee, 1948).

The disease was described by Güssow (*Gdnrs' Chron.* **42**, 1907, 180), and by Briton-Jones (*Rep. Long Ashton Res. Sta. for 1925*, 105–8). Corke (ibid. *for 1953*, 154–8) studied perennation and infection, found the perfect stage on dead leaves in winter and (*Plant Path.* **6**, 1957, 25–6) studied ascospore discharge. He also treated overwintered leaves to prevent maturation of the ascospores (*Rep. Long Ashton Res. Sta. for 1956*, 120–2). The perfect stage was first recorded with certainty in Hants in 1930 (*Report*, vii, 81). Clarke and Corke (*Rep. Long Ashton Res. Sta. for 1955*, 196–200) devised a pictorial diagram for the rapid assessment of the percentage leaf area affected.

CONTROL. Marsh and Maynard (*Rep. Long Ashton Res. Sta. for 1928*, 109–11; *for 1929*, 166–7; *J. Minist. Agric.* **37**, 1930, 255–9) carried out spraying trials with Bordeaux Mixture, Marsh (*Rep. Long Ashton Res. Sta. for 1932*, 86–9) with colloidal copper, Marsh and Dickenson (ibid. *for 1944*, 150–7) with dithiocarbamates, and Glasscock and English (*Plant Path.* **5**, 1956, 97–8) with oil Bordeaux; whilst Corke (*J. hort. Sci.* **30**, 1955, 197–200) carried out laboratory trials with fungicides with a view to preventing sporing on overwintered leaves. The effect of organic spray residues on canned fruit has been studied by Adam, Dowson & Marsh (*Rep. Long Ashton Res. Sta. for 1947*, 19–29).

Pseudopeziza trifolii (Fr.) Fuckel, *Symb. Myc.* 1869, 290; Sacc., *Syll.* viii, 723; Ramsbottom & Browne, *TBMS*, **34**, 1951, 107.

*Leaf spot of clover (*Trifolium*). Very common in England and Wales on red clover and locally abundant on white clover, particularly in late autumn. Occurs also on alsike, crimson clover and hop clover. Rarely harmful. In Scotland is found mainly on white clover, but also on red clover and alsike.

The species is biologically distinct from *Pseudopeziza medicaginis* on *Medicago*, but the morphological differences are slight. (See Ivy Massee, *J. econ. Biol.* **9**, 1914, 65–7.)

Pseudoplea trifolii (Rostr.) Petr. = *Sphaerulina trifolii*.

Puccinia aegra Grove, *J. Bot., Lond.*, **21**, 1883, 274; Sacc., *Syll.* vii, 614; Grove, *Brit. Rust Fungi*, 202; Wilson & Bisby, *TBMS*, **37**, 1954, 67.

*Viola rust. Common in many districts on *Viola* and pansy. For notes on the rust on *V. tricolor* and *V. cornuta* see *Gdnrs' Chron.* **6**, 1876, 49, 175 and 361.

Puccinia aethusae Mart., *Prodromus Florae mosquensis*, 2nd ed. 1817, 225; Sacc., *Syll.* vii, 634 (as *P. bullata* (Pers.) Schröt.); Grove, *Brit. Rust Fungi*, 190; Wilson & Bisby, *TBMS*, 37, 1954, 67.

*Rust of parsley (*Petroselinum*). Rare. Yorks, 1932 (*Report*, vii, 56).

Puccinia agropyrina Erikss., *Ann. Sci. nat.* 9, 1899, 273; Sacc., *Syll.* xvii, 384; Grove, *Brit. Rust Fungi*, 263; Wilson & Bisby, *TBMS*, 37, 1954, 67.

*Rust of *Agropyron*. Widely distributed in Scotland.

Puccinia agrostidis Plowr., *Gdnrs' Chron.* 8, 1890, 41 and 139; *Grevillea*, 21, 110; Sacc., *Syll.* xi, 202; Grove, *Brit. Rust Fungi*, 275; Wilson & Bisby, *TBMS*, 37, 1954, 67.

*Agrostis rust. Uncommon in England and Wales; widely distributed in Scotland. The aecidial stage occurs on *Aquilegia vulgaris*.

Puccinia airae Mayor & Cruchet, *Bull. Soc. vaud. Sci. nat.* 51, 1917, 628; Sacc., *Syll.* xxiii, 732; Grove, *Brit. Rust Fungi*, 265; Wilson & Bisby, *TBMS*, 37, 1954, 67.

*On *Deschampsia*. Uredo-stage occasional on *D. caespitosa* and *D. flexuosa*. Scarborough, Birmingham, Saltaire (Grove, loc. cit.).

Puccinia allii (DC.) Rud., *Linnaea*, 4, 1829, 392; Sacc., *Syll.* vii, 655.

*Rust of garlic (*Allium sativum*). Rare. On leaves attached to bulbs of garlic imported from Portugal in Jan. 1944; Wye, Kent, 1945. There was some doubt whether the fungus at Wye was *Puccinia allii* or *P. porri* (*Report*, ix, 47).

Puccinia anomala Rostr. = *P. hordei*.

Puccinia anthoxanthi Fuckel, *Symb. Myc.*, ii, *Nachtr.* 1873, 15; Sacc., *Syll.* vii, 665; Grove, *Brit. Rust Fungi*, 269; Wilson & Bisby, *TBMS*, 37, 1954, 67.

*Stem and leaf rust of sweet vernal grass (*Anthoxanthum*). Known since 1884 and now common. (See Grove, *J. Bot., Lond.*, **59**, 1921, 313, and Wilson, *TBMS*, **9**, 1923, 138.)

Puccinia antirrhini Diet. & Holw., *Hedwigia*, **36**, 1897, 298; Sacc., *Syll.* xiv, 327; Wilson & Bisby, *TBMS*, **37**, 1954, 68.

*Antirrhinum rust. General in the southern half of England and Wales and destructive in hot, dry summers; sporadic in the north and now becoming frequent in Scotland.

Before 1933 this rust was known only in North America and Bermuda. In July of that year—a hot, dry one—it appeared suddenly in Kent (Green, *Gdnrs' Chron.* **94**, 1933, 131), and by the end of October sixty-three outbreaks had been confirmed in south, east and south-west England (Pethybridge, *J. Minist. Agric.* **41**, 1934, 336–40). In an effort to stop the spread Chittenden (*J. R. hort. Soc.* **59**, 1934, 450–1; *Gdnrs' Chron.* **96**, 1934, 359) advocated the eradication of all antirrhinum plants, whether healthy or diseased, in November 1934, but during the next two years the rust proved very destructive south of a line from the Wash to the Severn, and it gradually spread north and into Wales. It was first seen in Scotland in 1935. As it became acclimatized its destructiveness waned, but it still does much damage in some years.

Green (*J. R. hort. Soc.* **59**, 1934, 119–26; **61**, 1936, 64–76; **62**, 1937, 214–19; **66**, 1941, 83–6) investigated the possibility of controlling it by spraying and the use of resistant varieties. He found it (*Gdnrs' Chron.* **95**, 1934, 81) on seedlings in January.

Puccinia apii Desm., *Catalogue des plantes omises*, 1823, 25; Sacc., *Syll.* xvii, 339; Grove, *Brit. Rust Fungi*, 184; Wilson & Bisby, *TBMS*, **37**, 1954, 68.

*Celery rust. Rare. Tay, Scotland (Trail, *Scot. Nat.*, n.s., **4**, 309). At one time (*circa* 1875–90) this rust was apparently not uncommon and W. G. Smith (*Gdnrs' Chron.* **26**, 1886, 756) referred to some severe attacks. (See also Cooke, *Fung. Pests Cult. Pl.* 89.)

297

Puccinia arenariae (Schum.) Wint., *Pilze*, 1884, 169; Sacc., *Syll.* vii, 683; Wilson & Bisby, *TBMS*, **37**, 1954, 68; syn. *P. lychnidearum* Link, *Sp. Pl.* **2**, 1825, 80; Grove, *Brit. Rust Fungi*, 218 p.p.

*On *Dianthus deltoides*, Kent, 1934; and on new seedling varieties of *Dianthus*, Berks, 1935 (as *Puccinia lychnidearum* Link f. *dianthi* Grove).

*Gypsophila rust. Rare. Corn, in same nursery, 1935 and 1938; on *Gypsophila elegans*, Som, 1929; Devon, 1949; Jersey, 1954.

*Rust of sweet william (*Dianthus barbatus*). Common and widespread, especially in the south. Described by Plowright (*Gdnrs' Chron.* **21**, 1884, 88–9) and W. G. Smith (ibid. p. 120).

Puccinia arrhenatheri (Kleb.) Erikss. in Cohn's *Beiträge zur Biologie der Pflanzen*, **8**, 1898, 1; Sacc., *Syll.* xvii, 384; Grove, *Brit. Rust Fungi*, 284; Wilson & Bisby, *TBMS*, **37**, 1954, 68.

*Rust of tall oat grass (*Arrhenatherum*). Uredo- and teleutostages not uncommon in the north of England; occasional elsewhere; common in Scotland.

The aecidial stage on *Berberis vulgaris* is very rare. Wooler, Northumb, 1944.

Puccinia asparagi DC., *Flor. Fr.* ii, 1805, 595; Sacc., *Syll.* vii, 601; Grove, *Brit. Rust Fungi*, 233; Wilson & Bisby, *TBMS*, **37**, 1954, 68.

*Asparagus rust. Localized and very sporadic. This rust has been known in Britain since 1824 (Greville, *Flora edensis*, 429) when it was found at Edinburgh. It was a matter of concern in many districts in 1895 (Abbey, *J. Hort.* **31**, 1895, 452–3) and in the Evesham area in 1904–6. It was then quiescent until 1933, when it was widely distributed in East Anglia and was also seen in Worcs (*J. Minist. Agric.* **40**, 1933, 800). Since then the rust has appeared sporadically in East Anglia and the Vale of Evesham, and was found in Lancs in 1956, but has not been recorded elsewhere and has not been seen in Scotland since the original discovery in 1824. (See *Report*, ix, 41.)

Puccinia bromina Erikss., *Ann. Sci. nat.* **9**, 1899, 268; Sacc., *Syll.* xvii, 382; Grove, *Brit. Rust Fungi*, 262; Wilson & Bisby, *TBMS*, **37**, 1954, 68.

*Brown rust of brome grasses (*Bromus*). Uredo- and teleuto-stages; aecidia unknown.

Freeman (*Ann. Bot., Lond.,* **14**, 1902, 487–94) and Marshall Ward (ibid. **14**, 1902, 233–315; *Ann. mycol., Berl.,* **1**, 1903, 132–51) carried out some early studies on the rusts of brome grasses, and it was on the basis of the results obtained that Ward developed his 'bridging-host' theory of rust behaviour. Subsequently Bean, Brian and Brooks (*Ann. Bot., Lond.,* **18**, 1954, 129–42) reinvestigated these diseases, found no evidence for the 'bridging-host' theory, and distinguished at least five physiologic races of brown rust on *Bromus* spp. They used material from abroad as well as from England and all the races fell within the group species *Puccinia dispersa* Erikss. & Henn. *sensu lata.* At least two races from English material belong in *P. bromina,* but it was not determined with certainty whether *P. symphyti-bromorum* Müller (with aecidia on *Symphytum officinale* in Switzerland) and *Puccinia bromi-maximi* Guyot were also represented in English material. Many attempts to infect *Symphytum* with teleutospores from material collected near Cambridge were unsuccessful.

Puccinia buxi DC., *Flor. Fr.* vi, 1805, 60; Sacc., *Syll.* vii, 688; Grove, *Brit. Rust Fungi,* 205; Wilson & Bisby, *TBMS,* **37**, 1954, 69.

*Rust of box (*Buxus*). Teleutospores rather common on *B. sempervirens* and widely distributed, notably in Scotland. A severe attack was reported from Glam in 1924. (See W. G. Smith, *Gdnrs' Chron.* **19**, 1883, 509.)

Puccinia campanulae Carm. ex Berk., *English Flora,* **5**, 365; Sacc., *Syll.* vii, 677; Grove, *Brit. Rust Fungi,* 159; Wilson & Bisby, *TBMS,* **37**, 1954, 69.

*Campanula rust. Not common. On *Campanula rotundifolia,* Clyde, Tay, Argyll, Dee; on *C. persicifolia,* Welshpool.

Puccinia caricina DC., see *P. pringsheimiana*.

Puccinia chrysanthemi Roze, *Bull. Soc. mycol. Fr.* **16**, 1900, 92; Sacc., *Syll.* xvi, 296; Grove, *Brit. Rust Fungi*, 131; Wilson & Bisby, *TBMS*, **37**, 1954, 70.

*Chrysanthemum rust. Common throughout Britain under glass, but much less prevalent than when first introduced, though now tending to get worse, notably in the west; occasional on outdoor plants.

This rust began to attract attention in Britain about 1895 and within a few years was causing much concern (M.C.C., *Gdnrs' Chron.* **22**, 1897, 256; Massee, ibid. **24**, 1898, 269; Abbey, *J. Hort.* **50**, 1898, 284–5), but it soon lost its virulence. Simeson (*Gdnrs' Chron.* **79**, 1926, 60) recommended a paraffin-oil treatment for its control, and Bailey (*J. R. hort. Soc.* **76**, 1951, 322–8) made some observations on varietal susceptibility.

Puccinia cichorii Bellynck. in Kickx, *Fl. Cr. Fl.* ii, 1867, 65; Sacc., *Syll.* xvii, 311; Grove, *Brit. Rust Fungi*, 148; Wilson & Bisby, *TBMS*, **37**, 1954, 70.

*Chicory rust. Uredo- and teleutospores on *Cichorium intybus*, Corn, Devon, Surrey, Suffolk, Norfolk. (See also *Puccinia endiviae*.)

Puccinia coronata Corda, *Ic. Fung.* i, 6; Sacc., *Syll.* vii, 623 p.p.; Grove, *Brit. Rust Fungi*, 253; Wilson & Bisby, *TBMS*, **37**, 1954, 70; syn. *P. lolii* Niels., *Ugeskr. Landm.* i, 549; *P. coronifera* Kleb., *Z. PflKrankh.* **4**, 1894, 132; Sacc., *Syll.* xi, 203.

*Crown rust of oats (*Avena*). Uredo- and teleuto-stages common and widely distributed in England and Wales; most prevalent in Wales and south-west England, with occasional heavy attacks in southern and northern districts, but usually slight in the east. Develops mainly in June–July. Crops usually free in Scotland but occasionally, as in 1943, epidemics occur in the south-west.

The fungus also attacks a number of grasses, including *Lolium perenne*, on which it is general throughout Britain, *Alopecurus pratensis* and *Arrhenatherum*.

The aecidial stage on *Rhamnus cathartica* and *R. frangula* is not uncommon.

Brown (*AAB*, **24**, 1937, 504–27; see also Brooks, ibid. **31**, 1944, 362) distinguished seven varieties on oats and grasses, but could not confirm Klebahn's division of Corda's species into *Puccinia coronata* Kleb. and *P. coronifera* Kleb. She also studied the aecidial hosts (ibid. **25**, 1938, 506–27). Griffiths (*Plant Path.* **2**, 1953, 73–7), who distinguished eight physiologic races from collections in Wales and south-west England, was concerned with varietal resistance and susceptibility in oats, and Davies and Jones (*Welsh J. Agric.* **2**, 1926, 212–21; **3**, 1927, 232–5) discussed the inheritance of resistance and susceptibility. Williams and Verma (*AAB*, **44**, 1956, 453–60) examined resistance in species of *Avena*.

Puccinia coronifera Kleb. = *P. coronata*.

Puccinia cyani Passer., in Rabenh., *Fung. Eur.* no. 1767; Sacc., *Syll.* vii, 634 (as *P. hieracii* (Schum.) Mart. p.p.); Grove, *Brit. Rust Fungi*, 140; Wilson & Bisby, *TBMS*, **37**, 1954, 70.
*Rust of cornflower. On cornflower (*Centaurea cyanus*) in gardens. Not uncommon in Britain.

Puccinia digraphidis Sopp. = *P. sessilis*.

Puccinia dispersa Erikss. & Henn. em. Erikss., *Getreideroste*, 1896, 210; Sacc., *Syll.* xi, 204; Grove, *Brit. Rust Fungi*, 260; Wilson & Bisby, *TBMS*, **37**, 1954, 71; syn. *P. secalina* Grove, loc. cit.
*Brown rust of rye (*Secale*). Uredo- and teleuto-stages common where the crop is grown, but the rust is unimportant.

The aecidial stage on *Lycopsis arvensis* has been seen only occasionally in England (Dorset, Kent, Norfolk, Yorks) and Scotland (Leuchars).

Puccinia endiviae Passer., *Hedwigia*, **12**, 1873, 113; Sacc., *Syll.* vii, 647; Wilson & Bisby, *TBMS*, **37**, 1954, 71.
*Rust of endive (*Cichorium endivia*). Uredo- and teleutospores.

Rare. The only British record was in the hot summer of 1947 when it was common and sometimes severe in the Penzance area of Corn. Wilson & Bisby (loc. cit.), who give Devon as a locality, thought the rust may be the same as *Puccinia cichorii.*

Puccinia festucae Plowr., *Gdnrs' Chron.* **8**, 1890, 42 and 140; Sacc., *Syll.* xi, 194; Grove, *Brit. Rust Fungi*, 257; Wilson & Bisby, *TBMS*, **37**, 1954, 71.

The species is doubtfully distinct from *P. coronata.*

*Rust of fescue grasses (*Festuca*). Uredo- and teleutospores in Scotland on *F. rubra* var. *arenaria* (Solway, Moray) and *F. ovina.*

The aecidial stage is seen occasionally on honeysuckle (*Lonicera*). Aberystwyth, 1945; Isles of Scilly, annually. (See Cooke, *Fung. Pests Cult. Pl.* 1906, 49.)

Puccinia gentianae Link, *Species Plantarum*, ii, 1825, 73; Sacc., *Syll.* vii, 604; Cooke, *Fung. Pests Cult. Pl.* 58; Grove, *Brit. Rust Fungi*, 178; Wilson & Bisby, *TBMS*, **37**, 1954, 71.

*Gentian rust. Rare. First British record, Horsham, Sussex (W. G. Smith, *Gdnrs' Chron.* **24**, 1885, 372) on *Gentiana acaulis*; on *G. amarella*, Salisbury, Wilts, 1904; Kew Gardens, 1905; on *G. verna* and *G. acaulis*, Cardiff, Glam, 1920 (Wilson, *TBMS*, **7**, 1921, 81); Ches, 1936; on *G. acaulis*, Devon, 1956.

Puccinia gladioli Cast., *Observations sur quelques plantes*, ii, 1843, 17; Sacc., *Syll.* vii, 728; Wilson & Bisby, *TBMS*, **37**, 1954, 71.

*Gladiolus rust. Very rare. The only British record is from Bohetherick, Corn, Nov. 1924 (*Rep. Dep. Pl. Path. Seale-Hayne agric. Coll. for 1933–4*, 55; *Report*, viii, 84).

Puccinia glumarum Erikss. & Henn., *Getreideroste*, 1896, 141; Sacc., *Syll.* xvii, 380; Grove, *Brit. Rust Fungi*, 258; Wilson & Bisby, *TBMS*, **37**, 1954, 71.

*Yellow rust of cereals. Uredo- and teleuto-stages. The most harmful rust on wheat in England and Wales, but seldom serious in Scotland. Not uncommon on barley, occasional on rye and some grasses, including cocksfoot (first seen on this host 1920—

Report, iv, 46). Occurs from January onwards and is usually noticeable by May. There is no known aecidial stage.

It was with this rust that Biffen (*J. agric. Sci.* 2, 1907, 109–28; *J. Bd Agric.* 15, 1908, 241–53; *TBMS*, 16, 1931, 19–37) proved that susceptibility to disease may be hereditary, and Armstrong (*J. agric. Sci.* 12, 1922, 57–96) followed this up, while Marryat (ibid. 2, 1907, 129–38) investigated the nature of immunity of some wheat varieties.

Mehta (*TBMS*, 8, 1923, 142–76) showed that the uredospores survive the winter but are killed by high summer temperatures, and Dillon-Weston (*AAB*, 16, 1929, 533–41) demonstrated that bunt (*Tilletia caries*) increases varietal susceptibility to yellow rust. Lists of resistant varieties have been issued (see, for example, *Agriculture, Lond.*, 50, 1943, 287), and Brooks (*AAB*, 31, 1944, 362–6) discussed physiologic specialization in this and other cereals. Manners (ibid. 37, 1950, 187–214) distinguished thirteen physiologic races in Britain, including eight on wheat and one on barley.

Batts (*J. nat. Inst. agric. Bot.* 6, 1953, 341–5) studied methods of testing varieties for resistance, and with Elliott (*Plant Path.* 1, 1952, 130–1) measured the effect of the rust on yields. Bell and Lupton (*Agriculture, Lond.*, 58, 1952, 572–6) discussed the significance of the rust in wheat cultivation and breeding in England.

Puccinia graminis Pers., *Tentamen Dispositionis Methodicae*, 1797, 39; Sacc., *Syll.* vii, 622; Grove, *Brit. Rust Fungi*, 250; Wilson & Bisby, *TBMS*, 37, 1954, 71.

*Black stem rust of cereals. The uredo- and teleuto-stages occur most years on wheat, especially in south-west Wales and locally elsewhere. Develops from July to September but is usually too late to do much harm. Occasionally it is generally epidemic, as in 1924 and 1940, with regional epidemics in the south-west more frequent (e.g. 1947, 1950, 1952, 1955—see *Plant Path.* 5, 1956, 38). Widely distributed but much less frequent on oats, rare on barley and rye. Occasional on a number of grasses, including *Agrostis, Lolium, Arrhenatherum, Dactylis* and *Agropyron*.

Black rust occurs sporadically on many of these hosts in Scot-

303

land and is not uncommon on oats in the Tweed Valley. It has also been recorded on the ornamental grass, *Bromus brizaeformis* (Perth, 1951); on *Poa trivialis* (south-east Scotland, 1948) and on *Deschampsia caespitosa* (Yorks, 1952).

The aecidial stage occurs on *Berberis vulgaris* and has very occasionally been seen on *Mahonia aquifolium* (e.g. by Plowright, *Gdnrs' Chron.* **19**, 1883, 736 and 768).

Plowright (*Gdnrs' Chron.* **18**, 1882, 231–4, 331–2) dealt with the heteroecism of this rust on wheat and barberry, and lengthy reports on the rust were issued by Carruthers (*J. R. agric. Soc.* **18**, 1882, 495–503) and Whitehead (*Report on Rust or Mildew on Wheat Plants 1892*, Bd Agric. Lond. 1893, 44 pp.). Later articles dealt with its distribution in Wales (Broadbent, *J. Minist. Agric.* **28**, 1921, 117–23) and Scotland (Maxwell & Wallace, *TBMS*, **11**, 1926, 138–45), and with varietal resistance (Jenkin & Sampson, *Bull. Welsh Pl. Breed. Sta.* no. 1, 1921, 41–9). Mehta (*TBMS*, **8**, 1923, 142–76) studied host specialization and found that winter temperatures in England kill the uredospores, while Waterhouse (*Ann. Bot., Lond.*, **35**, 1921, 557–64) investigated infection phenomena by the sporidia on *Berberis vulgaris*. Biffen (*TBMS*, **16**, 1931, 19–37) summarized the work that had been carried out to date on this rust and its control. It was nearly twenty years before any further serious study was made of the disease in Britain. Batts (*Nature, Lond.*, **163**, 1949, 107; *Trans. bot. Soc. Edinb.* **36**, 1952, 48–57) then investigated the distribution, host range and seasonal development of black rust in Scotland, as well as (*TBMS*, **34**, 1951, 533–8) the physiologic races occurring there and, with Elliott (*Plant Path.* **6**, 1957, 45–6), the effect on yield in southern England in 1955. Williams and Verma (*AAB*, **44**, 1956, 453–60) studied resistance in species of *Avena*.

Puccinia graminis Pers. var. **phlei-pratensis** (Erikss. & Henn.) Stakm. & Piem., *J. agric. Res.* **10**, 1917, 433; Wilson & Bisby, *TBMS*, **37**, 1954, 71; syn. *P. phlei-pratensis* Erikss. & Henn., *Z. PflKrankh.* **4**, 1894, 140; Sacc., *Syll.* xi, 204; Grove, *Brit. Rust Fungi*, 283.

*Stem rust of timothy (*Phleum*). The uredo- and teleuto-stages are frequent and destructive in Britain, especially in southern England. Occurs also on cocksfoot (*Dactylis*) and fescues (*Festuca*).

Puccinia holcina Erikss., *Ann. Sci. nat.* **9**, 1899, 274; Sacc., *Syll.* xvii, 379; Grove, *Brit. Rust Fungi*, 263; Wilson & Bisby, *TBMS*, **37**, 1954, 72.

*Holcus rust. Uredo- and teleuto-stages common throughout Britain on *Holcus lanatus* and *H. mollis*.

Puccinia hordei Otth, *Mitt. naturf. Ges. Bern 1870*, 1871, 114; Wilson & Bisby, *TBMS*, **37**, 1954, 72; syn. *P. anomala* Rostr., in Thümen, *Myc. Univ.* no. 831; *Flora*, **61**, 1878, 92; Sacc., *Syll.* vii, 625; *P. simplex* (Wint.) Erikss. & Henn., *Getreideroste*, 1896, 258; Sacc., *Syll.* xvii, 377; Grove, *Brit. Rust Fungi*, 264.

*Brown rust of barley (*Hordeum*). Uredo- and teleutospores common in May and June in England. Also Tweed, Forth and Sutherland areas of Scotland. The aecidial stage on *Ornithogalum* was first found in Britain by Dennis and Sandwith (*Nature, Lond.*, **162**, 1948, 461) on *O. pyrenaicum* in Wilts.

Five physiologic races of this rust have been distinguished in England and Wales (D'Oliviera, *AAB*, **26**, 1939, 56–82). Buchwald (*Ann. mycol., Berl.*, **41**, 1943, 306) pointed out that *Puccinia hordei* Otth is the valid name.

Puccinia hysterium Röhl., *Deutschlands Flora*, iii, 131; Sacc., *Syll.* vii, 668 (as *P. tragopogonis* (Pers.) Corda); Grove, *Brit. Rust Fungi*, 150 (as *P. tragopogi* Corda); Wilson & Bisby, *TBMS*, **37**, 1954, 72.

*Rust of salsify (*Tragopogon*). Said by Grove (loc. cit.) to be not uncommon, but it does not appear to have been reported for many years.

Puccinia iridis Rabenh., *Krypt. Fl.* i, 1844, 23; Sacc., *Syll.* vii, 657; Grove, *Brit. Rust Fungi*, 230; Wilson & Bisby, *TBMS*, **37**, 1954, 72.

*Iris rust. Uredo- and teleuto-stages. Not common on cultivated species, but has been found on *Iris tolmieana* (M.J.B., *Gdnrs'*

Chron. **16**, 1881, 502) and at times on varieties and hybrids of *I. germanica*. It is common on the native *I. pseudacorus* and *I. foetidissima*.

Jørstad and Roll-Hansen (*Nyt Mag. Naturv.* B, **87**, 1949) obtained aecidia on *Urtica dioica* with teleutospores from *Iris sibirica* in Norway.

Puccinia liliacearum Duby, *Botanicum Gallicum*, ii, 1830, 891; Sacc., *Syll.* vii, 668; Grove, *Brit. Rust Fungi*, 234; Wilson & Bisby, *TBMS*, **37**, 1954, 72; Wilson & Henderson, ibid. **37**, 1954, 251.

*Ornithogalum rust. On *Ornithogalum umbellatum*. Rare. Lytham, Lancs (W. G. Smith, *Gdnrs' Chron.* **4**, 1888, 104). The further record from near Carlisle, cited by Grove (loc. cit.) and others, really refers also to the Lytham specimens. Occasional in Scotland: East Lothian, 1935; Roxburgh, 1941; Blairgowrie, 1950. Also on *O. pyrenaicum*.

Puccinia leucanthemi Passer., *Hedwigia*, **13**, 1874, 47; Sacc., *Syll.* vii, 705; Grove, *Brit. Rust Fungi*, 133; Wilson & Bisby, *TBMS*, **37**, 1954, 72.

*Rust of ox-eye daisy (*Chrysanthemum leucanthemum*). Very rare. Lamorna Cove, Corn, Sept. 1906; possibly this species on *C. segetum*, Norfolk, 1949.

Puccinia lolii Niels. = *P. coronata*.

Puccinia lychnidearum Link = *P. arenariae*.

Puccinia malvacearum Mont., in Gay, *Historia fisica y politica de Chile*, viii, 1850–2, 43; Sacc., *Syll.* vii, 686; Grove, *Brit. Rust Fungi*, 206; Wilson & Bisby, *TBMS*, **37**, 1954, 72.

*Hollyhock rust. Common and widespread on *Althaea rosea* throughout Britain. Occurs also on *Malva* spp., including *M. moschata* and *M. sylvestris*, on *Sidalcea* and on *Lavatera arborea*. A slight attack was seen on *Brotex* in south Wales, 1931 (*Report*, vii, 96).

This rust spread rapidly and attracted very considerable attention after it was first observed in the south and east of England (M.J.B., *Gdnrs' Chron.* 33, 1873, 946). Early records are given in *Gdnrs' Chron.* 1, 1874, 127 and 767. W. G. Smith (ibid. 18, 1882, 22) found it was seed transmitted; Plowright (ibid. 18, 1882, 617) described the morphology of the fungus; and others who investigated it were Robinson (*Mem. Manchr lit. phil. Soc.* 57, 1913, 24 pp.; *Ann. Bot., Lond.*, 28, 1914, 331) and Bailey (ibid. 34, 1920, 173–200). Ashworth (*TBMS*, 16, 1931, 177–202) studied the fungus in monosporidial culture.

Puccinia menthae Pers., *Syn. Fung.* 1801, 227; Sacc., *Syll.* vii, 617; Grove, *Brit. Rust Fungi*, 170; Wilson & Bisby, *TBMS*, 37, 1954, 73.

*Mint rust. Common in all districts, both under glass and in the open. *Mentha villoso-nervata* and *M. spicata*, the mints usually grown in Britain, are very susceptible, while *M. cordifolia* is highly resistant. Reported occasionally on peppermint (*M. piperita*), Salop, Norfolk, Kent; on summer savory (*Satureja hortensis*), Herts, 1951.

The hot-water treatment for controlling it was described by Ogilvie and Brian (*Gdnrs' Chron.* 98, 1935, 65). See also *Reps. Long Ashton Res. Sta. for 1932, 1933 and 1935* and Shewell Cooper (*Gdnrs' Chron.* 108, 1940, 140). Treated beds soon become reinfected from nearby untreated plants (Keyworth & Howell, *Rep. Nat. Veg. Res. Sta. Wellesbourne for 1955*, 52).

Puccinia mirabilissima Peck = *Cumminsiella mirabilissima*.

Puccinia opizii Bubák, *Zbl. Bakt.* II, 9, 1902, 925; Sacc., *Syll.* xvii, 371; Wilson & Bisby, *TBMS*, 37, 1954, 73.

*Lettuce rust. Aecidia on *Lactuca sativa*. Not known to occur naturally on wild or cultivated lettuce in Britain until 1950, though previously intercepted occasionally between 1930 and 1946 on lettuces imported from Holland (W. C. Moore, *TBMS*, 29, 1946, 254). Was prevalent in market and private gardens near Norwich in 1950 (Ellis, *Trans. Norfolk Norw. Nat. Soc.* 17, 1951, 137–8), and seen also at Halesworth, Suffolk; Norwich, 1951.

The uredo- and teleuto-stages occur on species of *Carex* and have been found in Wheatfen Broad, Norwich on *C. appropinquata* and *C. paniculata*.

Puccinia pazschkei Diet., *Hedwigia*, 30, 1891, 103; Sacc., *Syll.* xi, 185; Grove, *Brit. Rust Fungi*, 213; Wilson & Bisby, *TBMS*, 37, 1954, 73.

*Saxifrage rust. Teleutospores only. Occasional on a number of cultivated species. On *Saxifraga longifolia*, Kew Gardens (Massee, *J. Bot., Lond.*, 46, 152); on *S. cotyledon* × *aizoon*, Sutton Coldfield, War (Grove, loc. cit); on *S. longifolia*, *S. hostii* and var. *rhaetica*, and *S. aizoon* var. *cultrata*, Royal Botanic Garden, Edinburgh, Sept. 1920 (Wilson, *TBMS*, 7, 1921, 81); on *S. cotyledon* var. *catheramensis* (Macdonald, *Trans. bot. Soc. Edinb.* 32, 1939, 556); on *S. obristii*; on *Saxifraga* ×, Harpenden, Herts, 1934; also in Scotland, Clyde, Tay.

Puccinia perplexans Plowr., *Quart. J. Micr. Soc.* 25, 164; *Hedwigia*, 25, 1886, 38; Sacc., *Syll.* vii, 632; Grove, *Brit. Rust Fungi*, 270; Wilson & Bisby, *TBMS*, 37, 1954, 73.

*Rust of meadow foxtail (*Alopecurus*). Uredo- and teleutospores common in England and Wales on *A. pratensis*; destructive at Aberystwyth, 1921.

Aecidia on *Ranunculus acris*; a single record in Scotland (Dee).

Puccinia phlei-pratensis Erikss. & Henn. = *P. graminis* var. *phlei-pratensis*.

Puccinia phragmitis (Schum.) Körn., *Hedwigia*, 15, 1876, 179; Sacc., *Syll.* vii, 630; Grove, *Brit. Rust Fungi*, 273; Wilson & Bisby, *TBMS*, 37, 1954, 73.

*Rhubarb rust. Aecidial stage on *Rheum rhaponticum*. Rarely reported, though there was an epidemic of it in Devon and Corn (Tamar Valley) in 1920 (*Report*, iv, 95). King's Lynn, Norfolk, 1883; Carns and Kent, 1927; Mon, 1933.

The uredo- and teleuto-stages occur on *Phragmites communis*.

Puccinia poae-nemoralis Otth, *Mitt. naturf. Ges. Bern 1870*, 1871, 113; Wilson & Bisby, *TBMS*, 37, 1954, 73.

*On *Anthoxanthum*. As *Uredo anthoxanthina* Bubák, widespread on *Anthoxanthum odoratum* (Grove, *J. Bot., Lond.*, **59**, 1921, 311–15).

*On *Poa*. Uredosori on *P. annua* and *P. nemoralis*, probably widespread. (See also *Puccinia poarum*.)

Puccinia poarum Niels., *Bot. Tidsskr.* 2, 1877, 26; Sacc., *Syll.* vii, 625; Grove, *Brit. Rust Fungi*, 278; Wilson & Bisby, *TBMS*, 37, 1954, 74.

*Rust of *Poa*. Uredo- and teleuto-stages very common throughout Britain on *P. annua*, *P. nemoralis*, *P. pratensis* and *P. trivialis*. The older records probably included *Puccinia poae-nemoralis* (q.v.).

The aecidial stage occurs on coltsfoot (*Tussilago farfara*).

Puccinia porri Wint., *Pilze*, 200; Sacc., *Syll.* vii, 605; Grove, *Brit. Rust Fungi*, 235; Wilson & Bisby, *TBMS*, 37, 1954, 74.

*Onion and leek rust. Common on leek in eastern Scotland, northern England, Worcs, and south Wales; occasional elsewhere. Localized but often intense and destructive on the seed crop and in dry seasons. Common in the south on chives, but not often seen on onion (Reading, south Wales, Hereford). Also in Scotland on chives (Dee, Orkney) and *Allium vineale*. W. G. Smith (*Gdnrs' Chron.* **22**, 1884, 593) described this rust on leeks.

Another rust, *Caeoma alliorum* Link, has been found once on leek in Clyde. (See also *Uromyces ambiguus*.)

Puccinia primulae Grev., *Flora edensis*, 1824, 432; Sacc., *Syll.* vii, 612; Grove, *Brit. Rust Fungi*, 179; Wilson & Bisby, *TBMS*, 37, 1954, 74.

*Primula rust. Widely distributed in Scotland on the native *Primula vulgaris*. Occurs, but not commonly, on *P. acaulis* in England.

Puccinia pringsheimiana Kleb., *Z. PflKrankh.* **5**, 1895, 76; Sacc., *Syll.* xvii, 468; Grove, *Brit. Rust Fungi*, 242; Wilson & Bisby, *TBMS*, **37**, 1954, 69 (as *P. caricina* DC.).

*Gooseberry cluster cup rust. Aecidia widely distributed in Britain, but varies greatly in amount from year to year. Found most years on the leaves and fruits in Wales and locally abundant elsewhere. Occasionally (1921, 1925, 1931, 1933, 1949) extremely abundant in many districts. A dry March, which delays teleuto-spore germination, seems to be the prelude to epidemics.

The rust is occasionally found on black and red currant and has been seen on *Ribes sanguineum* in Skye (July 1932) and Mull. The teleuto-stage occurs on certain species of *Carex*.

Saunderson and Cairns (*AAB*, **24**, 1937, 17–25) showed that the rust can be controlled by spraying with copper compounds. An early epidemic was described by W. G. Smith (*Gdnrs' Chron.* **16**, 1881, 76–7). (See also Soppitt, ibid. **24**, 1898, 145, and, for occurrence in Scotland, *Scot. J. Agric.* **9**, 1926, 308.)

Puccinia prostii Moug., Duby, *Botanicum Gallicum*, ii, 1830, 891; Sacc., *Syll.* vii, 732; Wilson & Bisby, *TBMS*, **37**, 1954, 74.

*Tulip rust. Rare. On *Tulipa sylvestris* in the Royal Botanic Garden, Edinburgh, from 1914, but it failed to spread to nearby cultivated varieties (Wilson, *Notes R. bot. Gdn Edinb.* **38**, 1914, 219–21); and once on cultivated tulips in England (1913). Morphology and cytology described by Lamb (*Trans. roy. Soc. Edinb.* **58**, 1934, 143–62).

Puccinia pruni-spinosae Pers. = *Tranzschelia pruni-spinosae.*

Puccinia purpurea Cooke, *Grevillea*, **5**, 1876, 15; Sacc., *Syll.* vii, 657.

*On sorghum. Brooks (*Plant Diseases*, 2nd ed., 1953, 279) stated that this species has been recorded on sorghum in England, but neither Wilson and Bisby (*TBMS*, **37**, 1954, 74) nor I have been able to find a published record or specimen of it.

Puccinia ribis DC., *Flor. Fr.* ii, 1805, 221; Sacc., *Syll.* vii, 679; Grove, *Brit. Rust Fungi*, 1913, 212; Wilson & Bisby, *TBMS*, **37**, 1954, 74.

*Red currant leaf rust. Rare. Teleutospores, Dallas, Forres, 16 July 1894 (Plowright, *Gdnrs' Chron.* **16**, 1894, 135; see also *TBMS*, **1**, 1897–8, 57); Forres (10 miles from the previous locality), 1946–7 (Noble & Gray, *Gdnrs' Chron.* **122**, 1947, 92); Inverness-shire, 1955.

Puccinia satyrii Sydow, *Monogr. Ured.* i, 1903, 594; Sacc., *Syll.* xvii, 363; Wilson & Bisby, *TBMS*, **37**, 1954, 74.

*On *Satyrium aureum* imported from Africa, Orchid House, Kew Gardens, 1929.

Puccinia saxifragae Sch., *Flora berolinensis*, ii, 1824, 134; Sacc., *Syll.* vii, 678; Grove, *Brit. Rust Fungi*, 212; Wilson & Bisby, *TBMS*, **37**, 1954, 74.

*On *Saxifraga umbrosa*. Teleutospores only. Rare. Also occurs on several native species of saxifrage.

Puccinia schroteri Passer., *N. Giorn. bot. Ital.* **7**, 1875, 255; Sacc., *Syll.* vii, 732; Grove, *Brit. Rust Fungi*, 232; Wilson & Bisby, *TBMS*, **37**, 1954, 75.

*Narcissus rust. Rare. On common double narcissus and *Narcissus poeticus*, Malpas, Ches (W. G. Smith, *Gdnrs' Chron.* **5**, 1889, 725; **7**, 1890, 558); on jonquil (Grove, loc. cit.); on *N. pseudonarcissus*, Kew Gardens, 1947–9.

Puccinia secalina Grove = *P. dispersa*.

Puccinia sessilis Schneid., in Schröt., *Die Brand- und Rostpilze Schlesien*, 19; Sacc., *Syll.* vii, 624; Wilson & Bisby, *TBMS*, **37**, 1954, 75; syn. *P. digraphidis* Sopp., *Gdnrs' Chron.* **7**, 1890, 643; *J. Bot., Lond.*, **28**, 1890, 213; Sacc., *Syll.* ix, 308; Grove, *Brit. Rust Fungi*, 267.

*Convallaria rust. Aecidia rare on *Convallaria majalis*. Has been called *Aecidium convallariae* Schum. (W. G. Smith, *Gdnrs' Chron.*

22, 1884, 12). The uredo- and teleuto-stages occur on *Phalaris arundinacea*.

Puccinia simplex (Wint.) Erikss. & Henn. = *P. hordei*.

Puccinia soldanellae Fuckel, *Symb. Myc.* iii, *Nachtr.* 1876, 14; Sacc., *Syll.* vii, 618; Grove, *Brit. Rust Fungi*, 180; Cooke, *Fung. Pests Cult. Pl.* 58; Wilson & Bisby, *TBMS*, **37**, 1954, 75.

*Rust of *Soldanella*. Aecidia only. Rare. On *S. alpina* imported to the Botanic Garden, Glasgow, before 1836; Botanic Garden, Cambridge, 1926.

Puccinia sorghi Schw., *Synopsis fungorum in America boreali*, 295; Sacc., *Syll.* vii, 659; Wilson & Bisby, *TBMS*, **37**, 1954, 75.

*Maize rust. Occasional. Norfolk, Cambs, Berks, Yorks, Cards; on sweet corn, Middx, Cards, Glam. The aecidial host is *Oxalis*.

Puccinia tragopogonis Corda = *P. hysterium*.

Puccinia triticina Erikss., *Ann. Sci. nat.* **9**, 1899, 270; Sacc., *Syll.* xvii, 376; Grove, *Brit. Rust Fungi*, 262; Wilson & Bisby, *TBMS*, **37**, 1954, 75.

*Brown rust of wheat (*Triticum*). Uredo- and teleuto-stages common in July and August in England and Wales, but relatively harmless. In Scotland first seen in 1945 (Clyde) and now widespread; sometimes severe in southern areas. The aecidial host is *Thalictrum*, but aecidia have not been found occurring naturally in Britain.

The susceptibility of some varieties is increased by the presence of mildew (Manners & Gandy, *AAB*, **41**, 1954, 393–404). Ten physiologic races have been found in England (Roberts, ibid. **23**, 1936, 271–301). Mehta (*TBMS*, **8**, 1923, 142–76) dealt with its overwintering.

Puccinia vincae Berk., *English Flora*, v, 1836, 364; Sacc., *Syll.* ix, 310; Grove, *Brit. Rust Fungi*, 176; Wilson & Bisby, *TBMS*, **37**, 1954, 76.

PUCCINIA

*Periwinkle rust. Rare. On *Vinca major* (Pim, *Gdnrs' Chron.* 2, 1887, 227), Devon, 1938, 1949; Dundry, Bristol, 1946, on *V. major* but not on *V. acutiloba* growing alongside; East Lothian, 1949; Tay. The mycelium is perennial in the rootstock.

Puccinia violae DC., *Flor. Fr.* vi, 1815, 62; Sacc., *Syll.* vii, 609; Grove, *Brit. Rust Fungi*, 200; Wilson & Bisby, *TBMS*, 37, 1954, 76.
*Violet rust. Not common on cultivated forms. Widely distributed and very common on wild violets.

Pucciniastrum epilobii Otth, *Mitt. naturf. Ges. Bern*, 1861, 72; Sacc., *Syll.* vii, 762; Wilson & Bisby, *TBMS*, 37, 1954, 76.
*On the needles of silver fir (*Abies*). Aecidia on *A. grandis*, south Scotland. The uredo- and teleuto-stages occur on willow herb (*Epilobium* spp.). (See also *Uredo fuchsiae*.)

Pullularia pullulans (de Bary) Berkh., see *Gloeotinia temulenta*.

Pycnostysanus azaleae (Peck) Mason, *Mycol. Pap.* 5, 1941, 130; syn. *Sporocybe azaleae* (Peck) Sacc., *Syll.* iv, 608; Wakefield & Bisby, *TBMS*, 25, 1941, 58.
*Leaf scorch and bud blast of *Rhododendron*. Widely distributed throughout Britain, and often damaging to many species. Bud blast is the important and common symptom.

The parasitism of this fungus and the etiology of the disease are still in doubt. Bud blast is often associated with infestation by the leaf hopper, *Graphocephala coccinea* Forst. (Bailey & Jepson, *J. R. hort. Soc.* 76, 1951, 355–65), which can be controlled by spraying with DDT. For notes on the disease see *J. R. hort. Soc.* 75, 1950, 230–2; *Gdnrs' Chron.* 127, 1950, 66, 98 and 238; and Street, *Rhododendron Yearb.* 5, 1950, 72–7.

Pycnothyrium gentianicola Grove, *J. Bot., Lond.*, 60, 1922, 143; *Coelomycetes*, ii, 197; syn. *Leptothyrium gentianaecolum* Bäuml., *Myk. Not.* 1; Sacc., *Syll.* x, 415.
*Leaf blight of *Gentiana*. On *G. macaulayi*, Forth, Clyde; on *G. acaulis*, Clyde; on *G. sino-ornata*, Corn, 1948.

313

Pyrenochaeta ferox (de Not.) Sacc., *Syll.* iii, 220; Grove, *Coelomycetes*, i, 153.

*On potato stems, Bridgwater, Som.

Pyrenochaeta phlogis Massee, *Kew Bull.* 1907, 241 (as *P. phloxidis*); Sacc., *Syll.* xxii, 932; Grove, *Coelomycetes*, i, 152.

*On stems of *Phlox*, Kew Gardens.

Pyrenophora avenae Ito & Kuribay., *Proc. imp. Acad. Japan*, 6, 1930, 352; Bisby & Mason, *TBMS*, 24, 1940, 192; stat. conid. *Helminthosporium avenae* Eidam, *Der Landwirt*, 27, 1891, 509; Wakefield & Bisby, *TBMS*, 25, 1941, 91.

*Leaf spot and seedling blight of oats (*Avena*). Present everywhere but less so than formerly. Despite the availability of effective seed treatment it is still serious in the colder and wetter districts of the north and west, especially Wales, northern England and Scotland. Seedling blight is still the worst disease of oats in Scotland but is becoming rarer in eastern England. The perfect stage of the fungus has been seen once only, in Ayrshire (Dennis, *TBMS*, 19, 1935, 288–90).

Considerable research has been carried out in Scotland on this disease. O'Brien, first with Prentice (*Scot. J. Agric.* 13, 1930, 272–84) and then with Dennis (ibid. 15, 1932, 39–45 and 406–10; *Trans. Highl. Agric. soc. Scot.* 46, 1934, 91–112), proved the value of Ceresan and other mercury dusts as effective seed dressings, and it was O'Brien's persuasion and enthusiasm that began the rapid swing towards the general use in Scotland, and later elsewhere, of organo-mercury products as cereal-seed disinfectants (see also *Tilletia caries*). Illuminating figures of seed disinfection in Scotland during that period were given by Dennis (*AAB*, 31, 1944, 370–4), who himself at the time was also studying the morphology and biology of *Pyrenophora avenae* (*TBMS*, 18, 1933, 223–8). A comprehensive account of the disease and its control was issued in 1933 (*Res. Bull. W. Scot. Coll. Agric. Dep. Pl. Husb.* no. 3, 74 pp.). The necessity for seed treatment annually has been stressed by Boyd (*Scot. J. Agric.* 24, 1943, 174–6).

Millard and Turner (*AAB*, **18**, 1931, 538–58) published an account of the disease in England. Dillon-Weston (*TBMS*, **20**, 1936, 112–15) described a method of making *Helminthosporium avenae* sporulate in culture, and Grainger (ibid. **37**, 1954, 412–19) also studied spore production and (*Phytopathology*, **46**, 1956, 445–56) the relation between host nutrition and attack. Dillon-Weston and Taylor (*J. agric. Sci.* **33**, 1943, 23–7) dealt with the control of the disease by seed treatment.

In Northern Ireland, Muskett (*Ann. Bot., Lond.*, n.s., **1**, 1937, 763–83) studied the epidemiology and control of the disease, and devised methods of evaluating seed disinfectants for use against it (ibid. **2**, 1938, 699–715). Later he became attracted by certain advantages of the short wet method of seed disinfection (*AAB*, **31**, 1944, 218–21). In Eire, McKay (*J. Dep. Agric. Dublin*, **32**, 1933, 234–56) and Carroll (ibid. **37**, 1940, 79–92) also carried out seed-disinfection trials.

Pyrenophora graminea Ito & Kuribay., *Proc. imp. Acad. Japan*, **6**, 1930, 352; Bisby & Mason, *TBMS*, **24**, 1940, 192; stat. conid. *Helminthosporium gramineum* Rabenh., *Herb. Myc.* 332; *Grevillea*, **17**, 1889, 67; Sacc., *Syll.* x, 615; Wakefield & Bisby, *TBMS*, **25**, 1941, 91.

The perfect stage has not yet been found in Britain.

*Leaf stripe of barley (*Hordeum*). Occurs throughout Britain and formerly caused much damage in the north and west. Following routine seed treatment it declined and is now uncommon. Until 1953 it had not been seen for many years in Scotland and even then was reintroduced with imported seed.

The fungus was first recorded in Britain about 1898 (*TBMS*, **1**, 1899–1900, 151; *J. R. agric. Soc.* **9**, 1898, 751). The disease caused by it was investigated by N. J. G. Smith (*AAB*, **16**, 1929, 236–60), and Dillon-Weston and Taylor (*J. agric. Sci.* **33**, 1943, 23–7) carried out seed-disinfection trials.

Pyrenophora lolii Dovaston, see *Helminthosporium siccans*.

Pyrenophora teres Drechsl., *J. agric. Res.* **24**, 1923, 656; Bisby & Mason, *TBMS*, **24**, 1940, 192; stat. conid. *Helminthosporium teres* Sacc., *Mich.* ii, 558; *Syll.* iv, 412; Wakefield & Bisby, *TBMS*, **25**, 1941, 92.

*Net blotch of barley (*Hordeum*). Widely distributed but unimportant. A method of measuring its intensity was described by Bescoby (*TBMS*, **18**, 1933, 180–2).

Webster (*TBMS*, **34**, 1951, 309–17) collected the perfect stage on barley stubble in Yorks, Notts and Norfolk in 1949–50, and followed Wehmeyer (*Mycologia*, **41**, 1949, 565–93) in regarding it as a form of *Pleospora trichostoma* (Fr.) Ces. & de Not., which is a 'species complex' comprising various groupings previously accorded specific rank, including *Pyrenophora teres*, the name retained here until the position has been further clarified.

Pythium aphanidermatum (Edson) Fitzpatr., *Mycologia*, **15**, 1923, 166.

*Stem rot of cucumber (*Cucumis*). Honeybourne, Worcs, 1942 (Hickman, *TBMS*, **28**, 1944, 63–7).

Stem rot and damping-off of cucumber due to *Pythium* sp. has been reported in England from time to time since 1900 (*Gdnrs' Chron.* **27**, 1900, 274). A fruit rot may be caused by *P. ultimum* (q.v.).

Pythium arrhenomanes Drechsl., *Phytopathology*, **18**, 1928, 874.

*Root rot of wheat (*Triticum*). Lincs, Herts, Kent, Sussex.

Most of the information about species of *Pythium* parasitic on cereals in England is given by Vanterpool (*AAB*, **25**, 1938, 528–43), who showed that root rot due to one or another species of *Pythium* occurred in Kent, Bucks, Herts, Cambs, Lincs and Yorks. In addition to *P. arrhenomanes* he found *P. graminicolum* Subraman. (*Bull. agric. Res. Inst. Pusa*, no. 177, 1928, 1) near Cambridge; *P. volutum* Vanterp. & Truscott (*Canad. J. Res.* **6**, 1932, 68) at Ramsgate, Kent and Slough, Bucks; *P. tardicrescens* Vanterp. (*AAB*, **25**, 1938, 533) at Slough; and *P. torulosum* Coker

& Patterson (*J. Elisha Mitchell sci. Soc.* **42**, 1927, 247) at Rothamsted, Herts and Reading, Berks.

Pythium de baryanum Hesse, *Pythium de Baryanum*, Halle, 1874, 34; Schröt., *Kr. Fl. Schles.* 232; Sacc., *Syll.* vii, 271.

For many years the species of *Pythium* commonly found associated with damping-off, stem rot and root injury of many plants was assigned without hesitation to this species. Drechsler (*Phytopathology*, **17**, 1927, 54) was the first to point out that more often than not it is *P. ultimum* (q.v.). Gupta (*Ann. Bot., Lond.*, **20**, 1956, 179–90) described the production of pectolytic enzymes by *P. de baryanum*.

*Damping-off of Japanese rape. Surrey, 1945.

*Root rot of pea (*Pisum*). Recorded by Carruthers (*J. R. agric. Soc.* **10**, 1899, 685) in Yorks; Cumb, 1938. (See also *P. ultimum.*)

*On *Gloxinia* (*Sinningia*), associated with a corm rot probably due primarily to non-parasitic causes (W. C. Moore & Tomlinson, *Plant Path.* **2**, 1953, 71).

*In *Impatiens sultani* (Pim, *Gdnrs' Chron.* **3**, 1888, 267).

Pythium graminicolum Subraman., see *P. arrhenomanes.*

Pythium intermedium de Bary, *Bot. Ztg*, **39**, 1881, 554; Sacc., *Syll.* xi, 244.

*In potato tubers, associated with symptoms resembling those of watery wound rot, Ches, Salop (1952). (See *P. ultimum.*)

*Black leg of sugar beet and mangold (*Beta*). (See *Pythium* sp.)

*Root rot of *Narcissus.* Clyde. *Pythium* sp. has been reported on this host from Staffs and Yorks.

Pythium mammilatum Meurs, Wortelrot, Thesis, Univ. Utrecht, Baarn, 1928, 44.

Barton (*Nature, Lond.*, **180**, 1957, 613–14) found that the germination of the oospores of this species was stimulated by exudates from living turnip seedlings.

*Damping-off of mustard seedlings (Duerden, *TBMS*, **26**, 1943, 15). First British record. *Pythium* sp. has been recorded from Lancs.

*Root rot of *Viola.* Described by Hickman (*TBMS*, **27**, 1944, 49–51). Only weakly parasitic.

Pythium megalacanthum de Bary, in Schröt. *Kr. Fl. Schles.* 232; Sacc., *Syll.* vii, 272.

*In watercress (*Nasturtium*), associated with stem rot, Sussex, 1937.
*Root rot of *Chrysanthemum*. Sometimes severe (e.g. 1947–8) in eastern Scotland, Forth, Tay. *Pythium* sp. has been recorded a number of times causing a foot rot of chrysanthemum cuttings in England.

Pythium oligandrum Drechsl., *J. Wash. Acad. Sci.* **20**, 1930, 398.
*Root rot of *Viola*. Frequent. (See *P. violae*.)

Pythium spinosum Sawada, in Sawada & Chen, *J. nat. Hist. Soc. Formosa*, **16**, 1926, 199.
*On *Primula*. Associated, perhaps secondarily, with a wilt of *P. sinensis*, Surrey, 1934 (Cook & Collins, *TBMS*, **21**, 1937, 29–33).
Pythium sp. has also been recorded associated with rotting of *Primula obconica*, Yorks, Herts (Williams, *Rep. exp. Res. Sta. Cheshunt for 1946*, 22), Hants; and on polyanthus, Perth, 1948.

Pythium tardicrescens Vanterp., see *P. arrhenomanes.*

Pythium torulosum Coker & Patterson, see *P. arrhenomanes.*

Pythium ultimum Trow, *Ann. Bot., Lond.*, **15**, 1901, 300; Sacc., *Syll.* xvii, 518.
*Watery wound rot of potato. Found in small amounts every year and apt to cause appreciable loss when the crop is lifted early in hot weather. Widely distributed throughout Britain. Described by Pethybridge and Smith (*J. Minist. Agric.* **37**, 1930, 335–40). *Pythium intermedium* occasionally causes similar symptoms.
*Black leg of sugar beet and mangold (*Beta*). (See *Pythium* sp.)
*Pre-emergence damping-off of pea (*Pisum*). A number of fungi, mainly species of *Pythium* (including *P. ultimum* and *P. de*

baryanum) but occasionally *Fusarium* spp. and possibly other fungi, are concerned in the so-called pre-emergence damping-off of peas (Baylis, *AAB*, **28**, 1941, 210–18), which is troublesome in early-sown varieties, especially if the soil is cold and wet soon after sowing. Hull (ibid. **24**, 1937, 681–9) studied the effect of soil moisture on this trouble, which can largely be prevented by seed treatment with organo-mercury dusts (Brett, Dillon-Weston & Booer, *J. agric. Sci.* **27**, 1937, 53–66) or cuprous oxide (Ogilvie, Croxall & Hickman, *Rep. Long Ashton Res. Sta. for 1939*, 88–99). These and other compounds were also used by Padwick (*AAB*, **25**, 1938, 100–14), Baylis *et al.* (ibid. **30**, 1943, 19–26), Croxall and Ogilvie (*Rep. Long Ashton Res. Sta. for 1942*, 65–76), and by Wain and Wilkinson (ibid. 59–64), who also studied the mode of action of copper compounds on pea seed (*AAB*, **32**, 1945, 240–7). Jacks (ibid. **38**, 1951, 135–68; see also *Prog. Rep. Field Expts 1945–9, Home Grown Threshed Peas Joint Cttee, 1950*, 9–12) used a wide range of materials and found thiram superior to most other chemicals.

P. ultimum has been recorded causing a distinct foot rot of peas in Wilts.

*Fruit rot of cucumber (*Cucumis*). Cambridge, 1943 (Brooks, *TBMS*, **27**, 1945, 134–6), Kent, Sussex, Essex. *Pythium* sp. has also been found causing a stem rot, Yorks, Ches, Essex, Devon, Glam, Flints, Caerns. (See also *P. aphanidermatum*.)

*Damping-off of tomato seedlings. Occasional, along with other species of *Pythium*. Can be controlled by the use of formalin dust (Montgomery, *J. hort. Sci.* **29**, 1954, 245–57).

*Corm rot of *Colchicum*. On *C. speciosum album* and *C. byzantinum*, Taplow, Bucks, 1940, but occurring there annually (W. C. Moore, *TBMS*, **24**, 1940, 349–50).

*Root rot of *Crocus*. Leeds, 1950 (Tomlinson, *Plant Path.* **1**, 1952, 50).

*Root rot of tulip. Not uncommon. Occurs both under glass (W. C. Moore & Buddin, *AAB*, **24**, 1937, 752–61) and in the open (W. C. Moore, *TBMS*, **24**, 1940, 61–3), leading to dwarfing and lack of vigour. Is one of the causes of chalking (chalkiness) in stored tulips. (See also *Pythium* sp.)

*Damping-off and foot rot of conifer seedlings. Common in forest nurseries. With other species of *Pythium* causing damping-off of Sitka spruce, Ampthill, Beds (Warcup, *Rep. For. Res., Lond. for year ending March 1952*, 108). Warcup (*TBMS*, **35**, 1952, 248–62) demonstrated that partial soil sterilization has a temporary alleviating effect on the disease in Sitka spruce.

Pythium undulatum Peterson, *Ann. mycol., Berl.*, **8**, 1910, 531; *Bot. Tidsskr.* **29**, 1909, 345; Sacc., *Syll.* xxi, 854.

*In leaves of water-lily (*Nymphaea*). Isolated from water-lilies in a pond, Royal Holloway College, Surrey, by Goldie Smith (*TBMS*, **33**, 1950, 92), who also described its morphology and life history (*J. Elisha Mitchell sci. Soc.* **68**, 1952, 273–92) and showed it was at most mildly pathogenic. Another *Pythium* sp. was parasitic.

Pythium violae Chesters & Hickman, *TBMS*, **27**, 1944, 60.

*Root and stem rot of *Viola*. Widely distributed in Britain. The disease is a complex caused by species of *Pythium*, *Corticium solani* and *Myrothecium roridum*, either alone or in combination, and was described by Chesters and Hickman (Rep. from *Nat. Viola and Pansy Soc. Yearb. 1938*, 8 pp.). Two of the species of *Pythium* concerned, *P. violae* and *P. oligandrum*, are described in *TBMS*, **27**, 1944, 55–62.

Pythium volutum Vanterp. & Truscott, see *P. arrhenomanes*.

Pythium sp. A common cause of damping-off and root rot in many plants. The following refer to damping-off except where stated.

Wright (*Plant and Soil*, **8**, 1956, 132–40) showed that dusting white mustard seed with spores of *Trichoderma viride* and other soil saprophytes partially controlled damping-off caused by a species of *Pythium* which was pathogenic to the seedlings of a wide range of hosts.

*Browning root rot of oats (*Avena*). Kent, Sussex, Ayrshire.

*Pre-emergence damping-off of swede seedlings, Belfast (Greeves & Muskett, *AAB*, **23**, 1936, 264–70).

*Black leg of mangold and sugar beet (*Beta*). Common on mangold in south-west England, occasional in the east (Gates & Hull, *AAB*, **41**, 1954, 541–61). *Pythium ultimum* and *P. intermedium* are the species most frequently implicated.

Frequent as the cause of black wood vessel disease of mangold in southern districts; occasional in red beet (Lincs); may cause a distinct black rot of mangold roots (Totnes, Devon, 1949).

*Root rot of *Vicia faba* (Yorks, Surrey) and dwarf bean (*Phaseolus*), Ches, 1950.

*Clover (*Trifolium*). War, 1947.

*Ryegrass (*Lolium*). Devon, 1944.

*In basal parts of *Asparagus* (Herefs, 1936) and on *A. plumosus*, Clyde.

*Brassicas, notably cauliflower. Occasional.

*Carrot (*Daucus*), and also associated on this host with a disease of uncertain origin called 'ringy' in Notts and 'watermark' or 'pit' in Lincs (*Report*, viii, 44).

*Damping-off and rusty root of celery (*Apium*). Occasional. At Bristol, 1952, the species was thought to be *Pythium artotrogus*. (See also *Centrospora acerina*.)

*Cress (*Lepidium*). Very common.

*Lettuce (*Lactuca*). Notably in northern England.

*Onion (*Allium*). Frequent.

*Vegetable marrow (*Cucurbita*). Sussex, 1951.

*Watercress (*Nasturtium*). Stem and root rot, Kent.

*Associated with wilting in hop (*Humulus*). Worcs, 1939.

Alyssum. Glam, 1955.

Antirrhinum. Occasional.

Begonia. Forth.

*Carnation (*Dianthus*). Foot rot of cuttings, Som, 1938; Hants, 1956; Corn, 1956.

*In *Cattleya* plant affected with black rot, along with *Phytophthora* sp., London, 1950.

Celosia argentea var. *cristata*. Root rot, Torquay, 1949.

*Wilt of cineraria (*Senecio*).

*Root rot of *Delphinium* (Clyde, Perth) and annual larkspur (Suffolk, War).

Gladiolus. Glam, 1949.

*Root rot of hyacinth. Occasional.

*Root rot of *Hydrangea*. Yorks, 1950.

*Root rot of *Iris*. Staffs, 1933; Yorks, 1952.

*Damping-off of *Kochia*. Lancs, 1956.

Lilium pardalinum, Salop, 1940; *L. martagon* and *L. tigrinum*, Angus, 1951; *Lilium* sp., Forth, Devon.

Lobelia. Common.

Nemesia. Aberdeen, 1953.

*Black leg of *Pelargonium*. Frequent. Described by Buddin and Wakefield (*Gdnrs' Chron.* **75**, 1924, 25) as due to *Pythium de baryanum*.

*Root rot of *Pteris wimsetii*. Dundee, 1952.

Salvia. Worcs, 1953–4, and occasionally elsewhere.

*Scabious. Cambs, 1937.

*Root rot of *Schizanthus*. Clyde.

*Sweet pea (*Lathyrus*). Salop, Glam, Scotland.

Tropaeolum. East Scotland, 1946.

*Tulip. A species of *Pythium*, closely related to, but not identical with, *P. de baryanum* var. *pelargonii* Braun, is sometimes associated with *P. ultimum* (q.v.) in root rot of tulips.

Ramularia agrestis Sacc., *Mich.* ii, 550; Sacc., *Syll.* iv, 202; Wakefield & Bisby, *TBMS*, **25**, 1941, 94.

*Leaf spot of viola and pansy. Not uncommon and at times troublesome.

Ramularia armoraciae Fuckel, *Symb. myc.* 361; Sacc., *Syll.* iv, 201; Wakefield & Bisby, *TBMS*, **25**, 1941, 94.

*Pale spot of horse-radish (*Armoracia*). Fairly common in England and Wales; also Clyde, Kincardine, Sutherland.

Ramularia bellunensis Speg., *Mich.* i, 475; Sacc., *Syll.* iv, 210.

*On Marguerite (*Chrysanthemum frutescens*). Rare. A single record (*Rep. exp. Res. Sta. Cheshunt for 1937*, 50).

Ramularia beticola Fautr. & Lamb., *Rev. mycol.* **19**, 1897, 54; Sacc., *Syll.* xiv, 1064; Wakefield & Bisby, *TBMS*, **25**, 1941, 94.

*Ramularia leaf spot of beet (*Beta*). Widely distributed and occasional on sugar beet, mangold, beetroot and spinach beet since 1914 (on spinach beet, Winchester, Hants), and in some seasons prevalent though of no economic importance.

Ramularia campanulae-latifoliae Allesch., *Hedwigia*, **34**, 1895, 283; Sacc., *Syll.* xiv, 1063; Wakefield & Bisby, *TBMS*, **25**, 1941, 94.

*Leaf spot of *Campanula*. On *C. latifolia*, Clyde, 1913 (Smith & Ramsbottom, *TBMS*, **5**, 1916, 242).

Ramularia deflectens Bresad., *Hedwigia*, **35**, 1896, 200; Sacc., *Syll.* xiv, 1059.

*On leaves of *Viola*, spreading in from the margins, Coventry, 1934 (Campbell, *AAB*, **25**, 1938, 115–21); on leaves of cultivated *Viola* hybrids, Scotland.

Ramularia deusta (Fuckel) Baker, Snyder & Davis, *Mycologia*, **42**, 1950, 416; syn. *Hyalodendron album* (Dowson) Diddens, *Zbl. Bakt.* II, **90**, 1934, 316; Wakefield & Bisby, *TBMS*, **25**, 1941, 73; *Cladosporium album* Dowson, *J. R. hort. Soc.* **49**, 1924, 221.

Erostrotheca multiformis Martin & Charles, *Phytopathology*, **18**, 1928, 844, was described as the perfect stage but is not now accepted as such (see Good, *Canad. J. Res.* C, **25**, 1947, 137–54 and Baker, Snyder & Davis (loc. cit.)).

*White mould of sweet pea (*Lathyrus*). Common in the south and west of England both under glass and in the open; occasional in the north and Scotland.

Ramularia hellebori Fuckel, *Symb. myc.* 361; Sacc., *Syll.* iv, 200; Wakefield & Bisby, *TBMS*, **25**, 1941, 95.

*Leaf spot of hellebore. Uncommon. Cooke (*J. R. hort. Soc.* **27**, 1902, 11).

Ramularia heteronema (Berk. & Br.) Wollenw., see *Cylindrocarpon album.*

Ramularia knautiae (Massal.) Bubák, *Öst. bot. Z.* **53**, 1903, 50; Sacc., *Syll.* x, 559 (as *R. succisae* Sacc. var. *knautiae* Massal.); Wakefield & Bisby, *TBMS*, **25**, 1941, 95.

*On scabious. A species of *Ramularia*, probably this, caused shot-hole lesions on *Scabiosa caucasica* in Yorks, 1935 (*Report*, viii, 89).

Ramularia lactea (Desm.) Sacc., *Mich.* ii, 549; Sacc., *Syll.* iv, 201; Wakefield & Bisby, *TBMS*, **25**, 1941, 95.

*White spot of viola and pansy. Occasional.
*White spot of violet. Uncommon. Kew, Herefs, Essex, 1887 (Cooke, *Grevillea*, **16**, 1887, 65); Herts, 1900; Devon, 1926; Surrey, 1929.

Ramularia macrospora Fresen., *Beitr.* 88; Sacc., *Syll.* iv, 211; Wakefield & Bisby, *TBMS*, **25**, 1941, 95.

*Leaf spot of campanula. Fairly common. Devon, Sussex, Kent, Herts, Forth, Clyde, Dee. Described by Chittenden (*J. R. hort. Soc.* **37**, 1911–12, 543–5).

Ramularia narcissi Chittenden = *R. vallisumbrosae.*

Ramularia nymphaearum (Allesch.) Ramsb. = *Ovularia nymphaearum.*

Ramularia onobrychidis Allesch., *Verzeichniss in Südbayern beobachteter Pilze*, iii, 1891, 104; Sacc., *Syll.* xi, 604.

*Leaf spot of sainfoin (*Onobrychis*). Vale of Glam, 1943 onwards. Described by Hughes (*TBMS*, **32**, 1949, 34–59).

Ramularia pastinacae Bubák, *S.B. böhm. Ges. Wiss.* 1903, 19; Sacc., *Syll.* xviii, 550; Wakefield & Bisby, *TBMS*, **25**, 1941, 95.

*Leaf spot of parsnip (*Pastinaca*). Few records but probably fairly common. First reported 1918 (Cotton, *Kew Bull.* 1918, 18). (See *Cercosporella pastinacae.*)

Ramularia petuniae Cooke, *Grevillea*, **20**, 1891, 8; Sacc., *Syll.* x, 561; Wakefield & Bisby, *TBMS*, **25**, 1941, 95.

*Leaf spot of *Petunia*. Very rare. Plymouth, Devon, 1891 (Cooke, *Gdnrs' Chron.* **10**, 1891, 114; *Fung. Pests Cult. Pl.* 61) and again in 1937 but apparently not in the same place (*Report*, viii, 87).

Ramularia primulae Thüm., *Öst. bot. Z.* **28**, 1878, 147; Sacc., *Syll.* iv, 214; Wakefield & Bisby, *TBMS*, **25**, 1941, 95.

*Leaf spot of *Primula*. Common everywhere in Britain, especially on polyanthus varieties, but rarely harmful.

Described as new to Britain by Massee (*Gdnrs' Chron.* **10**, 1891, 626), but Grove (*J. Bot., Lond.*, **50**, 1912, 13 and 154) considered *Ovularia primulana* Karst. (*Hedwigia*, **23**, 1884, 7; Sacc., *Syll.* iv, 143) to be a young state of this, as well as *O. interstitialis* Berk. & Br. on specimens from Scotland in 1875. (See *Cercosporella primulae*.)

Ramularia rhei Allesch., *Hedwigia*, **35**, 1896, 34; Sacc., *Syll.* xiv, 1063; Wakefield & Bisby, *TBMS*, **25**, 1941, 96.

*On rhubarb (*Rheum*), Middx, Essex, Glam, Argyll, Moray.

Ramularia tulasnei Sacc. = stat. conid. of *Mycosphaerella fragariae*.

Ramularia vallisumbrosae Cav., *Rev. mycol.* **21**, 1899, 101; Sacc. *Syll.* xvi, 1046; Wakefield & Bisby, *TBMS*, **25**, 1941, 96; syn. *R. narcissi* Chittend., *Gdnrs' Chron.* **39**, 1906, 277; Sacc., *Syll.* xxii, 1328.

Cercosporella narcissi Boud., *Bull. Soc. bot. Fr.* **48**, 1901, 110; Sacc., *Syll.* xviii, 563, and perhaps *Phyllosticta narcissicola* von Keissl., *Beih. bot. Zbl.* **34**, 1916, 97; Sacc., *Syll.* xxv, 20, are stages of this species.

*White mould of *Narcissus*. Widely distributed but rarely epidemic except in the Isles of Scilly and south-west England.

The disease was first reported in 1906 (Chittenden, loc. cit.), and the life history of the fungus was thoroughly investigated by Gregory (*TBMS*, **23**, 1939, 24–54), who also devised methods of controlling it (*J. Minist. Agric.* **43**, 1936, 865–9), and discussed its effect on the flower and bulb crop (*AAB*, **27**, 1940, 338, 472). W. C. Moore (*Diseases of Bulbs*, 1939, 61) included a full description of it. Beaumont (*Daffodil & Tulip Yearb.* **16**, 1950, 75–83) discussed varietal susceptibility to, and the effect of weather on, this and other foliage diseases of *Narcissus*.

Ramularia sp.

*On leaves of cyclamen, Som, 1937. (See *Cercosporella* sp.)

Ramulaspera holci-lanati (Cav.) Lindau, in Rabenh., *Krypt. Fl.* i, 8, 1907, 260; Griffiths, *TBMS*, **40**, 1957, 232–6.

*On *Holcus lanatus*, Aberystwyth, 1954–5, on eye-spot lesions. First British record.

Rehmiellopsis bohemica Bubák & Kabát, *Naturw. Z. Forst.- u. Landw.* **8**, 1910, 313; Sacc., *Syll.* xxii, 148; Bisby & Mason, *TBMS*, **24**, 1940, 171.

*Die back of silver fir (*Abies*). Observed in Mid- and East Lothian in 1910 and said by Wilson and Macdonald (*Trans. R. Scot. arb. Soc.* **38**, 1924, 114–18) to be spreading rapidly on *Abies alba* in south and west Scotland. Also attacks *A. nobilis*, *A. pinsapo* and *A. cephalonica*. Damaging to young trees. Often associated with attacks of *Chermes*.

Rhabdocline pseudotsugae Syd., *Ann. mycol., Berl.*, **20**, 1922, 194; Ramsbottom & Browne, *TBMS*, **34**, 1951, 107.

*Needle cast of Douglas fir (*Pseudotsuga*). Widely distributed in Britain, but of decreasing significance.

First seen in south Scotland (Dawyck) in 1921 causing serious defoliation of *Pseudotsuga taxifolia* var. *glauca* and var. *caesia* (Wilson & Wilson, *Trans. R. Scot. arb. Soc.* **40**, 1926, 37–40; *Gdnrs' Chron.* **81**, 1927, 323–4). Probably introduced from Dawyck to Exbury, Hants and Sevenoaks, Kent (1925); also occurs in Surrey, Oxon, Dorset, Berks. Its distribution was dealt with by Day (*Quart. J. For.* **21**, 1927, 193–9).

Peace (*Forestry*, **22**, 1948, 45) distinguished three main types of Douglas fir in its native habitat: Coastal, Colorado (Rocky Mountains south from British Columbia) and Intermediate (interior of British Columbia). In Britain the Coastal type is seldom seriously affected and often entirely free; damaging attacks have been recorded on the Intermediate type; on the Colorado type the disease is invariably serious, though individuals may escape.

Brown (*AAB*, **17**, 1930, 745–54) studied the abscission mechanism of infected leaves.

Rhabdospora ramealis (Rob. ex Desm.) Sacc., *Syll.* iii, 580; Grove, *Coelomycetes*, i, 443; syn. *Septocyta ramealis* (Rob. ex Desm.) Petrak, *Ann. mycol., Berl.*, **25**, 1927, 328.

*Purple blotch of blackberry (*Rubus*). Occasionally damaging in southern counties and widely distributed there. Also Yorks and Scotland. Recorded by Cooke (*Grevillea*, **14**, 1885, 34). Found on loganberry, Worcs, 1955.

Rhizina inflata (Schäff.) Quél., *Enchiridon fungorum*, 1886, 272; Sacc., *Syll.* viii, 57; Ramsbottom & Browne, *TBMS*, **34**, 1951, 51; syn. *R. undulata* Fr., *Observations mycologicae*, i, 1815, 161.

*Group dying of Sitka spruce (*Picea*). Known on *P. sitchensis* in Great Britain since 1936. Local, mainly in western districts and on 20–30-year-old trees. Occurs also in *P. abies*, *Larix decidua*, *L. leptolepis* and occasionally in *Pinus sylvestris*, *P. contorta* var. *latifolia* and *P. nigra* var. *poiretiana*.

Rhizina inflata was first found associated with group dying of spruce in Co. Wicklow, Ireland, in 1953 (McKay & Clear, *Irish For.* **10**, 1953, 58–9) and subsequently in Britain (see, for example, Murray, *Forestry*, **27**, 1954, 54–62), but the fungus had been reported long before by Brooks (*Quart. J. For.* **4**, 1910, 308–9 on Scots pine, Corsican pine and larch. It is thought that the fungus colonizes sites of fires and attacks the roots from these sites (Murray, *F.A.O. Pl. Prot. Bull.* **4**, 1955, 6).

Rhizina undulata Fr. = *R. inflata*.

Rhizoctonia crocorum Fr. = stat. mycel. of *Helicobasidium purpureum*.

Rhizoctonia monteithianum Bennett = stat. mycel. of *Sclerotinia homoeocarpa*.

Rhizoctonia solani Kühn = stat. mycel. of *Corticium solani*.

Rhizoctonia tuliparum (Kleb.) Whetz. & Arthur = *Sclerotium tuliparum*.

Rhizoctonia violacea Tul., see *Helicobasidium purpureum*.

Rhizoctonia sp.
*On wheat (*Triticum*), causing a root rot, Lancs, 1946.
**Anemone*. Root rot, Lanarkshire, 1953.
**Camellia*. Root rot, Renfrewshire, 1952.
**Coprosma*. Root and collar rot, Camborne, Corn, Oct. 1956.
*On *Cypripedium*, causing a root rot, Dorset, 1950.
*Damping-off of gentian. On *Gentiana sino-ornata*, Dorset, 1950.
*Poinsettia (*Euphorbia*). Causing death of many cuttings, south-east England, 1956.
*Damping-off of *Solanum capsicastrum*. Worcs, 1954.

Rhizopus necans Massee, *Kew Bull.* 1897, 89; Sacc., *Syll.* xiv, 435.
*Storage rot of lily bulbs. Caused rotting in bulbs of *Lilium speciosum* and *L. auratum* imported from Japan in 1896. See Massee (*J. R. hort. Soc.* **26**, 1901, 372–6), Grove (*Gdnrs' Chron.* **81**, 1927, 178–9, 197–8) and W. C. Moore (*Diseases of Bulbs*, 1939, 47–9).

Rhizopus nigricans Ehrenb. = *R. stolonifera*.

Rhizopus stolonifera (Fr.) Lind, *Danish Fungi*, 1913, 72; syn., *R. nigricans* Ehrenb., *Nova Acta Acad. Leop.* **10**, 1818, 198; Sacc., *Syll.* vii, 212.
*Mouldy fruit rot of tomato (*Lycopersicon*). Occasional, notably in the Lea Valley, Herts. (See *Report*, iii, 50.)

*Rotting apple fruits in storage. Occasional. Kidd and Beaumont (*TBMS*, **10**, 1924, 100), Herts, Worcs.
*Rotting peach fruits. Jersey, 1953.

Rhizopus sp.

*On strawberry (*Fragaria*). Not infrequent as a cause of fruit rot in wet seasons.

Rhizosphaera kalkhoffii Bubák, *Ber. dtsch. bot. Ges.* **32**, 1914, 190; Grove, *Coelomycetes*, i, 141.

*Needle cast of spruce (*Picea*). Frequent in Britain, but parasitism doubtful and probably secondary. Occurs mainly on trees of *P. pungens* and its varieties *argentea* and *glauca*, and also very common on *P. sitchensis* and *P. abies*. It has also been recorded on *Abies nobilis*, *A. alba*, *Pseudotsuga taxifolia*, *Pinus nigra austriaca* and *P. mugo* (Wilson & Waldie, *Trans. R. Scot. arb. Soc.* **40**, 1926, 34–6).

Rhynchosporium orthosporum Caldw., *J. agric. Res.* **55**, 1937, 184.
*Leaf blotch of cocksfoot (*Dactylis*). Frequent in Reading area of Berks; Norfolk (Owen, *Plant Path.* **1**, 1952, 122).

Rhynchosporium secalis (Oudem.) J. J. Davis, *Trans. Wis. Acad. Sci.* **20**, 1921, 420; Sacc., *Syll.* xiv, 1022; Grove, *Coelomycetes*, ii, 281; Wakefield & Bisby, *TBMS*, **25**, 1941, 86.
*Leaf blotch of barley (*Hordeum*) and rye (*Secale*). In England recognized in 1919 on rye in Devon and Corn (*Report*, iii, 31). Now common on barley and frequent on rye in East Anglia, widely distributed in the midlands, south and south-west, rare in the north. Some grasses, including *Hordeum murinum*, *Bromus mollis* and *Agropyron repens*, also affected.

In Scotland first recognized in 1944. Widely distributed there on barley; also occurs on timothy (*Phleum*) and *Bromus* spp. Brooks (*New Phytol.* **27**, 1928, 215–19) published observations on the disease. See also Owen (*TBMS*, **41**, 1958, 99–108).

The fungus on *Dactylis glomerata*, formerly referred to this species, is now distinguished as *Rhynchosporium orthosporum* (q.v.).

Rhytisma acerinum Fr., *Syst. Myc.* ii, 569; Sacc., *Syll.* viii, 753;
Ramsbottom & Browne, *TBMS*, **34**, 1951, 107; stat. conid.
Melasmia acerina Lév., *Ann. Sci. nat.* 1846, 276; Sacc., *Syll.* iii,
637; Grove, *Coelomycetes*, ii, 186.

*Tar spot of sycamore and maple (*Acer*). Common wherever
sycamore grows; less common on *Acer campestre* and *A.
platanoides*.

The life-history and cytology of the fungus was studied by S. G.
Jones (*Ann. Bot., Lond.*, **39**, 1925, 41–75) and by Bracher (*TBMS*,
9, 1924, 183–6). Maxwell (*Nature, Lond.*, **132**, 1933, 409 and 753)
drew attention to its relative absence in plantations 1200 ft. above
sea-level in Inverness-shire.

Rhytisma pseudoplatani Müller, *Zbl. Bakt.* II, **36**, 1913, 79;
Ramsbottom & Browne, *TBMS*, **34**, 1951, 108.

*Tar spot of sycamore (*Acer*). Bracher (*TBMS*, **9**, 1924, 183–6)
noted the slight differences between this species and *Rhytisma
acerinum*.

Rhytisma punctatum Fr., *Syst. myc.* ii, 569; Sacc., *Syll.* viii, 753;
Ramsbottom & Browne, *TBMS*, **34**, 1951, 108.

*Tar spot of sycamore (*Acer*). Less common than *Rhytisma
acerinum* but causes similar black blotches, though these are
formed of a number of small, distinct patches on a yellowish
ground.

Rhytisma salicinum Fr., *Syst. Myc.* ii, 568; Sacc., *Syll.* viii, 753;
Ramsbottom & Browne, *TBMS*, **34**, 1951, 108; stat. conid.
Melasmia salicina Lév. apud Tul., *Sel. Fung. Carp.* iii, 119;
Sacc., *Syll.* xxii, 1156; Grove, *Coelomycetes*, ii, 188.

*On leaves of willow (*Salix*). North Wootton, Norfolk; Ely,
Cambs; Birmingham; Ayrshire.

Rhytisma symmetricum J. Müller, *Pringsh. Jahrb. wiss. Bot.* **25**,
1893, 622; Sacc., *Syll.* xi, 433; Ramsbottom & Browne, *TBMS*,
34, 1951, 108.

*Leaf spot of willow (*Salix*). Rare. Great Yarmouth, Norfolk, Oct. 1926 (*Report*, vi, 65; *TBMS*, 29, 1946, 149).

Roesleria hypogaea Thüm. & Passer. = *R. pallida*.

Roesleria pallida Sacc., *Mich.* ii, 299; Sacc., *Syll.* viii, 826; syn. *Roesleria hypogaea* Thüm. & Passer., in Thüm., *Pilze de Weinstock*, 210.

The fungus was studied by Bayliss-Elliott and Grove (*Ann. Bot., Lond.*, 30, 1916, 407–14). Its pathogenicity is doubtful.

*On roots of dead pear trees. Frequent, Kent, Worcs, Herefs, Scotland; on young pear, plum and damson trees, Kent, 1924; on plum, Som, 1927; on peach, Berks, 1936.

*On roots of vine (*Vitis*). Painswick, Glos (*Gdnrs' Chron.* 32, 1872, 40).

*On roots of rose. Norfolk, 1932.

Rosellinia aquila (Fr.) de Not., *Sferiacei italici*, 1863, 21; Sacc., *Syll.* i, 252; Bisby & Mason, *TBMS*, 24, 1940, 154.

*Root rot of spruce (*Picea*). On 2-year-old plants of Norway spruce (*Picea abies*) in south Scotland (Wilson, *Trans. R. Scot. arb. Soc.* 36, 1922, 226–35).

Rosellinia necatrix Prill., *Maladies des Plantes agricoles*, ii, 1897, 133; Sacc., *Syll.* xvii, 595; Bisby & Mason, *TBMS*, 24, 1940, 155; stat. conid. *Dematophora necatrix* Hartig, *Untersuch. forstb. Inst.* iii, 1883, 95; Wakefield & Bisby, *TBMS*, 25, 1941, 72.

The perfect stage has not yet been seen in Britain.

White root rot. Common and destructive on a number of hosts in the Isles of Scilly and south-west England. Occasional in southern England and the Isle of Wight, a few records from East Anglia (mainly in hot summers) and once seen in Notts, but does not appear to occur farther north or in Scotland. For general notes on the disease see Massee (*Kew Bull.* 1896, 1–5; *J. Bd Agric.* 7, 1900, 10–16), Salmon and Wormald (*J. S.-E. agric.*

Coll. Wye, **22**, 1913, 450), and Nattrass (*Rep. Long Ashton Res. Sta. for 1926*, 66–72).

*Potato (*Solanum*). Uncommon. Isles of Scilly(*Gdnrs' Chron.* **80**, 1926, 359) and occasionally in south-western and southern counties; Notts, 1931.

*On sugar beet roots (*Beta*). Suffolk, 1936.

*Carrot (*Daucus*). Devon, 1937.

FRUIT

*Almond (*Prunus*). Devon, 1937.

*Apple (*Malus*). Fairly common in south and south-west England; Norfolk, 1948–9.

*Cherry (*Prunus*). Middx, 1909.

*Black currant (*Ribes*). Norfolk, 1947, 1953; Devon, 1950.

*Gooseberry (*Ribes*). St Ives, Corn, 1908; Kent, 1913; Devon, 1926; Norfolk, 1947, 1949.

*Nectarine (Devon) and peach (*Prunus*). Devon, Norfolk.

*Pear (*Pyrus*). Devon, 1937; Norfolk, 1947.

*Plum (*Prunus*). Norfolk, 1949.

*Raspberry (*Rubus*). Corn, 1949.

*Strawberry (*Fragaria*). Devon, 1937; Corn, 1948.

FLOWERS, ORNAMENTALS

*Arum lily (*Zantedeschia*). Isles of Scilly.

**Coprosma*. Isles of Scilly, 1956.

*On corms of *Cyclamen neapolitanum* imported from Italy, 1935.

*Elm (*Ulmus*). Isles of Scilly.

**Iris*. On *I. stylosa*, Devon, 1937; Corn; on *Iris* sp., Corn, 1949.

**Ixia*. Isles of Scilly. Troublesome.

*On *Jasminum officinale*. Devon, 1937.

*Lavender (*Lavandula*). Norfolk.

**Narcissus*. Not uncommon since 1926 in the Isles of Scilly and west Corn. Sometimes destructive (W. C. Moore, *Diseases of Bulbs*, 1939, 70).

*Peony. Lincs, 1949.

*Attacking hedges of *Pittosporum crassifolium*. Isles of Scilly.

*Privet (*Ligustrum*). Devon, 1937.

*Scabious. Suffolk, 1952.

Tritonia (*Montbretia*). Devon, 1937.
*Tulip. Corn, 1936.
*Violet (*Viola*). Corn, 1936; Devon, 1951.

Rosellinia quercina Hartig, *Lehrbuch der Baumkrankheiten*, 1882, 100; Sacc., *Syll.* ix, 496; Bisby & Mason, *TBMS*, **24**, 1940, 155.
*Rosellinia root rot of oak (*Quercus*). Uncommon. Known only on seedling trees in nursery (*J. Bot., Lond.*, **1**, 1912, 185); near Oxford, Forest of Dean (Waldie, *Forestry*, **4**, 1930, 1–6); Welshpool, 1947.

Schizothyrium ptarmicae Desm., *Ann. Sci. nat.* 3 Ser. **11**, 1849, 361; Sacc., *Syll.* ii, 725; Ramsbottom & Browne, *TBMS*, **34**, 1951, 108.
*On *Achillea*, causing stem and leaf spotting, Staffs, 1952.

Scleroderris fuliginosa (Pers.) Karst., *Mycol. fenn.* i, 1871, 216; Sacc., *Syll.* viii, 595; Dennis, *Mycol. Pap.* **62**, 1956, 6.
*On willow (*Salix*). A weak parasite on *S. fragilis*, following *Cryptomyces maximus* (q.v.), Scotland (Alcock & Maxwell, *Trans. R. Scot. arb. Soc.* **39**, 1925, 34–7; Alcock, *TBMS*, **11**, 1926, 161–7. (See also *Physalospora miyabeana*.)

Sclerophoma pithyophila Höhnel, *S.B. Akad. Wiss. Wien*, **98**, 1909, 1234; Sacc., *Syll.* iii, 101 (as *Phoma pithyophylla* (Corda) Sacc.); Grove, *Coelomycetes*, i, 157.
*On Scots pine (*Pinus sylvestris*). Occasional only, but caused severe defoliation in East Anglia in 1955.

Sclerotinia azaleae (Weiss) Dennis, see *Ovulinia azaleae*.

Sclerotinia bulborum (Wakk.) Rehm, in Rabenh., *Krypt. Fl.* 2nd ed. Abt. 3, 1896, 819; Sacc., *Syll.* viii, 197; Ramsbottom & Browne, *TBMS*, **34**, 1951, 79; Dennis, *Mycol. Pap.* **62**, 1956, 147.
*Black slime of hyacinth. Rare. Spalding, Lincs, 1938; Lanarkshire, 1955; and on one occasion (1938) on an imported bulb. Described by W. C. Moore (*Diseases of Bulbs*, 1939, 10).

333

*Black slime of *Muscari*. On plants in boxes, Hayle, Corn, 1955. Probably introduced with imported bulbs.
*Probably this species on *Scilla* bulbs, Lincs, 1934.

Sclerotinia candolleana (Lév.) Fuckel, *Symb. Myc.* 1870, 330; Sacc., *Syll.* viii, 198; Ramsbottom & Browne, *TBMS*, **34**, 1951, 79; Dennis, *Mycol. Pap.* **62**, 1956, 150.

*Leaf blotch of oak (*Quercus*). On leaves of *Q. robur* and *Q. petraea* in central and southern Scotland 1925–6 (Wilson & Waldie, *Trans. R. Scot. arb. Soc.* **39**, 1925, 206; *Gdnrs' Chron.* **80**, 1926, 106; *AAB*, **14**, 1927, 193–6); Surrey, Glos.

Sclerotinia convoluta Drayt., *Mycologia*, **29**, 1937, 314; Dennis, *Mycol. Pap.* **62**, 1956, 155; as *Botryotinia convoluta* (Drayt.) Whetz., *Mycologia*, **37**, 1945, 679; stat. conid. *Botrytis convoluta* Whetz. & Drayt., ibid. **24**, 1932, 469–76; Wakefield & Bisby, *TBMS*, **25**, 1941, 70.

*Botrytis rot of *Iris*. Whetzel and Drayton (loc. cit.) stated that this disease had been intercepted in shipments of rhizomatous irises from Europe, including England, but it has not yet been identified as occurring naturally in Britain. The apothecial stage was produced in culture and has not been seen in nature. (See also *Botrytis cinerea* and *Botrytis* sp.)

Sclerotinia crataegi Magn., *Ber. dtsch. bot. Ges.* **5**, 1905, 197; Sacc., *Syll.* xxii, 637; Ramsbottom & Browne, *TBMS*, **34**, 1951, 79; Dennis, *Mycol. Pap.* **62**, 1956, 142 (as *Monilinia johnsonii* (Ell. & Ev.) Honey).

The perfect stage has not been seen in Britain.

*Leaf blotch of hawthorn (*Crataegus*). Occasional. First recognized at Maidstone, Kent, 1926; prevalent and severe in Kent and eastern England in 1937 (Dowson & Dillon Weston, *Gdnrs' Chron.* **101**, 1937, 426; Wormald, ibid. **102**, 1937, 47).

Sclerotinia cydoniae Schellenb., *Zbl. Bakt.* II, **17**, 1907, 189; Sacc., *Syll.* xxii, 638; Ramsbottom & Browne, *TBMS*, **34**, 1951, 79; as *Monilinia cydoniae* (Schellenb.) Whetz., *Mycologia*,

37, 1945, 672; as *M. linhartiana* (Prill. & Delacr.) Dennis, *Mycol. Pap.* **62**, 1956, 141.

The perfect stage has not been seen in Britain.

*Leaf blotch of quince (*Cydonia*). Rare. East Malling, Kent, 1925 (Wormald, *TBMS*, **10**, 1926, 303–6); Wisley, Surrey, 1925; Kent, 1930; Oxon, 1932; Essex, 1952.

Sclerotinia draytoni Buddin & Wakef., *TBMS*, **29**, 1946, 150; Ramsbottom & Browne, *TBMS*, **34**, 1951, 80; Dennis, *Mycol. Pap.* **62**, 1956, 156.

*Botrytis rot of *Gladiolus*. One of the causes of this disease (Dennis & Wakefield, *TBMS*, **29**, 1946, 150) and probably not the common one, though McLennan, Baker and Gould (*Phytopathology*, **39**, 1949, 260–71) considered this species to be the perfect stage of *Botrytis gladiolorum* (q.v.).

Sclerotinia fructigena Aderh. & Ruhl., *Ber. dtsch. bot. Ges.* **22**, 1904, 262; Sacc., *Syll.* xviii, 40; Ramsbottom & Browne, *TBMS*, **34**, 1951, 81 (as *Stromatinia fructigena* (Schröt.) Boud.); Dennis, *Mycol. Pap.* **62**, 1956, 143 (as *Monilinia fructigena* (Aderh. & Ruhl.) Honey); stat. conid. *Monilia fructigena* Pers., *Syn. Fung.* 1801, 693; Sacc., *Syll.* iv, 34; Wakefield & Bisby, *TBMS*, **25**, 1941, 75.

The perfect stage has not been seen in Britain. (For general literature see *S. laxa.*)

*Apple brown rot, black rot and core rot. Brown rot is common in all districts and very severe about one year in three or four. A black form is frequently seen, especially in stored apples (Spinks, *Rep. Long Ashton Res. Sta. for 1915*, 94–6; *for 1916*, 24–6: Drummond, *TBMS*, **18**, 1934, 348–9; *J. Pomol.* **12**, 1934, 105–9), and the fungus may produce a core rot (Dowson, *Gdnrs' Chron.* **78**, 1925, 479–80; *TBMS*, **11**, 1926, 155–61).

BIOLOGY

Wormald (*J. Minist. Agric.* **34**, 1927, 552–4; *Gdnrs' Chron.* **92**, 1932, 200) emphasized the need for picking apples with the stalks left on. M. H. Moore (*Rep. E. Malling Res. Sta. for 1950*, 131) discussed infection through lenticel and hail bruises, and Croxall *et al.* (*AAB*, **38**, 1951, 833–43) the relation with earwig injury. The fungus not uncommonly causes a spur canker in some varieties, notably James Grieve and Laxton's Superb (Wormald, *TBMS*, **15**, 1930, 102–7; Wormald & M. H. Moore, *Rep. E. Malling Res. Sta. for 1944*, 124–6) and Merton Worcester (Jenkins & M. H. Moore, *Plant Path.* 3, 1954, 139). Byrde (*Rep. Long Ashton Res. Sta. for 1951*, 128–31) has studied the effect of the age of wounds and of weather on infection; made observations about sporulation on mummy apples and plums (ibid. *for 1953*, 163–6); discussed varietal susceptibility in dessert, culinary and cider apples (*J. hort. Sci.* **31**, 1956, 188–95); and (*J. hort. Sci.* **32**, 1957, 227–38) the nature of resistance in cider apples.

CONTROL

Byrde (*Rep. Long Ashton Res. Sta. for 1948*, 161–73) obtained promising results by spraying apples and plums with phenyl mercury chloride and calcium arsenite; and described the effect of spraying trials in summer (ibid. *for 1949*, 81–9; *J. hort. Sci.* **27**, 1952, 237–46) and in winter (ibid. **27**, 1952, 192–200; **29**, 1954, 226–9); as well as, with Fielding (*Rep. Long Ashton Res. Sta. for 1951*, 131–3), the use of eradicant fungicides.

M. H. Moore (*J. hort. Sci.* **25**, 1950, 225–34) also carried out trials with fungicides and studied the relative importance of various wound agents on infection.

*Plum brown rot. Common on the fruit in most districts. Its occurrence on fruit in cold storage was described by Wormald and Painter (*Rep. E. Malling Res. Sta. for 1934*, 148–50; *for 1936*, 198–200). (See also *Sclerotinia laxa*.)

*Pear brown rot. Common on fruit most years in England and Wales. Also Solway.

*Cherry brown rot. Frequent on fruit in England and Wales, especially on sweet cherries. As common on mature fruit as *S. laxa*, but less so on young fruits.

*Brown rot of peach and nectarine fruit. Uncommon but widely distributed in Britain under glass and in the open. M. H. Moore (*Rep. E. Malling Res. Sta. for 1951*, 148) described infection of the young shoots and leaves.

*Brown rot of almond fruit. Occasional. Surrey, 1934, and else-where in south-east England.
*Brown rot of apricot fruit. Unimportant.
*Brown rot of quince (*Cydonia*) fruit. Common.
*Brown rot of medlar (*Mespilus*) fruit. Very rare. West Malling, Kent, 1938.
*Nut drop of *Corylus*. On cob nuts and filberts in Kent (Wormald, *Gdnrs' Chron.* **115**, 1944, 60–1; M. H. Moore, *Rep. E. Malling Res. Sta. for 1946*, 120–1) it follows attacks by the nut weevil, *Balaninus nucum* (M. H. Moore, *J. hort. Sci.* **25**, 1950, 213–24).
*On blackberry (*Rubus*) fruits (M. H. Moore & Tallboys, *Rep. E. Malling Res. Sta. for 1952*, 136–7).
*Causing fruit rot of *Malus* (*Pyrus*) *purpurea*, Reading and East Malling, 1935; and of *Chaenomeles lagenaria* (*Cydonia japonica*), Bristol, 1947.

Sclerotinia fuckeliana (de Bary) Fuckel, *Symb. Myc.* 330; Sacc., *Syll.* viii, 196; Ramsbottom & Browne, *TBMS*, **34**, 1951, 80; Dennis, *Mycol. Pap.* **62**, 1956, 154; as *Botryotinia fuckeliana* (de Bary) Whetz., *Mycologia*, **37**, 1945, 679; stat. conid. *Botrytis cinerea* Fr., *Syst. Myc.* ii, 396; Sacc., *Syll.* iv, 129; Wakefield & Bisby, *TBMS*, **25**, 1941, 69.

See *Botrytis cinerea*.

For many years doubts were thrown on the views expressed by de Bary (*Bot. Ztg*, **44**, 1866, 377) and Istvanffi (*Ann. Inst. Centr. ampélologique Roy. Hongrois*, **3**, 1905, 183–360) that there was a genetic connexion between *Sclerotinia fuckeliana* and *Botrytis cinerea*. These were resolved as a result of research by Kharbush (*Bull. Soc. bot. Fr.* **74**, 1927, 257–62), by Groves and Drayton (*Mycologia*, **31**, 1939, 485–9), and by Gregory (*TBMS*, **32**, 1949, 1–10) who discovered the type material of *Sclerotinia fuckeliana* on which de Bary's description was based. Nevertheless, *Botrytis cinerea* is an aggregate species with a very wide host range, and some forms of it may ultimately be found to be referable to other species of *Sclerotinia*. For this reason the British records of *Botrytis cinerea* are listed under that species.

Sclerotinia gladioli Drayt., *Phytopathology*, **24**, 1934, 397; Rams-
bottom & Browne, *TBMS*, **34**, 1951, 80; as *Stromatinia gladioli*
(Drayt.) Whetz., *Mycologia*, **37**, 1945, 674; Dennis, *Mycol.
Pap.* **62**, 1956, 160; stat. mycel. *Sclerotium gladioli* Massey,
Phytopathology, **18**, 1928, 519.

*On *Acidanthera* sp., Middx, 1939.

*Dry rot of *Crocus*. Occasional on imported corms; on *C.
asturicus*, Bucks, 1931.

*Dry rot of *Freesia*. Has become prominent in southern districts
since 1952; also Yorks, Jersey, Dee, west Scotland, Edinburgh.

*Dry rot of *Gladiolus*. Known since 1926 to be widely distributed
in Britain, including the Isles of Scilly; common and sometimes
destructive; much less common than formerly on imported corms.
Fully described by W. C. Moore (*Diseases of Bulbs*, 1939, 108)
and investigated by Hawker, Bray and Burrows (*AAB*, **31**, 1944,
211–18).

*On *Tritonia* (*Montbretia*). Occasional on imported corms.

Sclerotinia homoeocarpa Bennett, *AAB*, **24**, 1937, 255; stat. mycel.
Rhizoctonia monteithianum Bennett, *Gdnrs' Chron.* **97**, 1935,
129.

*Dollar spot of turf. Common, widely distributed, and often
severe on sea-marsh fescue turf (*Festuca rubra*) used in lawns and
bowling greens. Occurs also, but much less commonly, on
Agrostis stolonifera. Most noticeable in early summer.

The disease was studied by Bennett (*AAB*, **24**, 1937, 236–57).
Drew Smith (*J. Sports Turf Res. Inst.* **8**, 1954, 439–44) described
preliminary fungicide trials and later (ibid. **9**, 1955, 35–59) fully
described the disease and its control. He combined fungicides
with nitrogen treatment (ibid. **9**, 1956, 235–43) and also found
griseofulvin to be effective (ibid. **9**, 1956, 203–9); see also *AAB*,
45, 1957, 206–8.

Sclerotinia laxa Aderh. & Ruhl., *Arb. biol. Abt., Berl.*, **4**, 1905,
427; Sacc., *Syll.* xxii, 637; Ramsbottom & Browne, *TBMS*, **34**,
1951, 80; Dennis, *Mycol. Pap.* **62**, 1956, 144 (as *Monilinia laxa*
(Aderh. & Ruhl.) Honey); stat. conid. *Monilia cinerea* Bon.,

Handb. 76; Sacc., *Syll.* iv, 34; Wakefield & Bisby, *TBMS*, **25**, 1941, 75.

Brown rot, blossom wilt, spur blight and wither-tip of fruit.

Until 1915 (Cayley, *Gdnrs' Chron.* **58**, 1915, 269–70) all the brown rot diseases in Britain were attributed to *Sclerotinia fructigena.* Berkeley (*English Flora*, v, Pt. II, 1836, 349) knew the fungus, and among early articles on brown rot may be mentioned those by W. G. Smith (*Gdnrs' Chron.* **24**, 1885, 51) and Salmon (*J. S.-E. agric. Coll. Wye*, **16**, 1907, 283–6; **19**, 1910, 355–7; *Gdnrs' Chron.* **47**, 1910, 327; **56**, 1914, 85). Between 1917 and 1946 Wormald published about thirty papers on the brown rot diseases, and it is to him we owe most of our knowledge about the various phases in which they are manifested, and the fungi that cause them. He dealt first with blossom wilt of apple (*AAB*, **3**, 1917, 159–204) and wither-tip of plum (ibid. **5**, 1918, 28–59) caused by *Monilia cinerea*, and then distinguished between *M. fructigena* and *M. cinerea* and demonstrated the existence of different physiologic races of *M. cinerea* on apple and plum (*Ann. Bot., Lond.*, **33**, 1919, 361–404; **34**, 1920, 143–71).

Some of his subsequent papers are cited under the particular host. Other, more general, ones were concerned with the world distribution of brown rot (*Ann. Bot., Lond.*, **41**, 1927, 287–99; *J. Minist. Agric.* **35**, 1928, 741–50); the nomenclature of the American brown rot fungi (*TBMS*, **13**, 1928, 194–204); and the ornamental host plants in Britain (ibid. **24**, 1940, 20–8). He found the ascigerous stage of *Sclerotinia laxa* once in Britain (*Ann. Bot., Lond.*, **35**, 1921, 125–35), that of *S. fructigena* not at all. In 1935 Wormald (*Bull. Minist. Agric., Lond.*, no. 88, 50 pp.) gathered together in popular form all the information available about brown rot diseases, and twenty years later (*Tech. Bull. Minist. Agric., Lond.*, no. 3, 1954, 113 pp.) produced a monograph on them for the technical specialist.

Among others who have studied these diseases are Curtis (*Ann. Bot., Lond.*, **42**, 1928, 39–68), who discussed the morphological aspect of resistance to brown rot; M. H. Moore (*Rep. E. Malling Res. Sta. for 1932*, 90–8) and Bennett, Kearns and Marsh (*Rep.*

Long Ashton Res. Sta. for 1944, 157–61), who dealt with the effect of spraying; Byrde (*J. hort. Sci.* **27**, 1952, 130–44), who studied the effect of over thirty fungicides in spore germination tests; and Cole (*Ann. Bot., Lond.*, **22**, 1956, 15–38), who considered the part played by pectolytic enzymes in the pathogenicity of *S. laxa* and *S. fructigena*.

The various diseases caused by *S. laxa* are given below under hosts.

FRUIT TREES

*Almond (*Prunus*). Wither tip. Moray, 1956.

*Apple (*Malus*). Fruit rot. Occasional in stored fruits, once in the black form (Wormald, *Gdnrs' Chron.* **117**, 1945, 115).

*Apricot (*Prunus*). Blossom wilt and brown rot. Uncommon (Denham & Wormald, *J. R. hort. Soc.* **67**, 1942, 261–3).

*Cherry (*Prunus*). Blossom wilt land brown rot. Blossom wilt seasonal, usually slight but now and then (e.g. 1939) very destructive locally, mostly in south-east England. More common on young fruits than *S. fructigena*. Occasional on ornamental flowering cherries. Wither tip is rare on this host. Blossom wilt was described by Wormald (*J. Pomol.* **9**, 1931, 232–7). (See also *Gdnrs' Chron.* **15**, 1894, 757.)

*Medlar (*Mespilus*). Very rare. Kent, 1920.

*Peach and nectarine (*Prunus*). Blossom wilt and brown rot. Not common.

*Pear (*Pyrus*). Blossom wilt and brown rot. Blossom wilt uncommon and known only on the variety Fertility (Wormald, *Ann. Bot., Lond.*, **44**, 1930, 965–74). Very occasionally causes brown rot of young fruits.

*Plum (*Prunus*). Blossom wilt, spur blight, wither tip and brown rot. Not common or severe except locally in some seasons. Widely distributed in Britain. Also on root-stocks. On the fruit less common than *S. fructigena* but the two species often occur together.

Wormald described wither-tip (*AAB*, **5**, 1918, 28–59); shoot wilt and canker (*Ann. Bot., Lond.*, **36**, 1922, 305–20); a shoot wilt in stools and layer beds of plum stocks (*J. Pomol.* **13**, 1935, 68–

77); and, with Painter (*Rep. E. Malling Res. Sta. for 1934*, 148–50; ibid. *for 1936*, 198–200), brown rot of the fruit in cold storage. Drummond (*J. Pomol.* **12**, 1934, 105–9; *TBMS*, **18**, 1934, 348–9) also dealt with wither tip.

*Quince (*Cydonia*). Fruit rot. Rare. East Malling, Kent, 1925.

ORNAMENTALS. Among the ornamental hosts on which *S. laxa* has been found in Britain are:

Chaenomeles lagenaria, syn. *Pyrus* (*Cydonia*) *japonica*. Blossom wilt not infrequent, Maidstone, Kent, 1917; East Malling, Kent, 1939; and since then in Norfolk, Cambs, Glos and Cards.

Malus (*Pyrus*) *floribunda*.

Malus (*Pyrus*) *purpurea*. Blossom wilt, Sussex, Kent, 1935; Hants, 1937.

Prunus besseyi. Blossom wilt and wither tip, Kent, Clyde.

Prunus padus (bird cherry). Blossom wilt and twig blight, East Malling, Kent, 1931–2; Scotland.

Prunus pumila. Blossom wilt and twig blight, East Malling, Kent, 1924.

Prunus serrulata. Twig die-back, Kent, 1925–6.

Prunus tenella (*P. nana*). Twig canker, Sussex, 1932.

Prunus tomentosa. Blossom wilt and twig blight, Kent, 1929, 1931.

Prunus triloba. Blossom wilt, Hants, 1949.

Pyrus elaeagrifolia. Blossom wilt and canker, East Malling, Kent, 1925.

Sorbus (*Pyrus*) *aria* (whitebeam). Blossom wilt, East Malling, 1924.

Sclerotinia laxa Aderh. & Ruhl. f. **mali** (Wormald) Harrison, *J. Proc. Roy. Soc. N.S. Wales*, **67**, 1933, 172; Ramsbottom & Browne, *TBMS*, **34**, 1951, 80; stat. conid. *Monilia cinerea* Bon., *Handb.* **76**; Sacc., *Syll.* iv, 34. (For general literature see *S. laxa*.)

*Brown rot, blossom wilt and spur canker of apple (*Malus*). Occurs in all districts but is usually slight, though locally severe some seasons (e.g. 1932, 1951). Occasional on wild and ornamental crab. Wormald (*AAB*, **3**, 1917, 159–204) described blossom wilt and spur canker.

Sclerotinia libertiana Fuckel = *S. sclerotiorum*.

Sclerotinia mespili Schellenb., *Zbl. Bakt.* III, **17**, 1907, 188; Sacc., *Syll.* xxii, 638; Ramsbottom & Browne, *TBMS*, **34**, 1951, 80; as *Monilinia mespili* (Schellenb.) Whetz., *Mycologia*, **37**, 1945, 673; Dennis, *Mycol. Pap.* **62**, 1956, 141.

*Leaf blotch of medlar (*Mespilus*). Not infrequent. Kent, Surrey, Som, Bristol. First recorded in the conidial stage in 1920 (Wormald, *AAB*, **7**, 1920, 173–7) and ascigerous stage found soon after (Wormald, *TBMS*, **7**, 1922, 287–93).

Sclerotinia minor Jagger, *J. agric. Res.* **20**, 1920, 333; Sacc., *Syll.* xxiv, 1178; Dennis, *Mycol. Pap.* **62**, 1956, 147.

*On artichoke (*Helianthus*). Isolated by Keay (*AAB*, **26**, 1939, 227) from blackened stems of *H. tuberosus*, Cambridge.

*On lettuce (*Lactuca*). This species is believed to play a part in the disease of lettuce commonly attributed to *Sclerotinia sclerotiorum* (q.v.).

*On tomato (*Lycopersicon*). What was presumed by cultural studies to be this species was found on a diseased tomato plant at Cambridge, 1953.

Sclerotinia narcissicola Greg., *TBMS*, **25**, 1941, 35; Ramsbottom & Browne, *TBMS*, **34**, 1951, 80; Dennis, *Mycol. Pap.* **26**, 1956, 156; as *Botryotinia narcissicola* (Greg.) Buchw., *K. VetHøjsk. Aarsskr.* no. 32, 1949, 148; stat. conid. *Botrytis narcissicola* Kleb., *Jb. hamburg. wiss. Anst.* **24**, 1906, beih. 3, 1907, 43; Sacc., *Syll.* xxv, 694; Wakefield & Bisby, *TBMS*, **25**, 1941, 70.

*Smoulder of *Narcissus*. Sporadic, though widely distributed. It may cause much loss to bulbs kept out of the ground until October and November. It is particularly troublesome in north-west Scotland. Described by W. C. Moore (*Daffodil Yearb.* **11**, 1940, 33).

*On snowdrop (*Galanthus*), destroying the foliage (Dowson, *Gdnrs' Chron.* **80**, 1926, 68).

Sclerotinia polyblastis Greg., *TBMS*, **22**, 1938, 201; Ramsbottom & Browne, *TBMS*, **34**, 1951, 80; Dennis, *Mycol. Pap.* **62**, 1956, 159; as *Botryotinia polyblastis* (Greg.) Buchw., *K. VetHøjsk. Aarsskr.* no. 32, 1949, 137; stat. conid. *Botrytis polyblastis* Dowson, *Gdnrs' Chron.* **80**, 1926, 68; Wakefield and Bisby, *TBMS*, **25**, 1941, 70.

*Narcissus fire. Causes much leaf and flower damage every year in Devon, Corn and the Isles of Scilly; slight elsewhere in Britain, including Forth and Clyde. The disease was described by Dowson (*TBMS*, **13**, 1928, 95–102) and W. C. Moore (*Diseases of Bulbs*, 1939, 69). The life cycle of the fungus was dealt with by Gregory (*TBMS*, **25**, 1941, 29–35), who, with Gibson (*AAB*, **33**, 1946, 40–5), studied its control. Small (*Gdnrs' Chron.* **97**, 1935, 304–5) reported the disease in Jersey.

Sclerotinia porri van Beyma, see *Botrytis byssoidea*.

Sclerotinia sclerotiorum (Lib.) de Bary, *Morphol.* 1884, 216; Dennis, *Mycol. Pap.* **62**, 1956, 146; as *S. libertiana* Fuckel, *Symb. myc.* 331; Sacc., *Syll.* viii, 196; Ramsbottom & Browne, *TBMS*, **34**, 1951, 80.

This fungus attacks a very wide range of hosts. Many new hosts, and notably hop, were added to the list in 1935, and it is not unlikely that there was a link between the many outbreaks of Sclerotinia disease and the occurrence of severe mid-May frosts in that year.

POTATO

*Stalk break of potato (*Solanum*). Destructive in the western parts of Ireland, frequent in north-west Scotland and the Western Isles (Hebrides), not common elsewhere in Britain. Seen twice in Jersey (1931, 1956). Short notes on the disease, following its first appearance in Ireland in 1880, will be found in *Gdnrs' Chron.* **14**, 1880, 264; **20**, 1883, 333; and **23**, 1885, 110. Pethybridge (*J. Dep. Agric. Ire.* **10**, 1910, 248–51; **11**, 1911, 429–35; **12**, 1912, 345–8; **13**, 1913, 454–7; **14**, 1914, 447–50) made a scientific investigation of it.

ROOTS AND FODDER CROPS

*On clamped swedes, Merioneth, 1937; Aberdeen, 1950.

*On turnip seed crops, Lincs, 1927, 1930; in decaying turnips, Cumb, 1931; sclerotia mixed with turnip seed from East Lothian crops, 1949.

*On ox cabbage, Staffs, Bristol, Corn, Sussex.

*On and in seed pods of kale, Sussex, 1943.

*Fairly common on sugar beet; occasional on mangold.

PULSE AND FORAGE CROPS

*On broad bean (*Vicia faba*), Staffs, 1950; on runner bean (*Phaseolus*) under glass, Sussex, 1951; Worcs, 1953; on *Phaseolus*, Surrey, 1922; Edinburgh, 1942; Jersey, 1955.

**Sclerotinia* sp. probably this species, on soya bean, Glam, 1936.

*Stalk break of topine, Bristol area, 1955; Argyll, 1957.

VEGETABLES

*Sclerotinia disease of Jerusalem artichoke (*Helianthus*). Not uncommon in southern counties in the field and in stored tubers. Also in Scotland, Forth, East Lothian, Dee. The disease is usually attributed to this species, but *Sclerotinia minor* Jagger (q.v.) may sometimes be concerned.

*On cauliflower (*Brassica*), Norfolk, 1949; Staffs, 1953.

*Sclerotinia rot of carrot (*Daucus*). Not uncommon in storage and occasionally very destructive.

*Causing damage to celery (*Apium*), Lincs, 1910 (*J. Bd Agric.* **17**, 301); in the field, Cambs, 1936; Lancs, 1948; on stored celery grown in Cambs and Lincs (Lowings, *Plant Path.* **4**, 1955, 106–7).

*Seedling rot of cress (*Lepidium*). Salop, 1944.

*Rotting chicory (*Cichorium*) roots. Occasional. Lincs, Surrey, Kent, Berks.

*On cucumber (*Cucumis*). Not uncommon, and sometimes also affects the fruits.

*On lettuce (*Lactuca*), causing a rot in the open from June onwards, or earlier in frames and under cloches. Fairly common in the south and widely distributed in England and Wales. There is some doubt as to what extent *Sclerotinia minor* (q.v.) is implicated in this disease.

*On cloche-grown melons (*Cucumis*), 1950; under Dutch lights, Kent, 1956.
*On mustard (*Brassica nigra*), Lincs, Norfolk. Sclerotia, presumed to be of this fungus, are occasionally found mixed with mustard seed.
*Storage rot of parsnip (*Pastinaca*). Occasional.
*On tomato (*Lycopersicon*) stems. Occasional.
*On vegetable marrow (*Cucurbita*). See Rees, *Welsh J. agric.* 1, 1925, 188.

FRUIT
*On raspberry (Bristol area, 1935), blackberry (Herefs, 1935), and on one strawberry fruit (Worcs, 1953).
Hop (*Humulus*). Rare. A single epidemic in the west midlands in 1935 (Ogilvie, *Rep. Long Ashton Res. Sta. for 1935*, 107–9), and affecting a few hills at Goudhurst, Kent (Salmon & Ware, *J. S.-E. agric. Coll. Wye*, 37, 1936, 25). *Sclerotinia* sp. was serious on layered hops at Newnham Bridge, Worcs, in 1952.

ORNAMENTALS
Alyssum. Salop, 1953, in boxes of a white variety.
Anchusa. On *A. italica*, Glos and Devon, 1935; Herefs, 1951.
Antirrhinum. Wisley, Surrey (Dowson, *Gdnrs' Chron.* 75, 1924, 62; *J. R. hort. Soc.* 51, 1926, 252–65); Glam, 1937; Solway.
*Aster (*Callistephus*). Westmorland, 1953.
Begonia. Leeds, 1923.
*Cactus (*Echinocereus* sp.). Herefs, 1935.
Campanula. Not infrequent on *C. pyramidalis* and occasional on Canterbury bells. Devon, Som, Glos, Glam, Oxon, Northumb. See Rees (*Welsh J. Agric.* 1, 1925, 188–90).
Chrysanthemum. Occasional in the south, west, and Scotland, notably in 1935. (See W. G. S[mith], *Gdnrs' Chron.* 24, 1885, 120.)
*Cornflower (*Centaurea*). Surrey, 1953.
Dahlia. Frequent in field and store.
Delphinium sp. Bristol area, 1935.
*Erysimum (*Cheiranthus*) allionii*. Dorset, 1934; on wallflower, Herefs, 1954.
*Twig die-back of *Forsythia*. Methven, Perthshire, 1952.

*Hollyhock (*Althaea*). Salop, 1935, 1937. See W. G. Smith, *Gdnrs' Chron.* **8**, 1890, 324; **9**, 1891, 791.

*Honesty (*Lunaria biennis*). West of England, 1935.

Lobelia. Causing damping-off, Som, 1929. Possibly this reported by Carruthers (*J. R. agric. Soc.* **63**, 1902, 279).

*Lupin. Herts, Mon, 1934; causing crown rot, Kelso, 1952.

Meconopsis. Forth.

*Michaelmas Daisy (*Aster*). Clyde, Cambs.

Nemesia. Bristol, 1950.

*Poppy (*Papaver*) and *P. somniferum.* Bristol area, Lincs.

Romneya. On *R. coulteri*, Dundry, Som, 1936.

Schizanthus. Wisley, Surrey (Dowson, *J. R. hort. Soc.* **51**, 1926, 257).

Solanum. On *S. capsicastrum*, Oxon, 1949; Kent, 1950.

*Stock (*Matthiola*). Northumb, 1927; Salop, 1935; Aberdeen, 1954; Surrey, 1956.

*Sunflower (*Helianthus*). Frequent and widely distributed.

Verbascum. On *V. olympicum*, Dundry, Som, 1933; Bristol area, 1935.

Zinnia. Berks, 1935.

Among the weed hosts recorded in Britain are *Conium maculatum, Convolvulus* sp., *Heracleum spondylium, Senecio vulgaris, Stellaria media* and *Urtica* sp.

Sclerotinia serica Keay, *J. Bot., Lond.*, **75**, 1937, 132; Ramsbottom & Browne, *TBMS*, **34**, 1951, 81; Dennis, *Mycol. Pap.* **62**, 1956, 148.

*Stem rot of *Gypsophila.* On *G. elegans*, Beds, 1933 (Keay, *AAB*, **26**, 1939, 227); in same field, 1945. Cronshey and Dillon-Weston (*Gdnrs' Chron.* **122**, 1947, 108) pointed out that the disease can be avoided by spring sowing.

Sclerotinia spermophila Noble, *TBMS*, **30**, 1948, 90; Dennis, *Mycol. Pap.* **62**, 1956, 158.

*On seed of white clover (*Trifolium*). Frequently parasitic on white clover seed; not detected in the field or on red clover seed.

In 1927 white clover seed imported into Britain from Europe

and New Zealand was found to be carrying a fungus that closely resembled *Sclerotinia trifoliorum* (Alcock, *C. R. Ass. Int. Essais Semences*, **6**, 1928, 31–6; Alcock & Martin, *Trans. bot. Soc. Edinb.* **30**, 1928, 13–18), and since then the same fungus has frequently been found in home-grown stocks of white clover seed. It was fully described by Noble (*TBMS*, **30**, 1948, 84–91; *Proc. int. Seed Test. Ass.* **14**, 1949, 182–6), and has as its imperfect stage a species of *Botrytis* which she thought may not be distinct from *B. anthophila* (q.v.).

Sclerotinia squamosa (Viennot-Bourgin) Dennis, see *Botrytis squamosa*.

Sclerotinia trifoliorum Erikss., *K. LantbrAkad. Handl. Stockh.* 1880, no. 1; Sacc., *Syll.* viii, 196; Ramsbottom & Browne, *TBMS*, **34**, 1951, 81; Dennis, *Mycol. Pap.* **62**, 1956, 148.

*Clover rot. Widespread and troublesome in East Anglia, in the chalk and limestone districts of Yorks, Lincs, the west midlands and south. Less common in the north, and virtually unknown in Wales and south-west England. Uncommon in Scotland. The disease usually begins in November–December, and develops from January to March, especially after a mild, wet autumn and winter. Most common on broad red clover, but late-flowering crimson, alsike and white clovers also affected. Not uncommon on lucerne and sainfoin, occasional on vetches (Norfolk, 1937; War, 1953). Trefoil appears to be losing the resistance formerly attributed to it and severe attacks have occurred on this host in recent years. There is some evidence that the strains on clover and lucerne are biologically distinct. The sclerotia of the fungus are sometimes carried in clover and lucerne seed samples.

Clover sickness was a familiar sight in England in the 1840's, but the name covered a number of things and it was not until about 1897 (Carruthers, *J. R. agric. Soc.* **8**, 1897, 735; **9**, 1898, 751) that *Sclerotinia trifoliorum* was recognized as one of the causes. Güssow (*J. R. hort. Soc.* **64**, 1903, 379–91) described clover sickness and Amos (*J. R. agric. Soc.* **79**, 1918, 68–88) sorted out several causes of it, including eelworm and *S. trifoliorum*, though other factors are sometimes involved (Mann, *J. agric. Sci.* **28**, 1938, 437–55). Wadham (*New Phytol.* **29**, 1925, 50–6) made observations on the spread of the fungus, and renewed attention was given to it after a lapse of many years by Keay (*AAB*, **26**, 1939, 227–46) and by Dillon-Weston, Loveless and Taylor (*J. agric. Sci.* **36**, 1946, 18–28), all of whom considered its relationship to Sclerotinia disease of beans; as well as by Monro and Ogilvie (*Rep. Long Ashton Res. Sta. for 1945*, 150–3) and by Dillon-Weston (*AAB*, **37**, 1950, 320–3). Carr and Davies (*Nature, Lond.*, **165**, 1950, 1023) devised a technique of artificial inoculation for use in selecting resistant clover seedlings; and Loveless (*AAB*, **38**, 1951, 642–64) traced the significance of ascospore infection in initiating outbreaks, and related the severity of the disease to rainfall in early winter. Tribe (*TBMS*, **40**, 1957, 489–99) showed that the sclerotia are frequently attacked in nature by *Coniothyrium minitans* Campb., *Mycologia*, **39**, 1947, 190.

*Apothecia of *S. trifoliorum* were present in an old-established lawn in which clover was attacked at Cambridge, December 1949.

*Sclerotinia disease of bean (*Vicia*). A cause, and probably a minor one, of this disease, which appears to be due mainly to *Sclerotinia trifoliorum* var. *fabae* (q.v.).

Sclerotinia trifoliorum Erikss. var. **fabae** Keay, *AAB*, **26**, 1939, 228.

*Sclerotinia disease of bean (*Vicia*). On field and, less often, broad beans. Very common in Suffolk and eastern districts generally (Dillon-Weston & Garner, *J. Minist. Agric. Lond.* **50**, 1943, 268–71); not uncommon in Wilts and neighbouring counties; occasional elsewhere in the south. Is most in evidence from April to June.

The *Sclerotinia* on this host was at one time somewhat arbitrarily named *S. trifoliorum* or *S. sclerotiorum*. Pathologically it was thought to be identical with *S. trifoliorum* on clover, but Keay (*AAB*, **26**, 1939, 227–46) considered there were slight but constant morphological differences between the forms on bean and clover, and she named the bean fungus *S. trifoliorum* var. *fabae*. Some years later Dillon-Weston, Loveless and Taylor

(*J. agric. Sci.* **36**, 1946, 18–28) could find no grounds for distinguishing the form on bean from that on clover, but after further study Loveless (*AAB*, **38**, 1951, 252–75) confirmed the distinction made by Keay. In fact he isolated both *S. trifoliorum* and its var. *fabae* from beans, though the variety was much the commoner. He failed to isolate the variety from other leguminous hosts.

A species of *Sclerotinia* of doubtful affinity has been seen on dwarf bean (Worcs, 1932; Sussex, 1940), and runner bean (Glam, 1929; Worcs, 1936).

Sclerotinia tuberosa (Fr.) Fuckel, *Symb. Myc.* 331; Sacc., *Syll.* viii, 195; Ramsbottom & Browne, *TBMS*, **34**, 91; Dennis, *Mycol. Pap.* **62**, 1956, 149.

*Black rot of *Anemone*. In Britain occurs mainly in gardens and rockeries on cultivated varieties of *A. nemorosa* and on *A. coronaria*, *A. apennina*, *A. blanda* and *A. ranunculoides*. It was at one time frequent (see, for example, Smith, *Gdnrs' Chron.* 3 Ser. **1**, 1887, 712–13), but has rarely been seen since about 1900. It is not known in commercial anemone plantations, but was found causing a rot in stored corms of *A. nemorosa* var. *robinsoniana* on a nursery in Bucks in 1941 (W. C. Moore, *TBMS*, **29**, 1946, 251) and again in 1953 on the same variety in the same nursery.

Sclerotinia sp.

*On *Calendula*, causing a foot rot in west Scotland, 1948.

Sclerotium cepivorum Berk., *Ann. Mag. nat. Hist.* **6**, 1841, 355; Sacc., *Syll.* xiv, 1151; Wakefield & Bisby, *TBMS*, **25**, 1941, 102; as *Stromatinia cepivorum* (Berk.) Whetz. in *Mycologia*, **37**, 1945, 674.

*White rot of onion, leek and shallot (*Allium*). Occurs on onion in most districts including south Scotland, but is most serious in Beds, Essex and the west midlands; frequent also on leeks and has been isolated from leek flower heads; less on shallots; reported occasionally on garlic (*A. sativum*) since 1920, on *A. carinatum* and *A. sphaerocephalum* (Swansea, 1926). *Sclero-*

tium cepivorum has been seen on the wild crow garlic (*Allium vineale*) in Worcs and Glam.

Special surveys carried out in 1943 and 1945 showed that about one-sixth of the total onion acreage in England and Wales was contaminated with this fungus.

The disease was fully described by Cotton and Owen (*J. Minist. Agric.* **26**, 1920, 1093–9). Nattrass (*Rep. Long Ashton Res. Sta. for 1926*, 65) and Holmes Smith (*Gdnrs' Chron.* **86**, 1929, 429–30) carried out variety resistant trials. Booer (*Nature, Lond.*, **155**, 1945, 241–2; *AAB*, **32**, 1945, 210–13; **33**, 1946, 413–19) obtained a measure of control by soil treatment with calomel, and Croxall, Sidwell and Jenkins (*AAB*, **40**, 1953, 166–75), as well as Wiggell (*Plant Path.* **5**, 1956, 60–1), obtained good control by seed treatment with pure calomel. Townsend and Willetts (*TBMS*, **37**, 1954, 213–21) followed the development of the sclerotia. Scott (*AAB*, **44**, 1956, 576–83 and 584–9) studied the biology of the fungus and concluded that it did not persist as growing mycelium in natural soils. It invades the bulb through the roots and spreads from plant to plant by root contact. According to Coley-Smith and Hickman (*Nature, Lond.*, **180**, 1957, 445) the presence of *Allium* roots has a marked stimulatory effect on the germination of the sclerotia.

Sclerotium delphinii Welch, *Phytopathology*, **14**, 1924, 31; Wakefield & Bisby, *TBMS*, **25**, 1941, 102.

*Crown rot of *Erythronium*. Causing much loss among tubers of *E. tuolumnense* stored in damp peat, Penryn, Corn, 1952.

*On English iris bulbs, Bucks, 1932 (*Report*, vii, 111); on var. Wedgwood imported from France, 1935 (W. C. Moore, *Diseases of Bulbs*, 1939, 137).

*On larkspur (*Delphinium*), Muir of Ord, Ross-shire, 1953.

*On lily bulbs imported from Japan, 1929 (*Report*, vii, 109).

*On violet, causing death of the plants, Penzance, Corn (*Report*, viii, 91).

Sclerotium gladioli Massey = stat. mycel. of *Sclerotinia gladioli*.

Sclerotium rhizodes Auersw., *Bot. Ztg*, **7**, 1849, 294; Sacc., *Syll.* xiv, 1154; Wakefield & Bisby, *TBMS*, **25**, 1941, 102.

*String of pearls disease of grasses. Rare. On *Agrostis tenuis*, Derby, 1931 (Stirrup, *TBMS*, **16**, 1932, 308); on *Holcus* sp., Forth.

Sclerotium rolfsii Sacc. = stat. mycel. of *Corticium rolfsii*.

Sclerotium tuliparum Kleb., *Jb. hamburg. wiss. Anst.* **24**, 1906, Beih. 3, 1907, 1; Wakefield & Bisby, *TBMS*, **25**, 1941, 103; syn. *Rhizoctonia tuliparum* (Kleb.) Whetz. & Arthur, *Mem. Cornell agric. Exp. Sta.* no. 89, 1925, 18.

*Grey bulb rot of flower bulbs. Widely distributed in Britain on tulip, and where it occurs, both under glass and in the open, it is very destructive. First recognized at Kew in 1922, and soon after substantial attacks were observed in England (Brooks, *Gdnrs' Chron.* **79**, 1926, 271) and Scotland (Cadman, *Garden*, **90**, 1926, 233). Since then the disease has been slowly spreading (W. C. Moore, *J. R. hort. Soc.* **75**, 1950, 113–17). *Iris reticulata* and Dutch irises are as susceptible as tulips. Slight attacks occasional on hyacinth, rare on colchicum (1931), narcissus (1933) and crocus (1938, 1947). The disease was fully described by W. C. Moore (*Diseases of Bulbs*, 1939, 25); its control with formalin and chloronitrobenzene was investigated by Buddin (*J. Minist. Agric.* **44**, 1937, 54–9; 1938, 1158–9).

Scolecotrichum graminis Fuckel = *Passalora graminis*.

Scolecotrichum melophthorum Prill. & Delacr., *Bull. Soc. mycol. Fr.* **7**, 1891, 219; Sacc., *Syll.* x, 599; Wakefield & Bisby, *TBMS*, **25**, 1941, 87.

*Reported new to Britain in 1897 (*J. R. hort. Soc.* **21**, clxxxvii) and found causing loss among cucumbers in Herts in 1918 (*Report*, ii, 50), but Höhnel (*Zbl. Bakt.* II, **60**, 1923, 8) considered the species a typical *Cladosporium* and identical with *C. cucumerinum* (q.v.).

Scopulariopsis fimicola (Cost. & Matruch.) Arn. & Barthelet, *Bull. Soc. mycol. Fr.* **52**, 1936, 65; syn. *Monilia fimicola* Cost. & Matruch., *Rev. gén. Bot.* **6**, 1894, 292; Sacc., *Syll.* xviii, 503; *Oospora fimicola* (Cost. & Matruch.) Cub. & Megl., *Atti Accad. Lincei, Rend. Cl. Sci. Fis. Mat. Nat.* **12**, 1903, 440; Wakefield & Bisby, *TBMS*, **25**, 1941, 57.

*White plaster mould of mushroom (*Agaricus*) beds. Widely distributed, fairly common and troublesome. Described by Ware (*Gdnrs' Chron.* **96**, 1934, 444–5) and by Atkins (*Mushroom Growers Ass. Midl. Group Publ. Yaxley*, 1946, 12 pp.). Williams (*Rep. exp. Res. Sta. Cheshunt for 1936*, 46–8) made observations on its growth, and Bewley, Harnett and Williams (*Gdnrs' Chron.* **102**, 1937, 130) on its control.

Selenophoma donacis (Passer.) Sprague & A. G. Johnson, *Mycologia*, **32**, 1940, 415; syn. *Septoria oxyspora* Penz. & Sacc., *Fung. Mortol.* 1884, 14; Sacc., *Syll.* iii, 565; Grove, *Coelomycetes*, i, 423.

*Halo spot of barley. On var. Herta, Boghall, Scotland, 1949. Since then seen in the Lothians and north Scotland on a number of crops, under conditions suggesting importation with seed barley from Denmark. In barley variety trial, Morpeth, Northumb, 1954; slight in 1955 in barley variety trials, Devon, Corn, Lincs. McKay (*Sci. Proc. R. Dublin Soc.* **24**, 1946, 99–110) has studied the fungus on barley in Ireland.

*Halo spot of grasses. Chiefly of importance on cocksfoot. (*Dactylis*) and timothy (*Phleum*), causing appreciable damage in wet seasons, especially to cocksfoot. Widely distributed in Britain. Has been seen at Aberystwyth also on *Poa trivialis*, tall oat grass (*Arrhenatherum*), meadow foxtail (*Alopecurus*) and *Agrostis* spp.

Sprague, in *Diseases of Cereals and Grasses in North America*, 1950, 203–9, regarded the forms on barley, cocksfoot, timothy and certain other grass hosts as a distinct variety, *Selenophoma donacis* var. *stomaticola* (Baüml.) Sprague & A. G. Johnson, syn. *Septoria culmifida* Lind in *Gartnertidende*, 1907, 112.

Selenophoma donacis var. **stomaticola** (Bäuml.) Sprague & A. G. Johns., see *S. donacis.*

Septocyta ramealis (Rob. ex Desm.) Petrak = *Rhabdospora ramealis.*

Septogloeum fragariae Höhnel = *Stagonospora fragariae.*

Septogloeum mori (Lév.) Briosi & Cav., *Fungh. Parass.* no. 21, 1888; Grove, *Coelomycetes*, ii, 290; syn. *Septoria mori* Lév., *Ann. Sci. nat.* 3 Ser. **5**, 1846, 279; Sacc., *Syll.* iii, 577 (as *Phleospora mori* (Lév.) Sacc.).
*Leaf spot of mulberry (*Morus*). Clevedon, Som (Cooke, *Gdnrs' Chron.* **8**, 1877, 599); Chepstow, Mon.

Septogloeum oxysporum Bomm., Rouss. & Sacc., *Contributions à la flore mycologique de Belgique*, iv, 294; Sacc., *Syll.* x, 497; syn. *Fusoma triseptatum* Sacc., *Syll.* x, 566.
*Blotch and char spot of timothy (*Phleum*). Edinburgh, Dee.

Septogloeum ulmi Died., 836; Grove, *Coelomycetes*, ii, 291; syn. *Phleospora ulmi* Wallr., *Flora Cryptogamica Germaniae*, no. 1545, 1833; Sacc., *Syll.* iii, 578; Cooke, *Fung. Pests Cult. Pl.* 204.
*On elm (*Ulmus*). On the leaves and occasionally the branches of *U. glabra* and *U. procera*. Very common.

Septoria aceris Berk. & Br. = *Phleospora aceris.*

Septoria affinis Sacc., *Mich.* i, 194; *Syll.* iii, 563; Grove, *Coelomycetes*, i, 425.
*Leaf spot of couch grass (*Agropyron*). Dee.

Septoria anemones Desm. var. **coronariae** Grove, *Coelomycetes*, ii, 1937, 359.
*Leaf spot of anemone. Occasional on *Anemone coronaria*, Gulval, Corn, 1936; Corn, 1938.

Septoria antirrhini Rob. & Desm., *Ann. Sci. nat.* **20**, 1853, 87; Sacc., *Syll.* iii, 535; Grove, *Coelomycetes*, i, 368.

*Leaf spot of *Antirrhinum*. Widely distributed but not usually damaging. Devon, Surrey, Kent, I.W., Essex, Suffolk, War, Salop, Staffs.

First reported by Chittenden (*J. R. hort. Soc.* **35**, 1909, 216–17). (See also Buddin and Wakefield, *Gdnrs' Chron.* **76**, 1924, 150–2.)

Septoria apii Chester, *Bull. Torrey bot. Cl.* **18**, 1891, 372; Grove, *Coelomycetes*, i, 368; syn. *S. petroselini* Desm. var. *apii* Briosi & Cav., *Fungh. Parass.* no. 144, 1891; Sacc., *Syll.* xiv, 972.

*Leaf spot of celery (*Apium*). Less common than *Septoria apiigraveolentis* (q.v.) and not as injurious.

Septoria apii-graveolentis Dorogin, *Mater. Mikol. Fitopat. Ross.* **1**, 1915, 57; Sacc., *Syll.* xxv, 454.

*Leaf spot of celery (*Apium*). Common everywhere in Britain and often destructive.

Until about 1940 this disease was always attributed in Britain to *Septoria apii* Chester. Two species are involved, however, and *S. apii-graveolentis*, which produces numerous small, green leaf spots bearing many pycnidia, is much more common and damaging than *S. apii*, which forms large brown spots with few pycnidia. 50 % or more of celery-seed samples grown in England and Wales are infected with *Septoria*.

The disease was described by Chittenden (*J. R. hort. Soc.* **37**, 1911, 115–22; *AAB*, **1**, 1914, 204–6) and Salmon (*Gdnrs' Chron.* **53**, 1913, 414–16; **54**, 1913, 3–4). Pethybridge dealt with its possible origin (*J. R. hort. Soc.* **40**, 1915, 476–80) as well as its spread and prevention (*J. Dep. Agric. Ire.* **14**, 1914, 687–94). Stirrup and Ewan (*Bull. Midl. agric. Coll.* **14**, 1927, 11 pp.; *Bull. Minist. Agric. Lond.* **25**, 1931, 34 pp.) summarized the results of a special investigation, and Holmes Smith (*Gdnrs' Chron.* **93**, 1933, 193–4) also wrote on its prevention. More recently Dillon-Weston and Cronshey (*Agriculture, Lond.,* **54**, 1947, 322–5) urged the need for growing clean seed; Taylor and Dillon-Weston (ibid. **55**,

1948, 201–3) provided figures showing the numbers of seed samples infected with the fungus; and Bant and Storey (*Plant Path.* 1, 1952, 81–3) showed how effective control can be obtained by hot-water seed treatment.

Septoria aristolochiae Sacc., *Mich.* i, 181; Sacc., *Syll.* iii, 558; Grove, *Coelomycetes*, i, 369.
*On leaves of *Aristolochia clematitis*, Kew Gardens.

Septoria armeriae Allesch., *Bibl. bot.* 8, 1897, 40; Sacc., *Syll.* xiv, 977.
*Leaf spot of *Armeria*. Rare. On *A.* 'cephalotes', London, 1935 (*Report*, viii, 80); Sheffield, 1946; Ches (Bond, *TBMS*, 35, 1952, 81–90).

Septoria armoraciae Sacc., *Mich.* i, 187; *Syll.* iii, 519; Grove, *Coelomycetes*, i, 377; syn. *Ascochyta armoraciae* Fuckel, *Symb. Myc.* 388; Sacc., *Syll.* iii, 397; Grove, *Coelomycetes*, i, 300.
*Leaf spot of horse radish (*Armoracia*). Widely distributed in Britain but not destructive.

Septoria avenae Frank = stat. conid. of *Leptosphaeria avenaria*.

Septoria azaleae Vogl., *Malpighia*, 13, 1899, 73; Sacc., *Syll.* xiv, 976; Grove, *Coelomycetes*, i, 371.
*Leaf scorch of *Azalea*. Occasional, becoming frequent. Kent, Herts, Oxon, Sussex, Surrey, Middx, Staffs, Yorks, Devon, east midlands. Usually on plants recently imported. First reported from Hythe, Kent (Salmon & Ware, *Gdnrs' Chron.* 81, 1927, 286–8). First recognized in Scotland in 1951, on imported plants.

Septoria badhami Berk. & Br., *Ann. nat. Hist.* 13, 1854, 460; Sacc., *Syll.* iii, 480; Cooke, *Fung. Pests Cult. Pl.* 152; Grove, *Coelomycetes*, i, 417.
*On leaves of vine (*Vitis*). First recognized in 1853 but rarely reported. Highgate, Shere, Twycross, East Bergholt.

Septoria berberidis Niessl., in *Rabenh. Fung. Eur.* no. 1080; Sacc., *Syll.* iii, 475; Grove, *Coelomycetes*, i, 371.

*On leaves of *Berberis*. On *B. vulgaris*, Darenth; Pembs.

Septoria brunneola (Fr.) Niessl, *Vorarbeiten Kryptogamenflora von Mähren, 1864,* 35; Sacc., *Syll.* iii, 573; Cooke, *Fung. Pests Cult. Pl.* 70.

*Leaf spot of *Convallaria*. Rare.

Septoria castanicola Desm., *Ann. Sci. nat.* **8,** 1847, 26; Sacc., *Syll.* iii, 504; Grove, *Coelomycetes*, i, 372.

*On leaves of chestnut (*Castanea*). On *C. sativa*. Common.

Septoria cercosporioides Trail, *Scot. Nat.* **9,** 1887, 89; Sacc., *Syll.* x, 370; Grove, *Coelomycetes*, i, 375.

*Leaf spot of *Chrysanthemum*. On *C. maximum*, Clyde.

Septoria chamaecisti Vestergr., *Bih. svensk. VetenskAkad. Handl.* **22,** 1896, 24; Sacc., *Syll.* xiv, 968; Moore, *TBMS,* **29,** 1946, 93.

*Leaf spot of *Helianthemum*. On garden hybrids of *H. nummularium*, Harpenden, Herts, July 1945.

Septoria cheiranthi Rob. & Desm., *Ann. Sci. nat.* **11,** 1847, 20; Sacc., *Syll.* iii, 521; Grove, *Coelomycetes*, i, 373.

*On leaves of *Erysimum*. On *E. (Cheiranthus) allionii*, Corn.

Septoria chrysanthemella Sacc., *Syll.* xi, 542; Grove, *Coelomycetes*, i, 374.

*Blotch of *Chrysanthemum*. Frequent in many districts throughout Britain on the cultivated varieties of *C. indicum* but seldom damaging. First recorded 1907 (Salmon, *Gdnrs' Chron.* **42,** 1907, 213). (See also *Septoria leucanthemi*.)

Further study is needed of the species of *Septoria* occurring on *Chrysanthemum indicum*, *C. leucanthemi* and *C. maximum* in Britain. *Cylindrosporium chrysanthemi* Ell. & Dearn., *Canad. Rec. Soc. 1893,* 271; Grove, *Coelomycetes*, ii, 293, which Trent (*Trans. Kans. Acad. Sci.* **42,** 1939, 203) regarded as synonymous with

Septoria chrysanthemella Sacc., is sometimes stated to occur in Britain, but I have been unable to confirm this.

Hemmi and Nakamura (*Mem. Coll. Agric. Kyoto*, no. 3, art. I, 1927, 24 pp.) made a careful study of the taxonomic side in Japan and concluded that two distinct species of *Septoria* are parasitic on the cultivated *Chrysanthemum* (*C. indicum*) in Japan, America and probably also other countries, viz. *Septoria chrysanthemella* Sacc. and *S. obesa* Syd. They also dealt with the synonymy involved.

Septoria clematidis Rob. & Desm., *Ann. Sci. nat.* **20**, 1853, 93; Sacc., *Syll.* iii, 524; Grove, *Coelomycetes*, i, 376.

*On leaves of *Clematis vitalba*, Dartford, Darenth, Tay.

Septoria consimilis Ell. & Martin = *Septoria lactucae.*

Septoria culmifida Lind, see *Selenophoma donacis.*

Septoria dianthi Desm., *Ann. Sci. nat.* **11**, 1849, 346; Sacc., *Syll.* iii, 516; Grove, *Coelomycetes*, i, 380.

*On *Dianthus*. Widely distributed and common on carnation. First record by Potter (*J. R. hort. Soc.* **27**, 1902, 428–30); on *Dianthus* 'Jordans', Som, 1938; on garden pink, Aberystwyth, 1943.

Septoria digitalis Passer., *Funghi parmensi enumerati*, no. 94; Sacc., *Syll.* iv, 534; Grove, *Coelomycetes*, i, 381.

*Leaf spot of *Digitalis*. On *D. purpurea*, Ches (*TBMS*, **4**, 1914, 125); on *D. lanata*, Kent, 1952 (Spilsbury, *TBMS*, **36**, 1953, 343–4).

Septoria divaricatae Ell. & Ev., see *S. drummondii.*

Septoria doronici Passer., *Funghi parmensi enumerati*, no. 75; Sacc., *Syll.* iii, 548; Grove, *Coelomycetes*, i, 381.

*On leaves of *Doronicum pardalianches*, Kew Gardens.

Septoria drummondii Ell. & Everh., *J. Mycol.* 7, 133; Sacc., *Syll.* xi, 544; Grove, *Coelomycetes*, i, 396.

*Leaf spot of *Phlox*. Widely distributed and not uncommon.

A number of species of *Septoria* have been recorded on *Phlox* at home and abroad, including *S. phlogis* Sacc. & Speg., *Mich.* i, 184; Sacc., *Syll.* iv, 533; *S. divaricatae* Ell. & Everh., *J. Mycol.* 5, 1889, 151; Sacc., *Syll.* x, 377; and *S. drummondii*. *S. phlogis* Sacc. & Speg. seems to be distinct and is probably not British. *S. phlogis* Ell. & Everh., *J. Mycol.* 3, 1887, 85 is considered to be synonymous with *S. drummondii*, and it is at least doubtful if *S. divaricatae* is different. All the British specimens examined fit *S. drummondii*.

Septoria euonymi Rabenh., *Flora*, 1848, 506; Sacc., *Syll.* iii, 482; Grove, *Coelomycetes*, i, 382.

*Leaf spot of *Euonymus*. On *E. japonicus*, Sefton Park, Liverpool, 1931 (Ellis, *TBMS*, 4, 1914, 294); Kew Gardens.

Septoria exotica Speg., *Fung. Arg.* ii, no. 107; Sacc., *Syll.* iii, 533; Grove, *Coelomycetes*, i, 415.

*Leaf spot of *Hebe*. Not uncommon in south-west England and Wales on shrubby veronicas grown as hedge plants; abundant in the Isles of Scilly; on *H. (Veronica) speciosa*, Som, 1937.

Septoria fragariae Desm. = *Stagonospora fragariae*.

Septoria fraxini Desm., *Exs.* no. 1086; Sacc., *Syll.* iii, 495; Grove, *Coelomycetes*, i, 383.

*On leaves of ash (*Fraxinus*). Rather common.

Septoria gladioli Passer., in Rabenh., *Fung. Eur.* no. 1956; Sacc., *Syll.* iii, 574; Grove, *Coelomycetes*, i, 421.

*On *Acidanthera murillae*, Cambs, 1950; Cumb, 1952.
*On *Crocus* corms, Ches, 1933.
*On *Freesia* corms.
*Hard rot and leaf spot of *Gladiolus*. The leaf-spot stage has been prominent at times in south-west England and the Isles of Scilly

since 1925; rare elsewhere and seen only once in Scotland (Clyde). The hard-rot stage on the corms is common wherever the plant is grown, and frequent on imported corms. The disease was fully described by W. C. Moore (*Diseases of Bulbs, 1939*, 104–8). Hawker (*AAB*, **31**, 1944, 204–10) attempted to control the disease by fungicidal treatment of the corms.

Septoria glumarum Passer. = *S. nodorum.*

Septoria graminum Desm., see *S. tritici.*

Septoria hederae Desm. = stat. conid. of *Mycosphaerella hedericola.*

Septoria helenii Ell. & Everh., *J. Mycol.* 3, 1887, 87; Sacc., *Syll.* x, 369; syn. *S. nubilosa* Ell. & Everh., *Proc. Acad. nat. Sci. Philad.* 1891, 76; Sacc., *Syll.* x, 369.

*Leaf spot of *Helenium*. Harpenden, Herts, 1941 (W. C. Moore, *TBMS*, **25**, 1941, 209); Luton, Beds, on plants originating from the infected plants at Harpenden; Twyford, Berks, 1949.

Septoria helleborina Höhn., *Ann. mycol., Berl.*, **3**, 1905, 333; Cooke, *Fung. Pests Cult. Pl.* 10 (as *S. hellebori* Thüm.).

*Leaf spot of Christmas rose (*Helleborus*). Occasional. Braunton, Devon, 1939.

Septoria hepaticae Desm., *Ann. Sci. nat.* **19**, 1843, 340; Sacc., *Syll.* iii, 522; Grove, *Coelomycetes*, i, 386.

*On leaves of *Hepatica americana* (*triloba*), Forden, Hants.

Septoria hippocastani Berk. & Br., *Ann. Mag. nat. Hist.* **5**, 1850, 379; Sacc., *Syll.* iii, 479; Grove, *Coelomycetes*, i, 367.

*On leaves of horse chestnut (*Aesculus*). Common.

Septoria humuli Westend., in Kickx, *Flor. Crypt. Flandr.* i, 433; Sacc., *Syll.* iii, 557; Grove, *Coelomycetes*, i, 386.

*On leaves of hop (*Humulus*), Highgate, Dartford.

Septoria hydrangeae Bizz., *Fungi Veneti novi v. critici, 1885*, 6; Sacc., *Syll.* x, 349; Grove, *Coelomycetes*, i, 387.

*Leaf spot of *Hydrangea.* Lincs, 1929.

Septoria lactucae Passer., *Erb. Critt. Ital.* **15**, 1878, no. 746; Sacc., *Syll.* iii, 551; syn. *S. lactucae* Peck, *Bot. Gaz.* **4**, 1879, 170; *S. consimilis* Ell. & Martin, *J. Mycol.* **1**, 1885, 100; Sacc., *Syll.* x, 368.

*Leaf spot of lettuce (*Lactuca*). Harpenden, Herts, 1940 (W. C. Moore, *TBMS*, **24**, 1940, 346–9).

Septoria lavandulae Desm., *Ann. Sci. nat.* **20**, 1853, 86; Sacc., *Syll.* iii, 537; Grove, *Coelomycetes*, i, 388.

*Leaf spot of lavender (*Lavandula*). Common but inconspicuous and innocuous.

Septoria leucanthemi Sacc. & Speg., *Mich.* i, 1878, 191; Sacc., *Syll.* iii, 549; Grove, *Coelomycetes*, i, 375.

*Leaf spot of ox-eye daisy (*Chrysanthemum leucanthemum*). Infrequent, Glam, 1934. Smith and Ramsbottom (*TBMS*, **4**, 1913, 177) recorded *Septoria chrysanthemella* on this host, Lanarkshire, 1912.

*Leaf spot of Shasta daisy (*Chrysanthemum maximum*). Frequent. First reported from Kent and Hants by Wormald (*Gdnrs' Chron.* **78**, 1925, 353–4; *Rep. E. Malling Res. Sta. for 1925*, II Suppl. **13**, 1927, 96–7). South Wales, west midlands, south and west of England.

Septoria ligustri (Desm.) Kickx, *Fl. Cr. Fl.* i, 1867, 354; Sacc., *Syll.* iii, 497; Grove, *Coelomycetes*, i, 389.

*On leaves of privet (*Ligustrum*). On *L. vulgare*, Kew Gardens, Highgate.

Septoria linicola (Speg.) Garassini = stat. conid. of *Sphaerella linorum* Wr.

Septoria lobeliae Peck, *Rep. N.Y. St. Mus. 1872*, **26**, 1874, 87; Sacc., *Syll.* iii, 532.

*Leaf blotch of *Lobelia*. On *L. syphilitica* var. *nana*, Maidenhead, Berks, 1939 (W. C. Moore, *TBMS*, **24**, 1940, 59); on *Lobelia* sp., Swansea, Glam, 1948.

Septoria lycopersici Speg., *Fung. Arg.* iv, no. 289; Sacc., *Syll.* iii, 535; Grove, *Coelomycetes*, i, 410.
*Leaf spot of tomato (*Lycopersicon*). Rare. First seen 1908 (Güssow, *J. Bd Agric.* **15**, 111; *Gdnrs' Chron.* **44**, 1908, 121); Glos, 1909; Yorks, 1913; Denbigh, 1926; Cumb, 1936.

Septoria mori Lév. = *Septogloeum mori*.

Septoria nigro-maculans Thüm., *Symbolae ad floram mycologicam austriacam*, iii, no. 66; Sacc., *Syll.* iii, 559.
*Leaf spot of walnut (*Juglans*). Beds (Carruthers, *J. R. agric. Soc.* **1**, 1890, 836).

Septoria nodorum Berk. = stat. conid. of *Leptosphaeria nodorum*.

Septoria nubilosa Ell. & Everh. = *S. helenii*.

Septoria obscura Trail, *Scot. Nat.* **10**, 1889, 73; Sacc., *Syll.* x, 377; Grove, *Coelomycetes*, i, 372.
*On *Campanula*. On *C. rotundifolia*, Clyde. A species closely resembling this has been recorded on *C. raineri* at Maidenhead, Berks (W. C. Moore, *TBMS*, **24**, 1940, 60).

Septoria oenotherae Westend., *Bull. Acad. Roy. Belg.* 2 Ser. **12**, no. 7; Sacc., *Syll.* iii, 513; Grove, *Coelomycetes*, i, 394.
*On leaves of *Oenothera*. Not common.

Septoria orobina Sacc., *Mich.* i, 187; *Syll.* iii, 509.
*Leaf spot of sainfoin (*Onobrychis*). Seen only at Cardiff, June 1942 (Hughes, *TBMS*, **32**, 1949, 60–2).

Septoria oxyspora Penz. & Sacc. = *Selenophoma donacis*.

PARASITES

Septoria paeoniae Westend., *Bull. Acad. Roy. Belg.* **19**, 1852, no. 9; Sacc., *Syll.* iii, 526; var. **berolinensis** Allesch., *Hedwigia*, **35**, 1896, 34; Sacc., *Syll.* xiv, 967; Grove, *Coelomycetes*, i, 394.

*Peony blotch. Widely distributed in Britain and fairly common. Early records were from Ayrshire, 1911 (Smith & Ramsbottom, *TBMS*, **4**, 1913, 177); Swansea, Mon, 1912; Langport, Som, 1913; and Wisley, Surrey, 1923. Since then reported at intervals from a number of districts, including Shetland. It was troublesome in Lincs in 1932 and is now common in northern Scotland. British specimens examined agree in spore measurements more closely with the variety than with the species.

Septoria passerinii Sacc., *Syll.* iii, 560.

*Leaf spot of barley (*Hordeum*). A fungus closely resembling this species was seen on barley awns, Wilts, 1953; on barley foliage, Essex, 1948; Berwick, 1955; Kirriemuir; Angus.

Septoria petroselini Desm., *Exs.* no. 674; *Mém. Soc. Sci. Lille*, 1842, 94; Sacc., *Syll.* iii, 530; Grove, *Coelomycetes*, i, 395.

*Leaf spot of parsley (*Petroselinum*). Frequent in most districts.

Septoria petroselini var. **apii** Briosi & Cav. = *S. apii*.

Septoria phlogis Ell. & Everh. = *S. drummondii*.

Septoria piricola Desm. = stat. conid. of *Mycosphaerella sentina*.

Septoria populi Desm., *Ann. Sci. nat.* **19**, 1843, 345; Sacc., *Syll.* iii, 502; Grove, *Coelomycetes*, i, 398.

*On leaves of poplar (*Populus*). Occurs especially on *P. nigra* and *P. canadensis serotina*, and probably not uncommon.

Septoria pseudoplatani Rob. & Desm. = *Phleospora pseudoplatani*.

Septoria rhododendri Cooke, *Grevillea*, **5**, 1877, 151; Sacc., *Syll.* iii, 494; Grove, *Coelomycetes*, i, 403.

*On leaves of *Rhododendron ponticum*, Looe, Corn.

Septoria ribis Desm. = stat. conid. of *Mycosphaerella ribis*.

Septoria rosae Desm. = stat. conid. of *Sphaerulina rehmiana*.

Septoria rubi Westend., *Exs.* no. 938; Sacc., *Syll.* iii, 486; Grove, *Coelomycetes*, i, 405.
*Septoria spot of blackberry (*Rubus*). Not uncommon, especially on Himalaya berry. Kent, Middx, Berks, Herefs, Dee. Also on loganberry, Haywards Heath, 1951 (as *Septoria* sp.) and honeyberry, Norfolk, 1953.

Septoria salicicola (Fr.) Sacc., *Mich.* i, 171; *Syll.* iii, 502; Grove, *Coelomycetes*, i, 405.
*On leaves of willow (*Salix*), Audley End, Aberdeen.

Septoria scillae Westend., in Kickx, *Fl. Cr. Fl.* i, 1867, 423; Sacc., *Syll.* iii, 571; Grove, *Coelomycetes*, i, 429.
*Leaf spot of *Scilla*. On *S. nonscripta*, Clyde, 1917 (Smith & Ramsbottom, *TBMS*, 6, 1918, 49).

Septoria sinarum Speg., *Nova addenda ad Mycologiam venetam*, no. 165; Sacc., *Syll.* iii, 517; Grove, *Coelomycetes*, i, 381.
*Leaf spot of sweet william (*Dianthus*). Fairly common in Scotland; on *D. chinensis* (Cooke, *Fung. Pests Cult. Pl.* 1906, 31).

Septoria tritici Rob. & Desm., *Ann. Sci. nat.* **17**, 1842, 107; Sacc., *Syll.* iii, 561; Grove, *Coelomycetes*, i, 423.
*Leaf spot of wheat (*Triticum*). Common but rarely harmful, mainly in the southern half of England in April. Not seen in Scotland until 1946 but since then frequent. Described by Grove (*Gdnrs' Chron.* **60**, 1916, 194 and 210) under the name *Septoria graminum* Desm., *Ann. Sci. nat.* **19**, 1843, 339; Sacc., *Syll.* iii, 565.

Septoria unedonis Rob. & Desm., *Ann. Sci. nat.* **8**, 1847, 20; Sacc., *Syll.* iii, 493; Grove, *Coelomycetes*, i, 369.
*On leaves of *Arbutus unedo*. Occasional, widely distributed.

Septoria violae Westend., *Exs.* II, no. 91; Sacc., *Syll.* iii, 518; Grove, *Coelomycetes*, i, 416.

*Leaf spot of violet (*Viola*). Not uncommon on cultivated hybrids as well as on wild species.

Septoria wistariae Brun., *Espèces et variétés nouvelles de Sphéropsidées trouvées aux environs de Saintes*, 1887, 3; Sacc., *Syll.* x, 351; Grove, *Coelomycetes*, i, 417.

*On leaves of *Wistaria*. On *W. sinensis*, Kew Gardens; Besford Court, Worcs.

Septoria sp.
*On *Viburnum davidi*, Devon, 1952.

Sorosporium panici-miliacei (Pers.) Takah. = *Sphacelotheca destruens*.

Sorosporium saponariae Rudolphi, *Linnaea*, iv, 1829, 116; Sacc., *Syll.* vii, 511; Cooke, *Fung. Pests Cult. Pl.* 32; Sampson, *TBMS*, **24**, 1940, 296.

*On *Dianthus*. In ovaries of *D. deltoides*, Norwich, 1881.

Ainsworth and Sampson (*Brit. Smut Fungi*, 113) pointed out that the description of the smut recorded by Southwell (*Grevillea*, **10**, 1881, 67) from a garden at Norwich as *Ustilago rudolphi* Tul., and compiled by Plowright (*Brit. Ured. Ustil.* 1889, 297) as *Sorosporium saponariae* Rudolphi, does not agree with this species, and in the absence of a specimen the identity of the smut on *Dianthus deltoides* must remain in doubt.

Sphacelia segetum Lév., see *Claviceps purpurea*.

Sphaceloma ampelinum de Bary = stat. conid. of *Elsinoe ampelina*.

Sphaceloma rosarum (Passer.) Jenkins, *J. agric. Res.* **45**, 1932, 330; syn. *Phyllosticta rosarum* Passer., *Erb. Critt. Ital.* no. 1092; Sacc., *Syll.* x, 109; Grove, *Coelomycetes*, i, 42; *Gloeosporium rosarum* (Passer.) Grove, *Coelomycetes*, ii, 224.

*Leaf spot of rose. First recognized in 1926 in England and not uncommon in the south. Forth, Clyde, Moray, Dee. Jenkins (*J. agric. Res.* **45**, 1932, 321–37) found it at various places in England.

Sphacelotheca destruens (Schlecht.) Stevens & A. G. Johns., *Phytopathology*, **34**, 1944, 613; Ainsworth & Sampson, *Brit. Smut Fungi*, 76; syn. *Sorosporium panici-miliacei* (Pers.) Takah., *Bot. Mag., Tokyo*, **16**, 1902, 184 and 247; Sacc., *Syll.* xx, 809, vii, 454.

*Millet smut. Rare. Slight in crops of *Panicum miliaceum*, Billing, Northants, Sept. 1944; Walton-on-the-Naze, Essex, 1949.

Dillon-Weston and Schofield (*Plant Path.* **1**, 1952, 29–30) obtained effective control by seed treatment with organo-mercury dusts and copper carbonate, and Dillon-Weston (ibid. **2**, 1953, 87) by steeping the seed in formaldehyde.

Sphacelotheca reiliana (Kühn) Clinton, *J. Mycol.* **8**, 1902, 140; Sacc., *Syll.* xvii, 487 and vii, 471; Cooke, *Fung. Pests Cult. Pl.* 236; Sampson, *TBMS*, **24**, 1940, 296.

*On male florets of maize (*Zea mays*). This was compiled by Cooke (loc. cit.), but there seems to be no British record of it.

Sphaerella linorum Wr., *Rev. Bot. Inst. 'Miguel Lillo'*, **2**, 1938, 483; stat. conid. *Septoria linicola* (Speg.) Garassini, *Rev. Fac. Agron. La Plata*, **20**, 1935, 170–201; syn. *Phlyctaena linicola* Speg., *An. Mus. nac. B. Aires*, **20**, 1910, 389; Sacc., *Syll.* xxii, 1135.

*Pasmo of flax (*Linum*). Not yet found in Britain, but seen on wild flax in Ireland (Lafferty & McKay, *Nature, Lond.*, **154**, 1944, 709), where Loughnane and McKay (*Sci. Proc. R. Dublin Soc.* **24**, 1946, 89–98) have studied the fungus.

Sphaerella sp.

*On *Ranunculus*. Associated with leaf spotting of cultivated varieties, Devon, 1955.

Sphaeropsis malorum Peck = stat. conid. of *Physalospora obtusa*.

365

Sphaerotheca euphorbiae (Castagne) Salm., *Bull. Torrey bot. Cl.* **29**, 1902, 95; *Monogr. Erysiph.* 1900, 65 (as *S. mors-uvae* (Schwein.) Berk. & Curt.); syn. *S. tomentosa* Otth, in *Jacz. Bull. l'Herb. Boiss.* **4**, 1896, 723; Sacc., *Syll.* xiv, 462.

*On *Euphorbia*, Tay.

Sphaerotheca fuliginea (Schlecht.) Salm., *Bull. Torrey bot. Cl.* **29**, 1902; Sacc., *Syll.* i, 4 (as *S. castagnei* Lév.); syn. *S. humuli* (DC.) Burr. var. *fuliginea* (Schlecht.) Salm., *Mem. Torrey bot. Cl.* **9**, 1900, 49 p.p.; Bisby & Mason, *TBMS*, **24**, 1940, 135.

*Powdery mildew of *Calendula*. In its *Oidium* stage common in south-west England and Worcs; occasional elsewhere. Perithecial stage first seen in Devon and Herts, 1949 (F. Joan Moore, *Plant Path.* **1**, 1952, 53) and common in 1953.

*On cultivated *Doronicum*, Kent, Beds, Herts, Devon and probably widely distributed.

*On *Veronica longiflora subsessilis*, Bucks, 1934 (*Report*, viii, 90).

Sphaerotheca humuli (DC.) Burr., *Bull. Ill. Lab. nat. Hist.* **2**, 1887, 400; Sacc., *Syll.* i, 4 (as *S. castagnei* Lév. p.p.); Salmon, *Monogr. Erysiph.* 1900, 45; Bisby & Mason, *TBMS*, **24**, 1940, 135.

*Raspberry mildew. Not common. Slight most years in the southern half of England, occasional in the north and Scotland (Dee).

*Strawberry mildew. Widely distributed in Britain and troublesome. In England most prominent in the south, south-west and west. Described in *J. Bd Agric.* **5**, 1898, 198–201 and by Salmon (*J. R. hort. Soc.* **25**, 1900–1, 132–8). M. H. Moore (*Rep. E. Malling Res. Sta. for 1936*, 276–9) dealt with its control by spraying or dusting with sulphur, and Ingram, Schofield and Taylor (*Plant Path.* **6**, 1957, 63–4) with dinitrocaprylphenyl-crotonate (karathane).

*Powdery mildew of hop (*Humulus*). Occurs every year in the hop districts, but is usually adequately controlled by dusting with sulphur.

It was seen many years ago by Berkeley (*Gdnrs' Chron.* 9, 1849, 467) and described at length by Whitehead (reprinted from *J. Bath W.S. Co. Ass.* 15, 1884, 24 pp.). (See also *J. Bd Agric.* 3, 1896, 291–3.) General articles on this mildew were published by Salmon (*J. Minist. Agric.* 28, 1921, 150–7, 260–3; *Misc. Publ. Minist. Agric. Fish. Lond.* 42, 1925, 41–9), who studied the disease (*J. agric. Sci.* 2, 1907, 327–32), varietal resistance to it (*AAB*, 8, 1921, 146–63), temporary loss of immunity from it (ibid. 14, 1927, 263–75) and, with Ware (ibid. 14, 1927, 276–89), carried out grafting experiments with resistant varieties. Beard (*Rep. E. Malling Res. Sta. for 1945*, 107–14) described a severe attack on new seedling hops.

Sphaerotheca humuli (DC.) Burr. var. **fuliginea** (Schlecht.) Salm. = *S. fuliginea.*

Sphaerotheca mors-uvae (Schw.) Berk., *Grevillea*, 4, 1876, 158; Sacc., *Syll.* i, 5; Salmon, *Monogr. Erysiph.* 1900, 70; Bisby & Mason, *TBMS*, 24, 1940, 136.

*American gooseberry mildew. Common throughout Britain. It sometimes appears in April but spreads chiefly in May or June, and is usually most severe in Wales, west and northern England, and north-west Scotland. Seasonal in intensity and of considerable economic importance.

When this mildew was first found in Northern Ireland in July 1900 (Massee, *Gdnrs' Chron.* 28, 1900, 143; Salmon, *J. R. hort. Soc.* 25, 1900, 139–42; 26, 1902, 778–9), considerable concern about the need to prevent its entry into England was expressed by Salmon (ibid. 27, 1902, 596–601; 29, 1904, 102–10; *Gdnrs' Chron.* 38, 1905, 305–6). It was, however, observed in 1906 in Worcs and Kent (Salmon, ibid. 40, 1906, 317; Carruthers, *J. R. agric. Soc.* 67, 1906, 262).

Early spraying trials to control it with lime sulphur were carried out by Salmon (*J. Bd Agric.* 19, 1912, 99–106; 19, 1913, 994–1004; *J. S.-E. agric. Coll. Wye*, 22, 1913, 403–31), who also studied the life history of the fungus (ibid. 22, 1913, 432–8, 439–45). A summary of the results appeared in *J. Bd Agric.* 20, 1914, 1057–79.

Brooks, Petherbridge and Spinks (ibid. **22**, 1915, 227–30) re-commended tipping the shoots and were dubious about the value of spraying. Other spraying trials about that time were done by Hector and Auld (*Gdnrs' Chron.* **58**, 1915, 79–80) with lime sulphur, borax and soda; by Petherbridge and Cole (*J. Bd Agric.* **23**, 1916, 750–5) with lime sulphur; by Barker and Lees (ibid. **22**, 1916, 1244–9) with a special liver of sulphur prepara-tion; and by Eyre and Salmon (ibid. **22**, 1916, 1118–25; **23**, 1917, 1098–1100) with ammonium polysulphide.

Later experiments were carried out by Nattrass (*J. Minist. Agric.* **33**, 1926, 265–8; **33**, 1927, 1017–22; **35**, 1928, 161–7; *Rep. Long Ashton Res. Sta. for 1927*, 101–3) using Burgundy Mixture, washing soda and various forms of sulphur, and comparing dusting with spraying. In Northern Ireland trials were done by Muskett and Turner (*J. Minist. Agric. N. Ire.* **1**, 1927, 45–67; **3**, 1931, 83–96) and in Eire by Murphy (*J. Dep. Agric. Dublin*, **29**, 1930, 188–204). Recently, Taylor and Schofield (*Plant Path.* **6**, 1957, 88–90) successfully used dinitrocaprylphenylcrotonate (karathane).

*On currant (*Ribes*). Not infrequent on red and black currant; on *Ribes fasciculatum chinense*, Kent, 1940.

Salmon reported it first on red currant (*Gdnrs' Chron.* **42**, 1907, 26) and the following year on black currant (ibid. **44**, 1908, 203). Later, with Wormald (ibid. **70**, 1921, 47), he discussed varietal resistance in red currants.

Sphaerotheca pannosa (Wallr.) Lév., *Ann. Sci. nat.* **15**, 1851, 138; Sacc., *Syll.* i, 3; Salmon, *Monogr. Erysiph.* 1900, 65; Bisby & Mason, *TBMS*, **24**, 1940, 136.

*Rose mildew. The commonest disease of roses, occurring annually in all districts. Ramblers, especially Dorothy Perkins and Crimson Rambler, are particularly susceptible.

Easlea (*J. R. hort. Soc.* **43**, 1918, 253–60) discussed varietal resistance and Corner (*New Phytol.* **34**, 1935, 180–200) studied the behaviour of the fungus on resistant varieties. Orchard (*Rep. exp. Res. Sta. Cheshunt for 1936*, 43) dealt with its control using a copper-in-oil emulsion, and Dillon-Weston and Taylor (*TBMS*, **27**, 1945, 119–20) with overwintering mycelium. (See also Green (*J. R. hort. Soc.* **59**, 1934, 470) for notes on the disease.)

Sphaerotheca pannosa var. **persicae** Woronich., *Bull. Soc. mycol. Fr.* **30**, 1914, 391; Sacc., *Syll.* xxiv, 1333; Bisby & Mason, *TBMS*, **24**, 1940, 136.

*Powdery mildew of peach and nectarine (*Prunus*). Common under glass, less frequent in the open. Occurs in all districts but most often in the south.

Sphaerotheca tomentosa Otth = *S. euphorbiae.*

Sphaerulina rehmiana Jaap, *Abh. Bot. Ver. Prov. Brandenb.* **52**, 1910, Extr. 10; Sacc., *Syll.* xxii, 190; Bisby & Mason, *TBMS*, **24**, 1940, 180; stat. conid. *Septoria rosae* Desm., *Exs.* no. 535; Sacc., *Syll.* iii, 485; Grove, *Coelomycetes*, i, 404.

The perfect stage has not been seen in Britain.

*Leaf scorch of rose. Occasionally reported on wild and culti-vated roses throughout Britain.

Sphaerulina taxi (Cooke) Massee, *Dis. Cult. Pl. Trees*, 1910, 220; Sacc., *Syll.* xxii, 191; Bisby & Mason, *TBMS*, **24**, 1940, 180; stat. conid. *Cytospora taxifolia* Cooke & Massee em. Pilát & Macal. apud Callen, *TBMS*, **22**, 1938, 99.

*Leaf scorch of yew (*Taxus*). Widespread in Scotland; reported in south-west England. Parasitism doubtful. Descriptions by Cooke (*Gdnrs' Chron.* **9**, 1878, 274; **12**, 1879, 800), by W. G. Smith (ibid. **21**, 1884, 827) and by Callen (*TBMS*, **22**, 1938, 94–106).

Sphaerulina trifolii Rostr., *Bot. Tidsskr.* **22**, 1899, 265; Sacc., *Syll.* xvi, 528; Bisby & Mason, *TBMS*, **24**, 1940, 180; syn. *Pseudoplea trifolii* (Rostr.) Petr., *Ann. mycol., Berl.*, **19**, 1921, 29.

*Burn of white clover (*Trifolium repens*). First recorded 1922 in small amount, Wales (Sampson, *Bull. Welsh Pl. Breed. Sta. Ser.*

H, **1**, 1922, 15). Now fairly common, especially in pure stands of the host. Not uncommon on white clover in lawns.

*On lucerne (*Medicago*). First British record, Aberystwyth, July 1956 (Carr, *Plant Path*. **6**, 1957, 38).

Spondylocladium atrovirens Harz, *Bull. Soc. Impér. Moscou*, 1872, 42; Sacc., *Syll*. iv, 483; Wakefield & Bisby, *TBMS*, **25**, 1941, 97; as *Helminthosporium atrovirens* (Harz) Mason & Hughes, *Canad. J. Bot*. **31**, 1953, 631.

*Silver scurf of potato (*Solanum*). A very common but slight blemish of the tubers.

The blemish was described in Ireland by Johnson (*Econ. Proc. R. Dublin Soc*. **1**, 1903, 161), but the fructifications of the parasite were not seen there until 1907 (Smith & Rea, *TBMS*, **3**, 1908, 37). The disease was first recognized in Britain two years later (Massee, *Kew Bull*. 1909, 16–18). See also Pethybridge (*J. Dep. Agric. Ire*. **15**, 1915, 517–20).

Spongospora subterranea (Wallr.) Lagerh., *J. Mycol*. **7**, 1892, 103; Sacc., *Syll*. vii, 513 (as *Sorosporium scabies* (Berk.) Fisch. v. Waldh.).

*Powdery scab of potato (*Solanum*). Widely distributed in Britain. In England and Wales not in general of great economic importance, but often damaging in the wetter districts of the west and north, especially in wet seasons. Widespread in Scotland. The disease is found to some extent every year on seed tubers from Scotland and Northern Ireland, but it does not develop appreciably in the south and east of England except in wet seasons or on low-lying wet soil where potatoes are grown too often.

The parasite not infrequently produces small galls on potato roots and sprouts (McKee & Webster, *Plant Path*. 3, 1954, 123–4).

This species was seen a century ago by Berkeley (*J. hort. Soc*. **1**, 1846, 33; *Ann. Mag. nat. Hist*. **5**, 1850, 464), who named it *Tubercinia scabies* Berk. In Ireland it was studied by Johnson (*Econ. Proc. R. Dublin Soc*. **1**, 1908, 453–64; *Sci. Proc. R. Dublin Soc*. **12**, 1909, 165–74) and by Pethybridge (*J. Dep. Agric. Ire*. **10–**

16, 1910–16). Pethybridge (*J. R. hort. Soc.* **38**, 1913, 524–30) also dealt with its nomenclature, and Osborn (*Ann. Bot., Lond.,* **25**, 1911, 327–41) with its cytology. General accounts of powdery scab were given in *J. Bd Agric.* **15**, 1908, 592–9; by Horne (*J. R. hort. Soc.* **37**, 1911, 362–89); and, in his monograph of the Plasmodiophorales, by Ivimey Cook (*Arch. Protistenk.* **80**, 1933, 209–15).

*On tomato (*Lycopersicon*). Producing small galls on the roots. Widely distributed in Scotland (where known since about 1930) and thought occasionally to cause a wilt. A few records in England in recent years, mainly from the north (e.g. Dillon-Weston, Taylor & Moore, *Plant Path.* **1**, 1952, 102). Cumb, Westmorland, Dur, Yorks, Essex, Glos, Notts, I. of Ely, I.W.

*On *Solanum* spp. Galls were induced artificially on the roots of *S. mineatum*, but not on certain other non-tuber-forming *Solanums*, by Ives (*Plant Path.* **2**, 1953, 106), and Boyd (*Nature, Lond.,* **167**, 1951, 412) artificially inoculated *S. curtilobum*.

Spongospora sp.

*Crook root of watercress (*Nasturtium*). Widely distributed and often severe in most places where the crop is grown. South-east England, Herts, Berks, Wilts, Dorset, Cambs, Lincs and the east midlands. The hybrid 'brown' cress is much more susceptible than green watercress. First recognized about 1949.

Spencer (*Rep. nat. Veg. Res. Sta. Warwick for 1950*, 52–6; *for 1951*, 61–8) studied the disease, and with Glasscock (*Plant Path.* **2**, 1953, 19–21) gave the results of a survey made in 1949–51. (See also Lyon and Howard (*J. Minist. Agric.* **59**, 1952, 123–8); Tomlinson (*Rep. nat. Veg. Res. Sta. Warwick for 1953*, 14–15; *for 1954*, 58; *for 1955*, 50–1; *for 1956*, 60).) Tomlinson (*Nature, Lond.,* **178**, 1956, 1301–2) showed that control is possible by the addition of traces of zinc to the water in the beds.

Sporendonema purpurascens (Bon.) Mason & Hughes in Brooks, *Plant Diseases*, 2nd ed. 1953, 382; Wood, *Nature, Lond.,* **179**, 1957, 328; Sacc., *Syll.* iv, 40 (as *Geotrichum purpurascens* (Bon.) Sacc.).

*In mushroom (*Agaricus*) beds. An occasional contaminant of the compost and casing soil of mushroom beds, called by growers 'Red Geotrichum' (*Mushroom News*, Darlington and Co., 1949, 12–13, 20–1). Not usually damaging. Can be controlled by calcium or sodium hypochlorite dust (Sinden, *Bull. Mushroom Growers' Ass. 1950*, 199).

Sporocybe azaleae (Peck) Sacc. = *Pycnostysanus azaleae.*

Sporodesmium putrefaciens Fuckel, *Symb. myc.* 350; Sacc., *Syll.* iv, 393 (as *Clasterosporium putrefaciens* (Fuckel) Sacc.); Wakefield & Bisby, *TBMS*, **25**, 1941, 100.

*On sugar beet (*Beta*). At one time this fungus was thought to be the cause of leaf scorch of sugar beet (*Report*, vi, 27), but is now regarded as a saprophyte on leaves showing advanced symptoms of virus yellows. It is an *Alternaria.*

Sporodesmium solani-varians Vanha = *Alternaria solani.*

Sporonema obturatum (Fr.) Sacc., *Syll.* iii, 678; syn. *Clinterium obturatum* Fr., *Summ. Veg. Scand.* 418.

*Associated with die-back of heather (*Calluna vulgaris*) in Scotland: Forth, Clyde, Argyll.

Sporotrichum poae Peck = *Fusarium tricinctum* f. *poae.*

Spumaria alba (Bull.) DC., *Flor. Fr.* ii, 1805, 261; Sacc., *Syll.* vii, 388; syn. *Mucilago spongiosa* Morgan, *Bot. Gaz.* **24**, 1897, 56.

A Myxomycete which causes smother of various crop plants.

*Not infrequent on turf in autumn.
*Around stems of sage (*Salvia*) in herb garden, Kent, 1938.
*Occasionally overruns strawberry (*Fragaria*) beds. Som, 1924; Dorset, 1946.
*On *Meconopsis.* Tweed.

Stagonospora bromi Smith & Ramsb., *TBMS*, **5**, 1914, 160; Sacc., *Syll.* xxv, 367.

*Purple brown blotch of *Bromus.* On *B. ramosus*, Dalry, 1913; on *B. carinatus*, Muchalls, Kincardinesh, 1954.

Stagonospora compta (Sacc.) Died. = *S. meliloti*.

Stagonospora curtisii (Berk.) Sacc., *Syll*. iii, 451; Grove, *Coelomycetes*, i, 359; syn. *S. narcissi* Hollós, *Ann. hist.-nat. Mus. hung*. 4, 1906, 354; Sacc., *Syll*. xxii, 1055.
*On *Amaryllis belladonna*. Occasional in Devon.
*On *Galanthus byzantinus*, Bucks, 1929.
*Leaf scorch of *Narcissus*. Widely distributed in Britain but of no economic importance except in Wales and south-west England, including the Isles of Scilly, where it began to be troublesome about 1927. Descriptions were given by W. C. Moore (*Diseases of Bulbs*, 1939, 64) and Gregory (*Daffodil Yearb. 1937*, 46–52). Beaumont (*AAB*, **37**, 1950, 591–6) discussed varietal susceptibility. The mycelium of the fungus can sometimes be detected within the bulb (Roberts, *Plant Path*. **2**, 1953, 105).

Stagonospora fragariae Briard. & Har., *Rev. mycol*. **13**, 1891, 17; Sacc., *Syll*. x, 333; syn. *Septogloeum fragariae* Höhnel, *Ann. mycol., Berl*., **1**, 1903, 524; Grove, *Coelomycetes*, ii, 289; *Septoria fragariae* Desm., *Observations botaniques et zoologiques*, 1826, 11; Sacc., *Syll*. iv, 511.
*Heart rot and leaf spot of strawberry (*Fragaria*). Uncommon. Vale of Evesham, Tay, Clyde. Also on wild plants.
This fungus, under the name *Septoria fragariae*, was recorded in Britain by Cooke (*Grevillea*, **14**, 1886, 101) on *Fragaria vesca* and by Smith (*TBMS*, **2**, 1903–4, 56) on a strawberry fruit. It was not seen again until 1930 (Ogilvie, *Rep. Long Ashton Res. Sta. for 1931*, 118; *for 1932*, 102). Reference to *Stagonospora* or *Septogloeum* depends on the presence or absence of a true pycnidial wall, and this requires further investigation (*Report*, vii, 85).

Stagonospora hortensis Sacc. & Malbr. = *Ascochyta boltshauseri*.

Stagonospora meliloti (Lasch) Petr., *Ann. mycol., Berl*., **17**, 1919, 64; syn. *S. compta* (Sacc.) Died., *Ann. mycol., Berl*., **10**, 1912, 482; Sacc., *Syll*. iii, 508 (as *Septoria compta* Sacc.); Grove, *Coelomycetes*, i, 349; *Stagonospora trifolii* Fautr., *Rev. mycol*. **12**, 1890, 167; Sacc., *Syll*. x, 333.

PARASITES

*Leaf spot of clover (*Trifolium*). On *T. repens*, Clyde, 1911 (Boyd, *Glasg. Nat.* **5**, 123; Smith and Ramsbottom, *TBMS*, **4**, 1913, 178), Keswick, Reading; on *T. minus*, Corn; on red clover, Dee, 1956. Similar symptoms have been recorded on red, white and subterranean clovers at Aberystwyth, and those on red clover (*T. pratense*) were attributed to *Ascochyta trifolii* (q.v.) by Grove (*Coelomycetes*, i, 317), who considered this to be merely an immature stage of *Stagonospora meliloti*. Jones and Weimer (*J. agric. Res.* **57**, 1938, 807) agreed it was a *Stagonospora*, but a distinct one which they named *S. recedens*.

Stagonospora narcissi Hollós = *S. curtisii*.

Stagonospora recedens (Massal.) Jones & Weimer, see *Ascochyta trifolii*.

Stagonospora trifolii Fautr. = *S. meliloti*.

Stagonosporopsis hortensis (Sacc. & Malb.) Petr. = *Ascochyta boltshauseri*.

Stemphylium botryosum Wallr. = stat. conid. of *Pleospora herbarum*.

Stemphylium graminis (Corda) Bon., *Handb.* 1851, 83; Sacc., *Syll.* iv, 522; Wakefield & Bisby, *TBMS*, **25**, 1941, 100.

*On apple (*Malus*). Said by Kidd and Beaumont (*TBMS*, **10**, 1924, 12) to be parasitic on stored apple fruits.

Stemphylium radicinum (Meier, Drechsl. & Eddy) Neerg., *Aarsberetn. Ohlsens Enkes plantepat. Lab. 1938–9*, 6; syn. *Alternaria radicina* Meier, Drechsl. & Eddy, *Phytopathology*, **12**, 1922, 157–66.

*Black rot of carrot (*Daucus*). Occasional. Devon, Glam, Glos, Kent, Nairn, Forth.

The disease was first suspected in 1934 (Salmon & Ware, *J. S.-E. agric. Coll. Wye*, **33**, 1934, 19), but was not definitely identified until 1938. (See Wormald (*Gdnrs' Chron.* **111**, 1942,

172).) It is seed-borne and was observed causing a seedling rot in Hants in 1954. It occurs frequently on young carrots imported to Scotland from western Europe. It can be controlled by hot-water treatment of the seed at 122° F. for 25 minutes (Roberts, *Plant Path.* 5, 1956, 122).

*On celery (*Apium*). Causing a seedling disease, Cambs, 1945 (W. C. Moore, *TBMS*, 28, 1945, 131–2). *Alternaria* sp. was found heavily contaminating celery seed that failed to germinate in Lancs, 1940.

Stemphylium sarciniiforme (Cav.) Wiltsh., *TBMS*, 21, 1938, 228; Wakefield & Bisby, ibid. 25, 1941, 100; syn. *Macrosporium sarciniiforme* Cav., *Difesa dei parassita*, 1890, no. 4; Sacc., *Syll.* x, 675.

*Ring spot of red clover (*Trifolium*). Occasional since 1915; unimportant. Hants, Glam, Glos, Som, Clyde. For taxonomy and formation of conidia see Wiltshire (loc. cit.).

Stemphylium sp.

*Leaf spot of hyacinth. On lesions near leaf tips, Shrewsbury, Jan. 1956.

**Ixia*. On leaf spots, Corn, 1948.

Stereum frustulosum (Pers.) Fr., *Epicr.* 552; Sacc., *Syll.* vi, 572; Rea, *Brit. Basid.* 665.

*Partridge wood of oak (*Quercus*). Rare.

Stereum gausapatum (Fr.) Fr., *Hymenomycetes Europaei*, 1874, 638; Sacc., *Syll.* vi, 560; syn. *S. spadiceum* sensu Fr. (non Pers.), *Epicr.* 549; Sacc., *Syll.* vi, 564; Rea, *Brit. Basid.* 663.

*Piped rot of oak (*Quercus*). Common on twigs and small branches; the principal cause of 'pipe rot' in the heartwood of living oak trees.

Stereum hirsutum (Fr.) Fr., *Epicr.* 549; *Hym. Eur.* 639; Sacc., *Syll.* vi, 563; Rea, *Brit. Basid.* 664.

*Piped rot of oak (*Quercus*). Occasional in standing oaks. The species is normally saprophytic.

Stereum purpureum (Fr.) Fr., *Hymenomycetes Europaei*, 1874, 639; Sacc., *Syll.* vi, 563; Rea, *Brit. Basid.* 664.

*Silver leaf of fruit trees. Occurs throughout Britain, but is naturally most prevalent and severe in the fruit-growing areas, and especially the plum areas of Kent, East Anglia and Worcs. Notes are given below on its incidence in various fruit trees.

Silver leaf was known a long time before Percival (*J. Linn. Soc. (Bot.)*, **35**, 1902, 390–5) proved its connexion with *Stereum purpureum*. His work was confirmed by Bedford and Pickering (*Rep. Woburn exp. Fruit Fm*, **6**, 1906, 210–24; **12**, 1910, 1–34), and soon after began the series of six papers by Brooks, and later with his colleagues, which summarized the results of an exhaustive investigation of the disease extending over twenty years (*J. agric. Sci.* **4**, 1911, 133–44; **5**, 1913, 288–308; **9**, 1919, 189–215; *J. Pomol.* **3**, 1923, 117–41; **5**, 1926, 61–97; **9**, 1931, 1–29). In addition to these papers Brooks and W. C. Moore (*Trans. Camb. Phil. Soc. Biol. Ser.* **1**, 1923, 56–8) dealt with the initiation of infection, while Brooks and Brenchley (*New Phytol.* **28**, 1929, 218–24; **30**, 1931, 128–35) induced silver leaf by injecting healthy plum trees with non-living extracts of *Stereum purpureum* and (*J. Pomol.* **13**, 1935, 135–9) reported on the effect of rootstock on resistance (see also Hatton, ibid. **14**, 1934, 97–136).

Particular aspects of the disease, and the behaviour of the fungus, have been investigated by other workers. Thus, Wakefield (*Naturw. Z. Forst- u. Landw.* **7**, 1909, Heft 11) and Salmon (*TBMS*, **3**, 1911, 310–24) discussed the possibility of physiologic races of the fungus, Exell (ibid. **10**, 1925, 207–15) studied its hymenial features, Mayo (*New Phytol.* **24**, 1925, 162–71) its enzymes, Tutin (*Biochem. J.* **19**, 1925, 414–15; *Rep. Long Ashton Res. Sta. for 1927*, 90–2) the pectin content of normal and silvered leaves, and Smolak (*AAB*, **2**, 1915, 138–57) and Tetley (*Ann. Bot., Lond.*, **46**, 1932, 633–52) the cytology of affected leaves.

Wormald (*J. Pomol.* **20**, 1943, 144–6) dealt with the relation of papery bark canker to silver leaf, and Hilton and Hoblyn (ibid. **19**, 1942, 168–85) and Garner (ibid. 186) the effect of framework grafting on the disease in apple and plum. Bintner (*Kew Bull.*

1919, 241–63), in a discussion of true and false silver leaf, included a host list of affected plants.

OCCURRENCE ON FRUIT TREES

*Plum (*Prunus*). Widely distributed and very common and destructive. Usually worse after a heavy cropping season, when many branches are broken, thus permitting ready access for the wound parasite.

*Apple (*Malus*). Not common. Chiefly on top-grafted trees, particularly Newton Wonder and Early Victoria.

*Pear (*Pyrus*). Rare. Corn, 1918 (*Report*, ii, 54), Kent, Surrey, Herefs (*Report*, iv, 74).

*Cherry (*Prunus*). Common, especially in Morellos.

*Peach, nectarine, apricot and almond (*Prunus*). Sporadic.

*Gooseberry (*Ribes*). Uncommon. Corn, Staffs, Worcs, Sussex.

*Red and black currant (*Ribes*). Not infrequent.

*Quince (*Cydonia*). Rare. On stool beds, East Malling, Kent, 1920–1.

*Medlar (*Mespilus*). Rare. Glam (*Report*, vii, 79).

*Walnut (*Juglans*). Rare.

*Raspberry (*Rubus*). Devon, 1947.

OTHER HOSTS. Silver leaf, or fructifications of *Stereum purpureum*, or both, have been reported in Britain on the following hosts other than fruit trees. Appropriate references are included, but these should not be taken to imply the only records on the particular host. In most instances there is insufficient evidence to permit an indication of prevalence.

Acer pseudoplatanus (sycamore).
Aesculus hippocastanum (horse chestnut) and *A. carnea*.
Azara microphylla. Som, 1955.
Betula sp. (birch).
Cedrus sp.
Corylus sp. Sussex.
Cotoneaster horizontalis.
Crataegus oxyacantha (hawthorn).
Cytisus (broom).
Escallonia exoniensis.

377

Eucalyptus gunnii. Growing out-of-doors, Biddenden, Kent, 1957.

Exochorda sp.

Fagus sylvatica (beech).

Genista sp.

Laburnum anagyroides (Güssow, *Gdnrs' Chron.* **39**, 1906, 332); and *L. alpinum.* Bangor, 1947.

Larix decidua (larch).

Lonicera nitida. Occasional.

Lupinus sp. (tree lupin).

Neviusia alabamensis.

Pernettya mucronata.

Philadelphus sp.

Pinus sp.

Platanus sp.

Populus sp. (poplar). Common on dead branches.

Prunus avium; *P. laurocerasus* (cherry laurel); *P. lusitanica* (portugal laurel), see *Gdnrs' Chron.* **9**, 1849, 71; *P. mahaleb*; *P. spinosa* (sloe) and *P. triloba.*

Rhododendron spp. On *R. griffithianum*, *R. barbatum*, *R. arboreum*, etc., Kew, 1919 (Cotton, *Gdnrs' Chron.* **77**, 1925, 112). (See also ibid. **128**, 1950, 58.)

Rosa sp. (rose).

Salix spp. (willow). Occasional. Silvered foliage, Worcs, 1923 (*Report*, v, 91); fructifications, Hereford, 1927 (ibid. vi, 65).

Sorbus aucuparia (mountain ash).

Spiraea arguta and *S. japonica.*

Syringa sp.

Ulmus sp. (elm).

Stereum rugosum (Fr.) Fr., *Hymenomycetes Europaei*, 1874, 643; Sacc., *Syll.* vi, 572; Rea, *Brit. Basid.* 663.

*Canker of oak (*Quercus*).* Causes decay of oak and beech wood in standing trees and may be a cause of canker in oak (Potter, *Trans. Engl. arb. Soc.* **5**, 1901–2, 105 as *Stereum quercinum*). Banerjee (*TBMS*, **39**, 1956, 267–77) described it as the cause of destructive cankers on oak in Scotland (Dawyck).

Stereum sanguinolentum (Fr.) Fr., *Hymenomycetes Europaei*, 1874, 640; Sacc., *Syll.* vi, 564; Rea, *Brit. Basid.* 663.

*On larch (*Larix*). Causing a soft brown or streaked rot of European larch (Peace, *Quart. J. For.* **32**, 1938, 81–104).

*Red stain of spruce (*Picea*). Cartwright (*AAB*, **25**, 1938, 430–2) described the association between this rot and certain wood wasps (Siricidae).

Stereum spadiceum sensu Fr., non Pers. = *S. gausapatum*.

Strasseria carpophila Bresad. & Sacc., in Strasser, *Verh. Kaiserl.-König.-zool. Ges. Wien*, **52**, 1902, 436; Sacc., *Syll.* xviii, 284.

*Black rot of apple (*Malus*). Edinburgh, 1942; ? west Ross-shire, 1940 (Dennis, *Gdnrs' Chron.* **114**, 1943, 221). Probably secondary to *Botrytis cinerea* which was also present in the fruit at Edinburgh.

Streptomyces scabies (Thaxt.) Waksm. & Henrici, in Bergey's *Manual of Determinative Bacteriology*, 6th ed. 1943 (1948), 957; syn. *Actinomyces scabies* (Thaxt.) Güssow, *Science*, **39**, 1914, 431; *Oospora scabies* Thaxt., *Rep. Conn. agric. Exp. Sta.* 1890; Sacc., *Syll.* xxii, 1240.

*Common scab of potato tubers. Troublesome and often severe in sandy or gravelly soils poor in organic matter, in alkaline soils and on newly ploughed up grassland, especially after hot dry summers. Widespread in Britain but rarely destructive, though it lowers the market value of the crop.

This disease was fully investigated by Millard (*Pamphl. Univ. Leeds and Yorks Comm. Agric. Educ.* **118**, 1921, 22 pp.; *AAB*, **9**, 1922, 156–64; **10**, 1923, 70–88). Later, with Burr (ibid. **13**, 1926, 580–644), he studied the relation between different strains of the parasite and different types of scab, and with Taylor (ibid. **14**, 1927, 202–16) discussed control by green manuring. A. P. Jones (ibid. **18**, 1931, 313–33) dealt with the histogeny of scab and Cairns, Greeves and Muskett (ibid. **23**, 1926, 718–42) with control by tuber disinfection. Varietal susceptibility in different soils in Ireland was studied by McKay (*Sci. Proc. R. Dublin Soc.* **25**,

1949, 65–84). Large and Honey (*Plant Path.* **4**, 1955, 1–8) have described a method of measuring the amount of scab, and by its use showed that 2–4 % of the national yield was substantially scabbed in 1952–3.

*Scab of turnip and swede (*Brassica*). Not uncommon on turnips in soils where potatoes become badly scabbed. Occasional on swede.

*On sugar beet, mangold and garden beet (*Beta*). Found most years, but has no economic significance except occasionally on red beetroot. Sometimes caused by *Streptomyces tumuli* (q.v.). The disease was described by Millard and Beeley (*AAB*, **14**, 1927, 296–311). The form known as girth scab is thought to be caused by other species.

*On parsnip (*Pastinaca*). A. P. Jones (*Nature, Lond.*, **171**, 1953, 574) claimed that this species causes a form of parsnip canker, which is normally a result of adverse soil conditions on growth or follows attacks of carrot fly.

*Radish scab. Not infrequent.

Streptomyces tumuli (Millard & Beeley); syn. *Actinomyces tumuli* Millard & Beeley, *AAB*, **14**, 1927, 308.

*Beet scab. Found along with *Streptomyces scabies* (q.v.), and now regarded as identical with it.

Stromatinia cepivorum (Berk.) Whetz., see *Sclerotium cepivorum*.

Stromatinia fructigena (Schröt.) Boud. = *Sclerotinia fructigena*.

Stromatinia gladioli (Drayt.) Whetz. = *Sclerotinia gladioli*.

Synchytrium endobioticum (Schilb.) Perc., *Zbl. Bakt.* II, **25**, 1909, 445; syn. *Chrysophlyctis endobiotica* Schilb., *Ber. dtsch. bot. Ges.* **14**, 1896, 36; Sacc., *Syll.* xiv, 447.

*Wart disease of potato (*Solanum*). Known in England with certainty since 1896 and first definitely recorded in Scotland in 1907. It became widely prevalent and caused considerable alarm in England and Wales during and immediately after the 1914–18 War, and in 1923 and 1929 was made the subject of special

Government Orders in an attempt to prevent its further spread. The effect of these Orders, coupled with concentration on the production and cultivation of new immune varieties, gradually led to a very marked change in the economic importance of wart disease. Now that about 75 % of the potato acreage is planted with immune varieties, crop losses due to the disease are negligible, though wart disease is still potentially a menace if only because soil once contaminated with the fungus remains so for many years, and there is always the possibility of new and more virulent strains appearing. The disease is still dealt with administratively under the Wart Disease of Potatoes Order of 1941, but the measures to be enforced are much less stringent than under the older Orders.

Wart disease has been found in every county in England and Wales, but the contaminated areas are mainly in the north-west and west midlands, and in allotments and gardens rather than in large farm fields. Most outbreaks have occurred in Lancs, Ches, Staffs, Salop and Worcs in England, and Carns, Flints, Glam and Mon in Wales. South of a line from the Wash to the Severn the disease is slight, and, except for parts of Som and around London, only isolated outbreaks have been seen. In Scotland it is most prevalent in the midlands and south. *Synchytrium endobioticum* has not been found occurring naturally on any plant other than potato, but under experimental conditions it will attack various wild and cultivated solanaceous plants. The disease occurs in the Isle of Man, but is unknown in the Channel Islands.

HISTORY. The first clear evidence of the existence of wart disease in Britain is contained in a reply by M. C. Cooke (*Gdnrs' Chron.* **20**, 1896, 227) to a correspondent, and the first scientific account of it was one by Potter (*J. Bd Agric.* **9**, 1902, 320–3). From then onwards there were many references in British literature to the disease. Most of them are cited in general historical accounts given by Horne (*J. R. hort. Soc.* **35**, 1909–10, 362–89), Taylor (*J. Minist. Agric.* **27**, 1920–1, 733–8, 863–7, 946–53) and Gough (*J. R. hort. Soc.* **45**, 1919–20, 301–12). The last-named also described the early discovery of immune varieties. The variety

381

tests that led to this discovery were dealt with fully in a Bulletin by Malthouse (*Wart Disease of Potatoes*, Synchytrium endobioticum *Percival*, Bull. no. 8, Shrewsbury, 1914, 58 pp.). An up-to-date map of the then distribution of the disease was included in a historical account published in *Scot. J. Agric.* 7, 1924, 72–82.

RESISTANCE AND IMMUNITY. The discovery of immunity opened up several fields of research (Brierley, *Rep. Int. Potato Conf. 1921*, 93–104). Of the nature of resistance Cartwright (*Ann. Bot., Lond.*, **40**, 1926, 391–5) could find no explanation on anatomical grounds, and Roach (*AAB*, **10**, 1923, 142–6; **14**, 1927, 181–92) concluded that it was not due to any chemical compound that might traverse the plant. Laboratory methods of testing for immunity were soon devised to supplement field trials, first in Germany by Spieckerman and Kotthof (*Dtsch. landw. Pr.* **51**, 1924, 114) and shortly afterwards in England by Glynne (*AAB*, **12**, 1925, 34–60), who showed that some varieties immune in the field were only highly resistant under conditions of the laboratory test (ibid. **13**, 1926, 358–9). Details of Government tests in England were given by Bryan (*J. agric. Sci.* **18**, 1928, 507–14; *Gdnrs' Chron.* **84**, 1928, 30) and Parker (*J. Minist. Agric.* **35**, 1928, 275), and in Scotland in *Scot. J. Agric.* **9**, 1926, 302–4; **10**, 1927, 333–7.

Inheritance of immunity was discussed by Collins (*Gdnrs' Chron.* **70**, 1921, 260 et seq.); by Salaman and Lesley (*Rep. Int. Potato Conf. of 1921*, 105–11; *J. Genet.* **13**, 1923, 177–86); by Black (ibid. **30**, 1935, 127–46); and by Collins (*Ann. Bot., Lond.*, **49**, 1935, 479–91).

Percival (*Zbl. Bakt.* II, **25**, 1909, 440–7) and Curtis (*Phil. Trans.* Ser. B, **210**, 1921, 409–78) dealt with the life history and cytology of *Synchytrium endobioticum*, and Glynne (*AAB*, **13**, 1926, 19–36) described a staining method for testing the viability of winter sporangia.

OTHER HOSTS. The susceptibility of certain solanaceous hosts other than potato was demonstrated artificially by Cotton (*Kew Bull. 1916*, 272–5), Glynne (*AAB*, **12**, 1925, 34–60) and Martin (ibid. **16**, 1929, 422–9).

CONTROL. Attempts have been made from time to time to control wart disease by suitable soil treatment. Potter (*TBMS*, **8**, 1923, 247–9) tried the effect of increasing soil alkalinity, and later Roach and his colleagues investigated the action of sulphur on the parasite (*AAB*, **12**, 1925, 152–90; **13**, 1926, 301–7; **14**, 1927, 422–7; **15**, 1928, 168–90: *J. agric. Sci.* **20**, 1930, 74–96).

Systremma ulmi (Duval ex Fr.) Theiss. & Syd., *Ann. mycol., Berl.*, **13**, 1915, 334; Bisby & Mason, *TBMS*, **24**, 1940, 207; syn. *Dothidella ulmi* (Fr.) Wint. in Rabenh., *Krypt. Fl.* ii, 904; Sacc., *Syll.* ii, 594 (as *Phyllachora ulmi* (Duv.) Fuckel.

*Tar spot of elm (*Ulmus*). Common on moribund elm leaves in autumn.

Taphrina alni-incanae (Kühn) Sadeb., *Jb. hamburg. wiss. Anst.* **8**, 1891, 61; Sacc., *Syll.* x, 69 (as *Exoascus alni-incanae* Kühn); syn. *E. alnitorquus* (Tul.) Sadeb., *Monogr. Exoasceen*, 115; Sacc., *Syll.* viii, 817.

*Leaf blister of alder (*Alnus*). On leaves of *A. glutinosa*; on female catkins of *A. glutinosa* and *A. incana*.

Taphrina athyrii Siemaszko, *Arch. Nauk biol. Warsaw*, **1**, 1923, 17; Dennis, *Notes R. bot. Gdn Edinb.* **22**, 1956, 130.

*On *Athyrium filix-femina*, West Ross, Scotland, Sept. 1946.

Taphrina aurea Fr. = *T. populina*.

Taphrina betulina Rostr., *Tidsskr. f. Skovbrug.* **6**, 1883, 246; *Bot. Zbl.* **15**, 1883, 149; syn. *T. turgida* Sadeb., *Jb. hamburg. wiss. Anst.* **8**, 1891, 61; *Exoascus turgidus* Sadeb., *Monogr. Exoasceen*, 116; Sacc., *Syll.* viii, 818; Ramsbottom & Browne, *TBMS*, **34**, 1951, 63.

*Witches' brooms of birch (*Betula*). Frequent. Massee (*Dis. Cult. Pl. Trees*, 140) reported it to be very abundant on *B. pendula*. On birch, as on other trees, mites sometimes cause similar witches' brooms.

PARASITES

Taphrina bullata (Berk.) Tul., *Ann. Sci. nat.* 5 Ser. **5**, 1866,
127 p.p.; syn. *Exoascus bullatus* (Berk. & Br.) Fuckel, *Symb.
myc.* App. II, 49; Sacc., *Syll.* viii, 817; Ramsbottom & Browne,
TBMS, **34**, 1951, 63.

*Pear leaf blister. Comparatively rare. Worcs, Bucks, Berks,
Herts, Kent, Tweed, Clyde, Moray. Notes on the disease were
made by Berkeley (*J. hort. Soc.* **9**, 1855, 48–51; *Gdnrs' Chron.* **21**,
1884, 804) and by Briton-Jones (*Rep. Long Ashton Res. Sta. for
1923*, 89–90).

Taphrina cerasi (Fuckel) Sadeb., *Jb. hamburg. wiss. Anst.* **8**, 1891,
61; syn. *Exoascus cerasi* Fuckel, *Symb. myc.* 252; Sacc., *Syll.* x,
69; Ramsbottom & Browne, *TBMS*, **34**, 1951, 63; *Taphrina
minor* Sadeb., *Jb. hamburg. wiss. Anst.* **8**, 1891, 61; *Exoascus
minor* Sadeb., *Monogr. Exoasceen*, 55; Sacc., *Syll.* x, 70;
Ramsbottom & Browne, *TBMS*, **34**, 1951, 63.

*Witches' brooms and leaf curl of cherry (*Prunus*). Fairly common, especially in southern counties; Clyde, Dee. Occasional
also on flowering cherries (Northumb, 1950) and *Prunus subhirtella* (Glos, 1953–4).

Described by Salmon (*J. S.-E. agric. Coll. Wye*, **17**, 1908,
320–3; *Gdnrs' Chron.* **43**, 1908, 209–10). Mix (*Kans. Univ. Sci.
Bull.* **33**, 1949, 3) considered *Taphrina cerasi* and *T. minor*
identical, and this was accepted by Henderson (*Notes R. bot.
Gdn Edinb.* **21**, 1954, 165–80) and Bond (*TBMS*, **39**, 1956, 65).

Taphrina deformans (Berk.) Tul., in Johans., *Om svampslägtet
Taphrina*, 1885, 34; syn. *Exoascus deformans* (Berk.) Fuckel,
Symb. myc. 252; Sacc., *Syll.* viii, 816; Ramsbottom & Browne,
TBMS, **34**, 1951, 63.

*Leaf curl of peach and almond (*Prunus*). Very common and
widely distributed in England and Wales on peach and flowering
almonds in the open; less on these and on nectarine under glass.
In Scotland much less common and up to 1950 known only from
the counties of Aberdeen, Moray and Banff. Also Argyll,
Edinburgh and Forfar in 1953.

The disease was described by W. G. Smith (*Gdnrs' Chron.* **4**,

1875, 136–7) and in *J. Bd Agric.* 4, 1897, 55–7. Horne (*J. R. hort. Soc.* 41, 1915, 110–14) obtained good control with Burgundy Mixture. Campbell (*Trans. bot. Soc. Edinb.* 29, 1925, 186–91) studied a rather abnormal form on *Prunus amygdalus* in the Botanic Garden, St Andrews.

Taphrina insititiae (Sadeb.) Johans., *Om svampslägtet Taphrina*, 1885, 33; syn. *Exoascus insititiae* Sadeb., *Monogr. Exoasceen*, 113; Sacc., *Syll.* viii, 817; Ramsbottom & Browne, *TBMS*, 34, 1951, 63.

*Witches' brooms of plum (*Prunus*). Occasional in western districts of England since 1921 (*Report*, iv, 80), notably on damsons; also Westmorland; on *Prunus subhirtella autumnalis*, Corn, 1948.

Taphrina minor Sadeb. = *T. cerasi.*

Taphrina populina Fr., *Observationes mycologicae*, i, 1815, 217; syn. *T. aurea* Fr., ibid. i, 217; Sacc., *Syll.* viii, 812; Ramsbottom & Browne, *TBMS*, 34, 1951, 63.

*Yellow leaf blister of poplar (*Populus*). Common on *P. nigra* and var. *italica*, and on other species, but seldom damaging.

Taphrina pruni (Fuckel) Tul., *Ann. Sci. nat.* 5 Ser. 5, 1866, 129; syn. *Exoascus pruni* Fuckel, *Enum. Fung. Nassoviae*, 1860, 29; Sacc., *Syll.* viii, 817; Ramsbottom & Browne, *TBMS*, 34, 1951, 63.

*Pocket plums. Not uncommon in Worcs and widely distributed throughout Britain, but now less frequent than formerly. There was an exceptionally widespread and severe attack in Wales and western counties in 1920 (*Report*, iv, 80). In Scotland reported also on *Prunus padus* (Clyde, Tay, Moray) and *P. spinosa* (Tweed).

Taphrina sadebeckii Johans., *Om svampslägtet Taphrina*, 1885, 38; Sacc., *Syll.* viii, 816; Ramsbottom & Browne, *TBMS*, 34, 1951, 64.

*Leaf blister of alder (*Alnus*). See Henderson (*Notes R. bot. Gdn Edinb.* 21, 1954, 165–80).

385

Taphrina tosquinetii (Westend.) Magn., *Hedwigia*, **29**, 1890, 23–5.
*Leaf blister of alder (*Alnus*). By far the commonest of the three species of *Taphrina* on *Alnus glutinosa*. Also causes stem distortion of suckers. Fully described by Bond (*TBMS*, **39**, 1956, 60–6). (See also Henderson, *Notes R. bot. Gdn Edinb*. **21**, 1954, 165–80.)

Taphrina turgida Sadeb. = *T. betulina*.

Taphrina ulmi (Fuckel) Johans., *Om svampslägtet Taphrina*, 1885, 43; syn. *Exoascus ulmi* Fuckel, *Symb. myc*. App. II, 49; Sacc., *Syll*. viii, 819.
*On elm (*Ulmus*). Attacks the foliage of elm suckers, and rather inconspicuous. Kent (see Mix, *Kans. Univ. Sci. Bull*. **33**, 1949, 3), Som (Bond, *TBMS*, **39**, 1956, 64), and perhaps elsewhere, but not known in Scotland.

Thekopsora areolata (Fr.) Magn., *Hedwigia*, **14**, 1875, 123; Sacc., *Syll*. vii, 764; Wilson & Bisby, *TBMS*, **37**, 1954, 76; syn. *T. padi* Grove, *Brit. Rust Fungi*, 368.
*Cone rust of spruce (*Picea*). Aecidia on cones of *P. abies*, Yorks, Scotland. The uredo- and teleuto-stages occur on *Prunus padus* and occasionally on planted *P. virginiana* and *P. serotina*. Not common.

Thekopsora padi Grove = *T. areolata*.

Thielaviopsis basicola (Berk. & Br.) Ferraris, *Fl. Ital. Crypt*. i (6), 1912, 233; Wakefield & Bisby, *TBMS*, **25**, 1941, 63. Brierley, *Ann. Bot., Lond.*, **29**, 1915, 483–93, studied the endoconidia of this fungus.
*Black root rot. Has a very wide host range: the following have been reported as hosts in Britain.

CEREALS. On oats, Lincs, 1936.

ROOTS. On beet seedlings, Lancs, 1936; on fodder beet, Isles of Scilly, 1951.

PULSE. On *Vicia faba*, occasional; on *Phaseolus* sp., occasional, including Scotland; on *Pisum sativum*, common, especially in soils with much lime, but does more damage in the north than the south.

PASTURE AND FORAGE. On clovers (*Trifolium*), Lancs, 1926 (*Report*, vi, 30), sainfoin (*Onobrychis*) and trefoil (*Medicago*); on sweet blue lupin grown for forage, Suffolk, 1949.

VEGETABLES. On broccoli (*Brassica*), Corn, 1948; on celery (*Apium*), Ches and Yorks; on cucumber (*Cucumis*), Yorks, Staffs, Ches; on lettuce (*Lactuca*), occasionally severe under glass in northern districts; on mint, Flints, Ches; on tomato (*Lycopersicon*), common every year in northern England from March to July and often responsible for serious reduction in yield: occasional in the south (Williams, *Rep. exp. Res. Sta. Cheshunt for 1926*, 26–9).

FRUIT. On raspberry (*Rubus*), Herefs, 1936; Devon and Corn, 1949.

ORNAMENTALS
Amaryllis (*Hippeastrum*). Herts, 1939.
Anemone. Som, 1936.
Antirrhinum. Ches, 1938; Argyll, 1952.
Arum lily (*Zantedeschia*). Yorks, 1936.
Aster (*Callistephus*). Flints, Yorks, north-west England. (See Massee, *Kew Bull.* 1912, 44–52.)
Begonia. Lancs.
Calceolaria. Yorks, Lancs.
Chrysanthemum. Frequent, especially in the north.
Cyclamen. Not infrequent.
Cypripedium. Surrey, 1939.
Dahlia. Ayrshire, 1956.
Daphne. On *D. mezereum*, Lancs, 1938.
Delphinium. Bath, 1933, troublesome.
Gloxinia (*Sinningia*). London, 1933; on imported corms, 1951.
Larch (*Larix*). 1955.
Lily. Ches, 1934.
Michaelmas Daisy (*Aster*). Ches, 1926.
Nemesia. West Lothian, 1954.
Pelargonium. Lancs, Yorks, Worcs.

Poinsettia (*Euphorbia pulcherrima*). Glam, 1937; Corn, 1954; Devon, 1955.
Primula. On *P. obconica*, Yorks, 1938; *Primula* spp., Forth.
Prunus. On *P. pissardi nigra*, Lancs, 1937.
Rose. Ches, 1933.
Scabious. Ches, 1938, 1948; Staffs, 1952.
Solanum. On *S. capsicastrum*, Glam, Middx.
Stock (*Matthiola*). Yorks, 1936.
Sweet pea (*Lathyrus*). Fairly common and rather destructive in northern England; Forth, Clyde, Dee. (See Massee, *Kew Bull.* 1912, 44–52.)
Tobacco (*Nicotiana*). See *Report*, iv, 96.
Verbena. Chiswick, Middx, July 1956.
Violet. Kent, Yorks; on viola, Devon, 1948.

Tilletia brevifaciens G. W. Fisch., see *T. caries.*

Tilletia caries (DC.) Tul., *Ann. Sci. nat.* 3 Ser. **7**, 1847, 113; Sampson, *TBMS*, **24**, 1940, 302; Ainsworth & Sampson, *Brit. Smut Fungi*, 83; syn. *T. tritici* (Bjerk.) Wolff, *Der Brand des Getreides*, 1874, 13; Sacc., *Syll.* vii, 481.

*Bunt of wheat (*Triticum*). In England and Wales severe for some years after the 1914–18 War but gradually declined, especially when seed dressing became the common practice. Some severe attacks seen towards the end of the 1939–45 War, and bunt can still be found in many districts though it is not often severe. The disease has long been rare in Scotland. Judged from naked-eye examination of seed samples at the National Institute of Agricultural Botany, Cambridge, for over 25 years up to 1942, bunt was at its peak in 1921–2, when 33 % of the samples examined were bunted. This figure dropped gradually from 5·1 % in 1932–3 to 1·2 % in 1940–1.

Dwarf bunt caused by *T. contraversa* Kühn, *Hedwigia*, **13**, 1874, 188, syn. *T. brevifaciens* G. W. Fisch., *Res. Stud. St. Coll. Wash.* **20**, 1952, 11 (see Conners, *Canad. J. Bot.* **32**, 1954, 426–31), and characterized by excessive dwarfing and tillering of affected plants, has not yet been recognized in Britain.

TILLETIA

*Bunt of rye (*Secale*). Rare. Salop, 1917; Cambs, 1929.

Certain aspects of bunt have been given attention by British workers, such as its effect on the growth of the wheat plant (Sampson & Davies, *AAB*, **14**, 1927, 83–104) and on the development of the ear (Dillon-Weston, *Phytopathology*, **19**, 1929, 681–5); varietal resistance (Sampson, *Welsh J. Agric.* **3**, 1927, 180–96); the possibility of physiologic races (Dillon-Weston, *Nature, Lond.*, **127**, 1931, 483; *AAB*, **19**, 1932, 35–54); and the financial side of bunt (Dillon-Weston & Engledow, reprint from *Essex County Farmers' Union Yearb. 1930*, 6 pp.).

It is to the control of bunt, however, that emphasis has been given in Britain. The earliest attempts at cereal-seed treatment were described by Buttress and Dennis (*Agric. Hist.* **21**, 1947, 93–103). Steeping the seed grain in copper-sulphate solution was the chief preventive from the time Samuel Taylor (*Gdnrs' Chron.* **6**, 1846, 242) described his experiments with it until Salmon and Wormald (*J. Minist. Agric.* **29**, 1922, 722–8; **30**, 1924, 918–25) showed that formaldehyde was a safe and sound wet treatment. Soon after attempts were made to find suitable dust treatments. Sampson and Davies (*Welsh J. Agric.* **2**, 1926, 188–212; *AAB*, **15**, 1928, 408–22) and Dillon-Weston (ibid. **16**, 1929, 86–92) carried out comparative trials with copper carbonate; Stirrup and Cranfield (ibid. **15**, 1928, 245–50) tried a formalin-gypsum dust; and Dillon-Weston (*Phytopathology*, **20**, 1930, 753–5) found iodine dust ineffective. This phase ended with official co-operative trials (Pethybridge & W. C. Moore, *J. Minist. Agric.* **37**, 1930, 429–39) in which copper-carbonate dust proved as effective as the wet treatments. But copper carbonate never became popular in this country; it prevented bunt and controlled covered smuts, but something was wanted that would deal at the same time with *Helminthosporium* on oats and barley. The attack was broadened, and the fundamental work on mercury-dust disinfectants carried out by Dillon-Weston and Booer (*J. agric. Sci.* **25**, 1935, 628–49; **27**, 1937, 43–52), together with the practical efforts of commercial interests and others (see, for example, Wakely & Mellor, *Nachr. SchädlBekämpf. Leverkusen*, **13**, 1938, 111–19 and under *Pyrenophora avenae*), led to general interest in the organo-mercury dusts.

These dusts were officially recommended from the beginning of the 1939–45 War (*J. Minist. Agric.* **45**, 1939, 1264) and are now predominantly used, either alone or combined with an insecticide, for cereal-seed disinfection. Brett and Dillon-Weston (*J. agric. Sci.* **31**, 1941, 500–17) drew attention to certain dangers that must be avoided when using these dusts, and Templeman and Marmoy (*AAB*, **27**, 1940, 453–71) showed that the addition of growth-promoting substances gives no added benefit. Dobson (*TBMS*, **11**, 1926, 82–91) had shown that bunted grain may be fed without injury to animals, and it was later established that grain pickled with formaldehyde (Dillon-Weston, *AAB*, **16**, 1929, 86–92) or treated with organo-mercury dusts could under certain conditions be fed to poultry without risk.

The most recent development in the history of cereal-seed treatment was the production of combined insecticide-fungicide seed dressings that control wireworm as well as seed-borne diseases (Holmes, *Search*, **2**, 1951, 21–6). Jameson and McC. Callan (*Nature, Lond.*, **167**, 1951, 490) studied the effect of these dressings on germination in seed testing and in the field. They sometimes cause abnormalities in seed-germination tests done in sand (Wellington, *J. nat. Inst. agric. Bot.* **6**, 1953, 538–44).

Tilletia contraversa Kühn, see *T. caries*.

Tilletia decipiens (Pers.) Körn., *Hedwigia*, **16**, 1877, 30; Sacc., *Syll.* vii, 482; Sampson, *TBMS*, **24**, 1940, 303; Ainsworth & Sampson, *Brit. Smut Fungi*, 86.

*Smut of *Agrostis*. Occurs on *A. canina*, *A. stolonifera* and *A. tenuis*. Common in Scotland, where it was also found in 1948 on seed imported from New Zealand; Yorks, 1933.

Tilletia foetida (Wallr.) Liro, *Maanv.-taloud. Koelaitos, Vuosi-kirja, Helsinki*, 1915–16, 27, 1920; syn. *T. laevis* Kühn, in Rabenh., *Fung. Eur.* 1873, no. 1697; Sacc., *Syll.* vii, 485.

*Bunt of wheat (*Triticum*). There is no authentic record of this species occurring in growing crops in Britain, but it was seen at Liverpool in 1929, along with *T. caries* (q.v.) in 'chicken corn' believed to have been imported.

Tilletia holci (Westend.) Schröt., in Cohn, *Beiträge zur Biologie der Pflanzen*, 2, 1877, 365; Sacc., *Syll.* vii, 484 (as *T. rauwenhoffii* Fisch. de Waldh.); Sampson, *TBMS*, 24, 1940, 303; Ainsworth & Sampson, *Brit. Smut Fungi*, 86.

*Smut of *Holcus*. Not common. On *H. mollis* and *H. lanatus*, Doncaster, Yorks, 1891 (*TBMS*, 1, 1897–8, 60); on hedgerow Yorkshire fog, Cards, 1945; Forth, Moray, Dee.

Tilletia laevis Kühn = *T. foetida*.

Tilletia lolii Auersw., in Klotzsch-Rabenh., *Herbarium vivum mycologicum*, 1854, no. 1899; Sacc., *Syll.* vii, 483; Ainsworth & Sampson, *Brit. Smut Fungi*, 86.

*Smut of *Lolium*. Very rare. Aberystwyth. The fungus was introduced into the Welsh Plant Breeding Station in seed of *L. temulentum* from Portugal, and infected also *L. multiflorum*, *L. perenne* and *L. remotum* in a pot experiment, 1937–8.

Tilletia olida (Riess) Wint. = *Ustilago olida*.

Tilletia tritici (Bjerk.) Wolff = *T. caries*.

Trametes pini (Fr.) Fr., *Syst. Myc.* i, 336; Sacc., *Syll.* vi, 345; Rea, *Brit. Basid.* 615.

*Heart rot of Scots pine (*Pinus*). Rare.

Trametes radiciperda Hartig = *Fomes annosus*.

Trametes suaveolens (L.) Fr., *Epicr.* 491; Sacc., *Syll.* vi, 338; Rea, *Brit. Basid.* 615.

*Heart rot of willow (*Salix*). Common in older trees and pollards of cricket-bat willow (*S. alba calva*). Rare in young unpollarded trees.

Tranzschelia pruni-spinosae (Pers.) Diet., *Ann. mycol., Berl.*, 20, 1922, 30; Wilson & Bisby, *TBMS*, 37, 1954, 77; syn. *Puccinia pruni-spinosae* Pers., *Syn. Fung.* 1801, 226; Sacc., *Syll.* vii, 648 (as *P. pruni* Pers.); Grove, *Brit. Rust Fungi*, 207.

On *Prunus* and *Anemone*.

*Plum rust. Uredo- and teleuto-stages common in southern England on plum, but usually develops too late (August onwards) to do much damage. Sometimes causes early defoliation, especially near anemones bearing the aecidial stage. Uncommon in Scotland. Life-history studied by Brooks (*New Phytol.* **10**, 1911, 207–8).

*Rust of peach, nectarine and apricot (*Prunus*). Rare. On apricot, Sussex, Kent, Devon; on peach and nectarine, Devon, 1933 (Salmon & Ware, *Gdnrs' Chron.* **94**, 1933, 490–2). Regarded at the time as new host records, but a note by Berkeley (ibid. **24**, 1864, 1130, 480—see also under 'Notices to Correspondents' on p. 992) almost certainly refers to a British specimen.

*Cluster cup rust of *Anemone*. The aecidial stage. In south-west England and the Isles of Scilly rare in commercial plantings of *Anemone coronaria* grown as annuals, but common in second-year beds. Often severe in *A. fulgens* grown as a perennial. Occasional in other districts, including Scotland.

Trichoderma lignorum (Tode) Harz = *T. viride*.

Trichoderma viride Pers. ex Fr., *Syst. Myc.* iii, 1829, 215; syn. *T. lignorum* (Tode) Harz, *Bull. Soc. Impér. Moscou*, **44**, 1871, 116; Sacc., *Syll.* iv, 59.

*In mushroom (*Agaricus*) beds as an invader. Northants, Herts, Derby.

*On narcissus bulbs. Not uncommon in stored bulbs late in the season, but doubtfully parasitic. It often occurs on bulbs that decay after hot-water treatment (*Rep. Dep. Pl. Path. Seale-Hayne agric. Coll. for 1938*, 22–8).

The taxonomy of this and related forms was dealt with by Bisby (*TBMS*, **23**, 1939, 149–68). Brian and Hemming (ibid. **33**, 1950, 132–41) studied the effect of nutrition on spore production.

Trichoscyphella resinaria (Cooke & Phill.) Dennis, *Mycol. Pap.* **32**, 1949, 93; Ramsbottom & Browne, *TBMS*, **34**, 1951, 89 (as *Trichoscypha resinaria* (Phill.) Boud.).

*Canker of spruce (*Picea*). Massee (*Dis. Cult. Pl. Trees*, 284) said that this fungus (as *Dasyscypha resinaria* Rehm, *Ascomycetes Lojkani*, i, 1882, 11; Sacc., *Syll.* viii, 438) was frequent on *Picea abies*, had proved destructive to the Bhotan pine (*Pinus griffithii*) in Wilts, and occurred on larch (*Larix*).

Trichoscyphella willkommii (Hart.) Nannf., *Nova Acta Soc. Sci. upsal.* 4 Ser. **8**, 1932, 299; syn. *Dasyscypha willkommii* (Hartig) Rehm, *26ste Ber. Naturh. Ver. Augsberg*, 1881, 19; Sacc., *Syll.* viii, 437 (as *D. calycina* (Schum.) Fuckel); Ramsbottom & Browne, *TBMS*, **34**, 1951, 89 (as *Trichoscypha calycina* (Schum. ex Fr.) Boud.).

*Larch canker. Extremely common everywhere on *Larix europaea*, occasional on *L. leptolepis*. There are big differences in susceptibility among European larch from different sources, those of high alpine origin being particularly badly affected. Terrell (*Quart. J. For.* **30**, 1936, 158–60) referred to rapid recovery after certain cultural measures had been taken on larch plantations in Hants, and MacDonald (*J. For. Comm.* 1949, 192–8) discussed the history of the disease in Britain.

Damage by frost and other agencies often play a part (Day, *Forestry*, **5**, 1931, 41–56), and there is still some doubt about the degree of pathogenicity of the fungus, but the work of Hahn and Ayers (*J. For.* **41**, 1943, 483–95) made it clear that it can be pathogenic. Manners (*TBMS*, **36**, 1953, 362–74) discussed the taxonomy of the parasite and distinguished it from the related saprophyte, *Trichoscypha hahniana* (Seaver) Manners, commonly found on the same host. He also (ibid. **40**, 1957, 500–8) satisfied himself that *Trichoscyphella willkommii* as well as frost is involved in the production of cankers in the field.

Trichosphaeria parasitica Hartig = *Acanthostigma parasiticum*.

Trichothecium roseum Fr., *Syst. Myc.* iii, 427; Sacc., *Syll.* iv, 178; Wakefield & Bisby, *TBMS*, **25**, 1941, 87. Ingold, ibid. **39**, 1956, 460–4, re-described its conidial apparatus.

*On cucumber (*Cucumis*), associated with leaf spotting, but doubtfully parasitic, Hants, 1946; Surrey, 1954.

*On tomato fruits (*Lycopersicon*), causing stem end rot, Baldock, Herts, 1951 (Taylor, *Plant Path.* 1, 1952, 102); Worcs, 1953.
*Pink mould of apple (*Malus*). Frequent on scab spots in stored apples.

Trochila buxi Capron apud Cooke, *Handb.* ii, 1871, 786; Sacc., *Syll.* viii, 729; Ramsbottom & Browne, *TBMS*, **34**, 1951, 102.
*Leaf fall of box (*Buxus*). Clyde, Tay, Moray.

Trochila laurocerasi (Desm.) Fr., *Summ. Veg. Scand.* 367; Sacc., *Syll.* viii, 729; Ramsbottom & Browne, *TBMS*, **34**, 1951, 102; stat. conid. *Gloeosporium phacidiellum* Grove, *J. Bot., Lond.*, **50**, 1912, 53, *Coelomycetes*, ii, 220; Sacc., *Syll.* xxv, 554.
*Leaf spot of cherry laurel (*Prunus*). Widely distributed in Scotland; Studley Castle, War. Studied by Gregor (*AAB*, **23**, 1936, 700–4).

Tuburcinia primulicola (Magn.) Bref., *Untersuch. Ges. Mykol.* **12**, 1895, 180; Sampson, *TBMS*, **24**, 1940, 303; Ainsworth & Sampson, *Brit. Smut Fungi*, 90; syn. *Urocystis primulicola* Magn., *Verh. bot. Ver. Brandenb.* **20**, 1878, 53; Sacc., *Syll.* vii, 517; Cooke, *Fung. Pests Cult. Pl.* 54.
*Ovary smut of *Primula*. On *P. vulgaris* and *P. farinosa*. Uncommon. Malpas, Salop, 1884, on *P. farinosa* (Dod, *Gdnrs' Chron.* **22**, 1884, 248 and 308; W. G. Smith, ibid. 268–9); Teesdale; Dee. For the life history and cytology see Wilson (*Rep. Brit. Ass. Adv. Sci. Manchester 1915*, 730–1).

Typhula graminum Karst., see *T. incarnata*.

Typhula gyrans (Batsch) Fr., *Epicr.* 585; Sacc., *Syll.* vi, 746.
*On turnip and swede (*Brassica*). Macdonald (*AAB*, **21**, 1934, 590–613) studied the life history and pathogenicity of this fungus on turnip and swede in Scotland, and concluded it was only saprophytic.

Typhula incarnata Lasch ex Fr., *Epicr.* 585; Sacc., *Syll.* vi, 745; syn. *T. itoana* Imai, *Jap. J. Bot.* **8**, 1936, 5; Corner, *A Monograph of Clavaria and allied genera*, 1950, 673.

*Snow rot of cereals. Rare. On wheat, Herts, 1945 (W. C. Moore, *TBMS*, **28**, 1945, 132 as *T. graminum* Karst.); on barley, Durham, 1953 (Croxall & F. Joan Moore, *Plant Path.* **2**, 1953, 140); on oats, Pewsey, Wilts, 1953.

Typhula itoana Imai = *T. incarnata.*

Typhula trifolii Rostr., *Ugestr. Land.* **35**, 1890, 72; Sacc., *Syll.* xxiii, 497.

*On clover (*Trifolium*). Sclerotia of this species mixed with clover seed imported from Poland were detected by Noble (*Ann. Bot., Lond.*, n.s., **1**, 1937, 67–98), who studied the morphology and cytology of the fungus.

Uncinula aceris (DC.) Sacc., *Syll.* i, 8.
*Powdery mildew of sycamore (*Acer*). Common.

Uncinula necator (Schw.) Burr., in Magnus, *Beitrag zur Pilzflora Franken*, iii, 1900, 35; Sacc., *Syll.* i, 8 (as *U. americana* How.).
*Vine mildew. Very common under glass wherever vines are grown; occurs also, sometimes epidemically (Cardiff, 1920–1), in the open.

This mildew was first seen at Margate by M. J. B[erkeley] (*Gdnrs' Chron.* **7**, 1847, 779) and named *Oidium tuckeri* Berk. The early history of vine mildew in England and elsewhere was given by Cooke (ibid. **9**, 1878, 74) and W. G. Smith (ibid. **25**, 1886, 619–22, 660–1). Wilson (ibid. **78**, 1925, 432) controlled it with sulphur, but Durham (ibid. **96**, 1934, 13) found Shirlan ineffective.

Uredo anthoxanthina Bubák = *Puccinia poae-nemoralis.*

Uredo behnickiana P. Henn. = *Hemileia oncidii.*

Uredo festucae DC., *Flor. Fr.* vi, 1805, 82; Sacc., *Syll.* vii, 847; Wilson & Bisby, *TBMS*, **37**, 1954, 77.

*On *Festuca ovina*, Scotland. Is best kept distinct from *Uromyces dactylidis* (*festucae*), q.v., until aecidia and teleutospores are found.

Uredo fuchsiae Arth. & Holw., *Amer. J. Bot.* **5**, 1918, 538; Sacc., *Syll.* xxiii, 943; Wilson & Bisby, *TBMS*, **37**, 1954, 77.

*Fuchsia rust. Rare. Cardiff, 1932 (Smith & Rees, *TBMS*, **16**, 1932, 308); Swansea, 1932; Cheltenham, 1938; Glam, 1954; Middx, 1956; Leics, 1957.

According to Gäumann (*Phytopath. Z.* **14**, 1942, 189–91) this is the uredo-stage of *Pucciniastrum epilobii* Otth f.sp. *palustris* Gäum., *Ber. schweiz. bot. Ges.* **51**, 1941, 338.

Uredo lynckii (Berk.) Plowr., *Brit. Ured. Ustil.* 259; Sacc., *Syll.* vii, 852; Grove, *Brit. Rust Fungi*, 385; Wilson & Bisby, *TBMS*, **37**, 1954, 77.

*On *Spiranthes.* Kew Gardens, on a plant recently imported from Trinidad (M.J.B., *Gdnrs' Chron.* **8**, 1877, 242); Kelvinside, Glasgow, 1890.

Uredo oncidii P. Henn., *Hedwigia*, **41**, 1902, 15; Sacc., *Syll.* xvii, 453; Wilson & Henderson, *TBMS*, **37**, 1954, 253.

*On *Oncidium.* On *O. cavendishianum*, Royal Botanic Garden, Edinburgh, 1938.

According to Wilson and Bisby (*TBMS*, **37**, 1954, 77) the species of *Hemileia* recorded on *Oncidium* at Enfield, Middx, in 1934 (*Report*, viii, 87), with uredospores larger than those of *Hemileia oncidii* Griff. & Maubl., *Bull. Soc. mycol. Fr.* **25**, 1909, 138; Sacc., *Syll.* xxi, 599, may have belonged here.

Uredo phaji Racib., *Parasitische Algen und Pilze Javas*, ii, 1900, 32; Sacc., *Syll.* xvi, 357; Wilson & Bisby, *TBMS*, **37**, 1954, 77; Grove, *Brit. Rust Fungi*, 382 (as *Hemileia phaji* Syd.).

*On *Phaius*. Uredospores on *Phaius* sp., Kew; on imported *P. wallichi*, Dublin. According to Wilson and Bisby (loc. cit.) possibly *Uredo behnickiana* (q.v.).

Uredo quercus Brond., Duby, *Botanicum Gallicum*, ii, 1830, 893; Sacc., *Syll.* vii, 851 (as *Uredo ?ilicis* Cast.); Grove, *Brit. Rust Fungi*, 315; Wilson & Bisby, *TBMS*, **37**, 1954, 77.

*Oak rust. Uredospores on *Quercus robur* in southern England. Rare. Hastings, St Leonards, Shere, Hurstmonceaux, Salisbury; on *Q. ilex*, Devon, Guernsey.

This rust is sometimes listed as *Cronartium quercuum* Miyabe, *Bot. Mag.*, *Tokyo*, **13**, 1899, 74, but the aecidial stage on branches of *Pinus* and the teleutospores are not known in Britain.

Uredo scolopendrii Schröt., see *Milesia blechni* and *M. scolopendrii*.

Uredo tropaeoli Desm. = *Coleosporium tropaeoli*.

Urocystis agropyri (Preuss) Schröt., *Die Brand- und Rostpilze Schlesiens*, 1869–72, 7; Sacc., *Syll.* vii, 516; Sampson, *TBMS*, **24**, 1940, 303; Ainsworth & Sampson, *Brit. Smut Fungi*, 92.

*On *Agropyron repens*, *A. pungens* and *Arrhenatherum elatius*. Occasional.

Urocystis anemones (Pers.) Wint., in Rabenh., *Krypt. Fl.* i, 1881, 123; Sacc., *Syll.* vii, 518; Sampson, *TBMS*, **24**, 1940, 304; Ainsworth & Sampson, *Brit. Smut Fungi*, 94.

*On *Anemone pulsatilla*, Sussex, 1931.
*On *Trollius*, Kent, Herts, Worcs, Dur, Scotland.

Urocystis cepulae Frost, *Rep. Sec. Mass. St. Bd Agric.* **24**, 1877, 175; Sacc., *Syll.* vii, 517; Sampson, *TBMS*, **24**, 1940, 304; Ainsworth & Sampson, *Brit. Smut Fungi*, 95.

*Onion and leek (*Allium*) smut. In England first recorded from Northumb and Northants in 1918 (Cotton, *J. Bd Agric.* **26**, 1919, 168–74) on onion, and between then and 1942 a relatively small number of outbreaks on onion or leek were confirmed in widely separated districts. Up to the end of 1932 there were eighteen known outbreaks and by the end of 1942 this number had increased to eighty-seven. The fungus was treated as a dangerous parasite, subject to the provisions of the Onion Smut Order of 1921. Special surveys carried out in 1942 and 1943 revealed that this smut was more widely distributed and less harmful than previously thought, and the Order was withdrawn. The chief centres of infection are in Northumb and Dur, Beds and Hunts, and the Vale of Evesham, and there have been outbreaks in about a dozen other counties (*Report*, ix, 51).

In Scotland, where the disease was first recorded in 1912, it occurs in Tweed, Forth and Clyde and is sometimes severe in the Lothians.

*In 1936 a species of *Urocystis* with spores larger than those of *U. cepulae* was found on *Allium vineale* in Som.

Whitehead (*TBMS*, **7**, 1920, 65–71) described the life-history and cytology of *Urocystis cepulae*, and both he (*J. Minist. Agric.* **28**, 1921, 443–50) and Alcock, McIntosh and Wallace (*Scot. J. Agric.* **9**, 1926, 65–70) tested the formalin drip treatment. Ogilvie and Hickman (*Rep. Long Ashton Res. Sta. for 1937*, 96–109) had no success with seed treatment, but later, using modern organic fungicides, Croxall and Hickman (*AAB*, **40**, 1953, 176–83) obtained good control by seed treatment with ferbam and thiram.

The fungus can persist in the soil for twenty years in the absence of a susceptible host (*Report*, ix, 52).

Urocystis colchici (Schlecht.) Rabenh., *Fung. Eur.* 1861, no. 396; Sacc., *Syll.* vii, 516; Sampson, *TBMS*, **24**, 1940, 304; Ainsworth & Sampson, *Brit. Smut Fungi*, 96.

*Colchicum smut. Occasional. Wilts, Bucks, Yorks. Was reported by Berkeley (*Outl. 1860*, 335) and has sometimes been seen on imported corms, including those of *Colchicum* (*Bulbocodium*) *vernum* in 1929 and 1937.

Urocystis eranthidis (Passer.) Ainsw. & Samps., *Brit. Smut Fungi*, 96.

*Smut of winter aconite (*Eranthis*). Frequent on *E. hyemalis* and widely distributed in southern England. Lincs, Norfolk, Cambs, Berks, Glos, Som, Dorset.

Urocystis floccosa (Wallr.) Henderson, *Notes R. bot. Gdn Edinb.* **21**, 1955, 241.

*Smut of *Helleborus*. Rare. Nr Bath, Som (see Henderson, loc. cit.).

Urocystis gladiolicola Ainsw., *TBMS*, **32**, 1950, 257; Ainsworth & Sampson, *Brit. Smut Fungi*, 98.

*Gladiolus smut. Occasional. Kew, 1906; Guernsey, 1911; Som, 1923; Yorks, 1933; Devon, 1936, 1955; Corn, 1944; Lincs, 1951; Herts, 1952, 1954.

The fungus on which W. G. Smith (*Monthly Micros. J.* **16**, 1876, 304) based *Urocystis gladioli* was a species of *Papulaspora* (*P. gladioli* (Requien) Dodge & Laskaris, *Bull. Torrey bot. Cl.* **68**, 1941, 293). See *Report*, viii, 84; W. C. Moore, *Diseases of Bulbs*, 1939, 121; and Ainsworth (*TBMS*, **32**, 1950, 255–7).

Urocystis hepaticae-trilobae (DC.) Ainsw. & Samps., *Brit. Smut Fungi*, 98.

*On *Hepatica pennsylvanica*, Kew Gardens, 1890 (as *Urocystis pompholygodes*).

Urocystis occulta (Wallr.) Rabenh., in Klotzsch, *Herbarium vivum mycologicum*, ii, 1856, no. 393; Sacc., *Syll.* vii, 515; Sampson, *TBMS.* **24**, 1940, 304; Ainsworth & Sampson, *Brit. Smut Fungi*, 98.

*Stripe smut of rye (*Secale*). Of long standing but rare. First recorded 1800 (Berkeley in Smith's *English Flora*, v, Pt. II, 375); Romsey, Hants, 1920; Rothamsted, Herts, 1932; Askham Bryan, Yorks, 1936; a few attacks in Norfolk (Dillon-Weston & Taylor, *Nature, Lond.*, **152**, 1943, 160), 1949 and 1952; Ches, 1943; Bristol, 1957. Taylor and Dillon-Weston (*J. agric. Sci.* **35**, 1945, 116–18) showed that it can be controlled by seed treatment with organo-mercury dusts.

Urocystis primulicola P. Magn. = *Tuburcinia primulicola*.

Urocystis violae (Sow.) Fisch. v. Waldh., *Bull. Soc. Nat. Moscou*, **40**, 1867, 258; Sacc., *Syll.* vii, 519; Sampson, *TBMS*, **24**, 1940, 305; Ainsworth & Sampson, *Brit. Smut Fungi*, 100.

*Violet smut. Rather common and widely distributed. Ches, Oxon, Devon, Som, Hants, Herts; on viola, Lincs, 1930.

Uromyces airae-flexuosae Ferd. & Winge, *Bull. Soc. mycol. Fr.* **36**, 1920, 162; Sacc., *Syll.* xxiii, 643; Wilson & Bisby, *TBMS*, **37**, 1954, 77.

*On *Deschampsia*. Uredo- and teleutospores on *D. flexuosa*. Yorks, Surrey, widely distributed in Scotland.

Uromyces aloes (Cooke) Magn., *Ber. dtsch. bot. Ges.* **10**, 1892, 48; Sacc., *Syll.* xi, 227 (as *Uredo aloes* Cooke); Wilson & Bisby, *TBMS*, **37**, 1954, 77; Wilson & Henderson, ibid. **37**, 1954, 253.

*Aloe rust. Teleutospores on *Aloes glauca* Mill. and three other unnamed species of *Aloes* in greenhouses, Roy. Bot. Gdn, Edinburgh, 1938. Plants imported from Africa.

Uromyces ambiguus (DC.) Lév., *Ann. Sci. nat.* 3 Ser. **8**, 1847, 375; Sacc., *Syll.* vii, 543; Grove, *Brit. Rust Fungi*, 121; Wilson & Bisby, *TBMS*, **37**, 1954, 77.

*Rust of chives (*Allium*). Rare. Uredo- and teleutospores on *A. schoenoprasum*, *A. babingtonii* and *A. scorodoprasum*; on *A. ampeloprasum*, Corn. The species is very close to *Puccinia porri* (q.v.).

Uromyces anthyllidis Schröt., *Hedwigia*, **14**, 1875, 162; Sacc., *Syll.* vii, 551; Grove, *Brit. Rust Fungi*, 95; Wilson & Bisby, *TBMS*, **37**, 1954, 77.

*On lupin. This rust occurs on the native *Anthyllis vulneraria*, and Grove (loc. cit.) thought it to be confined to that genus, but Cooke (*Fung. Pests Cult. Pl.* 41) recorded it 'also on cultivated lupins, as *Lupinus luteus* and *Lupinus albus* in Great Britain. . .'.

Uromyces appendiculatus (Pers.) Unger, *Die Exantheme der Pflanzen, Wien*, 1833, 279; Sacc., *Syll.* vii, 535; Wilson & Bisby, *TBMS*, **37**, 1954, 77; syn. *U. phaseolorum* de Bary, *Ann. Sci. nat.* **20**, 1863, 80; Grove, *Brit. Rust Fungi*, 101.

*Rust of dwarf and runner bean (*Phaseolus*). Until about 1952 rare; Devon, 1939; Beds, 1941; now frequent in the east and south-east and occasionally damaging; also Corn, Glam; on *Phaseolus coccineus*, Devon and Corn.

Uromyces ari-triphylli (Schw.) Seeler, *Rhodora*, **44**, 1942, 174; Wilson & Bisby, *TBMS*, **37**, 1954, 77; syn. *Aecidium dracontii* Schw., *Synopsis fungorum in America boreali*, 291; Sacc., *Syll.* vii, 831; Cooke, *Fung. Pests Cult. Pl.* 77.

*On *Arisaema*. On *A. triphyllum*. Very rare. Aecidia only, in garden, Melbury, 1863; Kew Gardens, 1945. Only the aecidium of this American rust develops on plants imported to Europe.

Uromyces armeriae Lév., *Ann. Sci. nat.* 3 Ser. **8**, 1847, 375; Sacc., *Syll.* vii, 533 (as *U. limonii* (DC.) Lév. p.p.); Grove, *Brit. Rust Fungi*, 89; Wilson & Bisby, *TBMS*, **37**, 1954, 78.

*Armeria rust. Widely distributed on garden varieties of *Armeria maritima*, Som, Herts, Forth, Clyde, Tay; on cultivated *A. pseudarmeria* and *A. plantaginea*, Scotland; on *A. maritima* var. *alpina* and var. *elongata*, and on *A. plantaginea* var. *plantaginea*, Roy. Bot. Gdn, Edinburgh.

Uromyces betae (Pers.) Lév., *Ann. Sci. nat.* 3 Ser. **8**, 1847, 375; Sacc., *Syll.* vii, 536; Grove, *Brit. Rust Fungi*, 113; Wilson & Bisby, *TBMS*, **37**, 1954, 78.

*Beet and mangold rust. Occurs regularly in the midlands, east and south on sugar beet, mangold, garden beet and spinach beet. Occasional on seakale beet. Rarely harmful. In Scotland on red beet (Dee, Moray) and sugar beet (Forth).

Uromyces caryophyllinus Wint. = *U. dianthi*.

Uromyces colchici Mass., *Grevillea*, **22**, 1892, 6; Sacc., *Syll.* xi, 180; Grove, *Brit. Rust Fungi*, 122; Wilson & Bisby, *TBMS*, **37**, 1954, 78.

*Colchicum rust. Very rare. On *Colchicum speciosum*, *C. autumnale* and *C. variegatum*, Kew Gardens, 1892. It developed for three years on *C. speciosum*, and in the third year spread to the other two species. Yorks.

Uromyces dactylidis Otth, *Mitt. naturf. Ges. Bern*, 1861, 85; Sacc., *Syll.* vii, 540 p.p.; Grove, *Brit. Rust Fungi*, 125; *J. Bot., Lond.*, **72**, 1934, 265–6; Wilson & Bisby, *TBMS*, **37**, 1954, 78; syn. *U. festucae* Syd., *Hedwigia*, **39**, 1900, 117; Sacc., *Syll.* xvi, 269; *U. poae* Rabenh., in Marcucci, *Unio itineraria*, 1866, no. 38; Grove, *Brit. Rust Fungi*, 127.

*Rust of cocksfoot (*Dactylis*). Uredo- and teleuto-stages. Common throughout Britain from midsummer onwards.

*Rust of *Poa*. Uredo- and teleuto-stages. On *P. annua*, *P. nemoralis*, *P. pratensis* and *P. trivialis* widely distributed in Scotland; Glam, 1935.

*Rust of fescue (*Festuca*). Uredo- and teleuto-stages. First seen in Britain at Ely, Glam, 1933, on *Festuca rubra* and chewing's fescue. Since then seen occasionally in south Wales on red and creeping fescue and once on sheep's fescue. Not seen elsewhere.

The aecidial stage of this species occurs on *Ranunculus* spp.

Uromyces dianthi (Pers.) Niessl, *Beiträge zur Kenntnis der Pilze*, 1872, 12; Wilson & Bisby, *TBMS*, **37**, 1954, 78; syn. *U. caryophyllinus* Wint., *Pilze*, 149; Sacc., *Syll.* vii, 545; Grove, *Brit. Rust Fungi*, 108.

*Rust of carnation (*Dianthus*). Uredo- and teleutospores common and widespread in England and Wales; listed from Clyde, Ayrshire, East Lothian, Moray, Roxburgh.

This rust first became troublesome about 1890 (W. G. Smith, *Gdnrs' Chron.* **3**, 1888, 151; *J. Hort.* 13 July 1893, 34–6: Douglas, *Gdnrs' Chron.* **15**, 1894, 410–11).

Uromyces erythronii (DC.) Lév., *Ann. Sci. nat.* 3 Ser. **8**, 1847, 371; Sacc., *Syll.* vii, 564; Wilson & Bisby, *TBMS*, **37**, 1954, 78.

*This species was found on the foliage of *Erythronium dens-canis* imported into Dorset from the French Pyrenees in April 1936 (*Report*, viii, 84).

Uromyces fabae de Bary, *Ann. Sci. nat.* 4 Ser. **20**, 1863, 72; Sacc., *Syll.* vii, 531 p.p.; Grove, *Brit. Rust Fungi*, 97; Wilson & Bisby, *TBMS*, **37**, 1954, 78, who suggest the correct name may be *U. viciae-fabae* (Pers.) Schröt., *Hedwigia*, **14**, 98.

*Bean rust. On broad and field beans (*Vicia faba*). Usually appears late and rarely does much damage. Common in the south, less frequent in the midlands and north. Widespread in Scotland. Not uncommon on vetches. The aecidial stage is not often seen on bean (Steven, *J. Bot., Lond.*, **76**, 1936, 79); Cullompton, Devon, 1948.

Uromyces festucae Syd. = *U. dactylidis.*

Uromyces flectens Lagerh., *Svensk bot. Tidskr.* **3**, 1909, 36; Sacc., *Syll.* xxi, 541; Grove, *Brit. Rust Fungi*, 92; Wilson & Bisby, *TBMS*, **37**, 1954, 78.

*On white clover (*Trifolium repens*). Teleuto-stage frequent on wild and cultivated forms throughout Britain.

Uromyces genistae-tinctoriae Wint. = *U. laburni.*

Uromyces geranii (DC.) Lév., *Ann. Sci. nat.* 3 Ser. **8**, 1847, 371; Sacc., *Syll.* vii, 535; Grove, *Brit. Rust Fungi*, 103; Wilson & Bisby, *TBMS*, **37**, 1954, 78.

*Pelargonium rust. Occasional on a number of wild species of *Geranium*; on cultivated plants, Som, 1933.

Uromyces holwayi Lagerh., *Hedwigia*, **28**, 1889, 108; Sacc., *Syll.* ix, 294; Wilson & Bisby, *TBMS*, **37**, 1954, 78.

*On lily. A specimen of *Uredo prostii* Duby in Herb. Kew., recorded on *Lilium columbianum* in *Gdnrs' Chron.* **11**, 1879, 820, was identified by Wilson and Bisby (loc. cit.) as this species, which is widespread in North America.

Uromyces jaapianus Kleb., *Krypt. Fl. Brandenb.* **5**, 1913, 239; Sacc., *Syll.* xxiii, 653; Wilson & Bisby, *TBMS*, **37**, 1954, 79.

*Trifolium rust. On *Trifolium dubium* (*minus*), not common, War, Worcs, Som, Dee; on *T. campestre*, Som, Glos. (See also *Uromyces striatus*.)

Uromyces laburni (DC.) Fuckel, *Symb. myc.* 62; Wilson & Bisby, *TBMS*, **37**, 1954, 79; syn. *U. genistae-tinctoriae* Wint., *Hedwigia*, **19**, 1880, 36; Sacc., *Syll.* vii, 550; Sydow, *Monogr. Ured.* ii, 90.

*Genista rust. Rare. Uredo- and teleuto-stages on *Genista tinctoria* growing wild, Worcs, 1933; on a double garden form, Hailsham, Sussex, 1935; on *G. anglica*, Inverness, 1934; on imported *G. sagittalis*, Yorks, 1955.

*On broom (*Cytisus*). Recorded in Scotland on *C. scoparius* and *Ulex europaeus* by Macdonald (*TBMS*, **29**, 1946, 64–7), who distinguished the forms on *Genista*, *Cytisus* and *Ulex* as f. *anglicae*, f. *scopari* and f. *ulicis* resp. of *Uromyces genistae-tinctoriae*. According to Wilson and Bisby (loc. cit.) there is a specimen in Herb. Kew. on broom, Kent, 1865.

*On *Laburnum*. Very rare. Uredo- and teleutospores, Thornton-le-Dale, Yorks, 1946. The aecidia of this rust occur on *Euphorbia*, but have not been seen in Britain.

Uromyces lilii (Link) J. Kunze, *Hedwigia*, **12**, 1873, 144; Grove, *Brit. Rust Fungi*, 118; Wilson & Bisby, *TBMS*, **37**, 1954, 79.

*Lily rust. Occasional on *Lilium candidum*, Kew, War, Middx, Norfolk.

Uromyces limonii (DC.) Lév., *Dict. Hist. Art. Uréd.* 19; Sacc., *Syll.* vii, 532 p.p.; Grove, *Brit. Rust Fungi*, 88; Wilson & Bisby, *TBMS*, **37**, 1954, 79.

*Rust of *Limonium*. Not uncommon locally on *L. latifolium*, Worcs, Cambs, Mon, Sussex, Jersey; on *L. tataricum* var. *angustifolium*, Kent, 1947.

Uromyces loti Blytt, *Christ. Vidensk.-SelskabsForhandl.* 1896, 37; Sacc., *Syll.* xxi, 541; Grove, *Brit. Rust Fungi*, 94; Wilson & Bisby, *TBMS*, **37**, 1954, 79.

*Lotus rust. Uredo- and teleuto-stages occasional on *Lotus angustissimus*, *L. corniculatus*, *L. hispidus* and *L. uliginosus*. Suffolk, Norfolk, Corn, Tweed. The aecidia on *Euphorbia* are unknown in Britain.

Uromyces onobrychidis (Desm.) Lév., *Ann. Sci. nat.* 3 Ser. **8**, 1847, 371; Sacc., *Syll.* xxi, 544; Wilson & Bisby, *TBMS*, **37**, 1954, 79.

*Rust of sainfoin (*Onobrychis*). Uredo- and teleutospores common on *O. viciifolia*. First British record by Hadden (*TBMS*, **5**, 1917, 438).

Uromyces phaseolorum de Bary = *U. appendiculatus*.

Uromyces pisi (Pers. ex DC.) Wint. in Rabenh., *Krypt. Fl.* i, 163; Sacc., *Syll.* vii, 542 p.p.; Grove, *Brit. Rust Fungi*, 99; Wilson & Bisby, *TBMS*, **37**, 1954, 79.

*Rust of pea (*Pisum*). Rare. Uredo- and teleuto-stages on *P. sativum*, Pincoed, Glam, Sept. 1933, and a few other places in England, including Yorks, 1947, and Herts, 1953. Occurs also on *Lathyrus pratensis* in Yorks and Scotland.

The aecidia occur on *Euphorbia cyparissias*. Reported (as *Aecidium cyparissiae* DC.) in garden, Kent, 1948.

Uromyces poae Rabenh. = *U. dactylidis*.

Uromyces scillarum (Grev.) Wint., *Pilze Deutschl.* 142; Sacc., *Syll.* vii, 567; Grove, *Brit. Rust Fungi*, 120; Wilson & Bisby, *TBMS*, **37**, 1954, 79.

*Scilla rust. Frequent in England and Wales on *Scilla hispanica*; occurs also on *S. nonscripta* and *S. bifolia*; on *S. verna*, Corn; widely distributed in Scotland on *S. nonscripta*.

*Muscari rust. Teleuto-stage. Rare. On *Muscari polyanthum*, Edinburgh (Wilson, *Notes R. bot. Gdn Edinb.* **8**, 1914, 220).

Uromyces striatus Schröt., *Die Brand- und Rostpilze Schlesiens*, 1869, 11; Sacc., *Syll.* vii, 542; Wilson & Bisby, *TBMS*, 37, 1954, 80.

*Rust of trefoil (*Medicago*). Uncommon. First collected in Kent, 1903; Cardiff, 1923; Faversham, Kent, 1945; Bridgwater, Som 1945; Glos, 1949; West Suffolk, 1950. On *M. arabica* and *M. denticulata*, Som, 1948. The aecidial stage on *Euphorbia cyparissias* is not known in Britain.

Glasscock and Ware (*TBMS*, 29, 1946, 167–9) dealt with the status of *Uromyces striatus* in Britain. Several early records on species of *Trifolium* growing wild were wrongly named, and the fungus concerned was *Uromyces jaapianus* Kleb., but *U. striatus* was found on wild *Medicago arabica* at Minehead in 1920, and it has been seen on native plants in Scotland (Macdonald, *Trans. bot. Soc. Edinb.* 32, 1939, 557). The rust seen on trefoil in a variety trial at Cardiff in 1923 (*Report*, v, 40) was probably *Uromyces striatus*, though attributed at the time to *U. trifolii-repentis* Liro.

Uromyces trifolii (DC.) Lév., *Ann. Sci. nat.* 3 Ser. 8, 1847, 371; Sacc., *Syll.* vii, 534 p.p.; Grove, *Brit. Rust Fungi*, 90; Wilson & Bisby, *TBMS*, 37, 1954, 80.

*Clover rust. Widely distributed and common on red clover in England and Wales, especially in the wetter districts; Moray, Argyll. Wilson and Bisby (loc. cit.) list it on *Trifolium pratense*, *T. hybridum*, *T. incarnatum* and *T. medium*. The aecidial stage is rare; and was first observed in 1931 (*Report*, vii, 49).

Uromyces trifolii-repentis Liro, *Acta Soc. Fauna Flora fenn.* 29, 1906, 15; Sacc., *Syll.* vii, 534 p.p.; Grove, *Brit. Rust Fungi*, 91; Wilson & Bisby, *TBMS*, 37, 1954, 80.

*White clover rust. Occurs only on *Trifolium repens*. Uncommon in England and Wales; Solway, Tay, Forth, Clyde. The aecidial stage is rare.

Uromyces valerianae Fuckel, *Symb. Myc.* 63; Sacc., *Syll.* vii, 536; Grove, *Brit. Rust Fungi*, 86; Wilson & Bisby, *TBMS*, 37, 1954, 80.

*Valerian rust. On *Valeriana officinalis* on commercial herb farm, Little Ness, Salop, April 1941.

Uromyces viciae-fabae (Pers.) Schröt., see *U. fabae*.

Urophlyctis alfalfae (Lagerh.) Magn., *Ber. dtsch. bot. Ges.* **20**, 1902, 291; Sacc., *Syll.* xvii, 515.

*Crown wart of lucerne (*Medicago*). Not uncommon in southern and eastern England; Notts; War; Worcs.

The disease was first recognized in south-east England (Salmon, *Gdnrs' Chron.* **39**, 1906, 122–3; *J. S.-E. agric. Coll. Wye*, **15**, 1906, 229–30; **16**, 1907, 296–7). Line (*Proc. Camb. phil. Soc. biol. Sci.* **20**, 1921, 360–5) studied the biology of the fungus.

Urophlyctis potteri Bartlett, *TBMS*, **11**, 1926, 279.

*Lotus gall. Rare. On *Lotus corniculatus*, Newcastle, Northumb. Full description by Bartlett (*TBMS*, **11**, 1926, 266–81).

Ustilago avenae (Pers.) Rostr., *Overs. danske Vidensk. Selsk. Forh.* 1890, 13; Sacc., *Syll.* ix, 283; Sampson, *TBMS*, **24**, 1940, 296; Ainsworth & Sampson, *Brit. Smut Fungi*, 60; syn. *U. perennans* Rostr., *Ustilagineae Daniae*, 1890, 25; Sacc., *Syll.* ix, 283; Sampson, *TBMS*, **24**, 1940, 298.

The seedling-infecting black loose smut of barley, *Ustilago nigra* Tapke, 1932, belongs here. It is widely distributed in the United States but has not yet been recognized in Britain.

*Loose smut of oats (*Avena*). Commoner than *U. hordei* (q.v.) and occurs regularly in small quantity almost everywhere in England and Wales; occasionally severe. Now rare in Scotland except perhaps in the north (Dennis, *AAB*, **31**, 1944, 372) and Hebrides, on the older varieties (Gray, *Plant Path.* **3**, 1954, 60).

The biology of the two oat smuts was dealt with in a series of contributions from Aberystwyth. The existence of physiologic races was indicated by Sampson (*AAB*, **12**, 1925, 314–25; **20**, 1933, 258–71), who also discussed the viability of the chlamydospores (ibid. **15**, 1928, 586–612) and varietal resistance (ibid. **16**, 1929, 65–85). Western (ibid. **23**, 1936, 245–63) studied the mode of infection of some susceptible and resistant varieties, and later with Sampson (ibid. 25, 1938, 490–505) summarized experience over ten years on the behaviour of three physiologic races each of *U. avenae* and *U. hordei (kolleri)*. Subsequently, in unpublished work, Radcliffe (see Ainsworth & Sampson, *Brit. Smut Fungi*, 61) detected seven races of *U. avenae* and five of *U. hordei*, including those distinguished earlier. Williams and Verma (*AAB*, 41, 1954, 405–16) have examined resistance in *Avena* species.

McKay described the incidence and control of loose smut in Eire (*J. Dep. Agric. Dublin*, 32, 1933, 234–56) and investigated the method of infection and dependence on external conditions (*Sci. Proc. R. Dublin Soc.* 21, 1936, 297–307), while in Northern Ireland Muskett and Cairns (*AAB*, 19, 1932, 462–74; *J. Minist. Agric. N. Ire.* 4, 1933, 105–15) compared the relative merits of seed treatment with copper sulphate, formaldehyde, the then new organo-mercury dusts, and other chemicals.

*Smut of tall oat grass (*Arrhenatherum*). On *A. elatius*. Common everywhere.

Ustilago bromivora (Tul.) Fisch. v. Waldh. = *U. bullata*.

Ustilago bullata Berk., in Hooker, *Flora of New Zealand*, 1855, 196; Sacc., *Syll.* vii, 468; Ainsworth & Sampson, *Brit. Smut Fungi*, 65; Reid, *TBMS*, 40, 1957, 193; syn. *U. bromivora* (Tul.) Fisch. de Waldh., *Aperçu*, 1867, 22; Sacc., *Syll.* vii, 461; Sampson, *TBMS*, 24, 1940, 297.

*Ear smut of *Bromus*. Frequent in England and occasionally spoils hay crops. Found on *B. briziformis*, *B. madritensis*, *B. maximus*, *B. mollis*, *B. secalinus*, *B. sterilis* and *B. uniloides*.
*Ear smut of *Agropyron*. On *A. trachycaulum* (= *A. pauciflorum*), Kincardinesh, 1955. New British host record.

Ustilago hordei (Pers.) Lagerh., *Mitt. bad. bot. Ver.* 1889, 70; Sacc., *Syll.* ix, 283; Sampson, *TBMS*, 24, 1940, 297; Ainsworth & Sampson, *Brit. Smut Fungi*, 58; syn. *U. kolleri* Wille, *Bot. Notiser*, 1893, 9; Sampson, *TBMS*, 24, 1940, 298; *U. levis* (Kellerm. & Swing.) Magn., *Verh. bot. Ver. Brandenb.* 37, 1895, 69.

*Covered smut of barley (*Hordeum*). Fairly common in England but less so than formerly, and rarely serious. Rare in the newer varieties in northern Scotland, but persists in Orkney, Shetland and the outer Hebrides (Gray, *Plant Path.* 3, 1954, 60).

Carruthers (*J. R. agric. Soc.* 7, 1896, 143–6) and Percival (*J. S.-E. agric. Coll. Wye*, 11, 1902, 81–3) published notes on the disease, and Salmon and Wormald (*J. Bd Agric.* 24, 1918, 1388–94) carried out seed-disinfection trials.

*Covered smut of oats. Rarely distinguished from *Ustilago avenae* (q.v.) but less common than that species, though regular on *Avena strigosa* varieties in mid-Wales, north-west Scotland and the Hebrides. Williams and Verma (*AAB*, 41, 1954, 405–16) studied the resistance of different *Avena* spp.

Ustilago hypodytes (Schlecht.) Fr., *Syst. Myc.* iii, 518; Sacc., *Syll.* vii, 453; Ainsworth & Sampson, *Brit. Smut Fungi*, 56.

*Sheath smut of grasses. Widely distributed, but usually found on grasses of little agricultural value, such as *Agropyron repens*, *A. acutum*, *Bromus erectus*, *Elymus arenarius* and *Festuca gigantea*. Angus (*TBMS*, 39, 1956, 115–24) discussed the taxonomy of the species and its allies.

Ustilago kolleri Wille = *U. hordei.*

Ustilago levis (Kellerm. & Swing.) Magn. = *U. hordei.*

Ustilago macrospora Desm., *Pl. Crypt. franc.* 1850, no. 2127; Ainsworth & Sampson, *Brit. Smut Fungi*, 69.

*On *Agropyron*, *Bromus* and *Calamagrostis*. Few records but probably confused with *Ustilago striiformis* (q.v.). First record on *Agropyron junceum*, Norfolk (*Trans. Norfolk Norw. Nat. Soc.* 13, 1932, 302); on *A. repens*, east Scotland, 1947, Surrey, Guernsey.

Ustilago maydis (DC.) Corda, *Ic. Fung.* 5, 1842, 3; Sacc., *Syll.* vii, 472; Ainsworth & Sampson, *Brit. Smut Fungi*, 66; syn. *U. zeae* (Beckm.) Unger, *Einfluss des Bodens*, 1836, 211; Sampson, *TBMS*, 24, 1940, 300.

*Maize smut. Occasional in south and east England. Surrey, Berks, Middx, Cambs, Suffolk, Norfolk. First recorded from Little Canford (M. J. B[erkeley], *Gdnrs' Chron.* **10**, 1850, 675).

Ustilago nigra Tapke, see *U. avenae.*

Ustilago nuda (Jens.) Rostr., *Tidsskr. Landøkon.* **8**, 1889, 745; Sacc., *Syll.* ix, 283; Sampson, *TBMS*, **24**, 1940, 298; Ainsworth & Sampson, *Brit. Smut Fungi*, 63; syn. *U. tritici* (Pers.) Rostr., *Ustilagineae Daniae*, 1890, 140; Sacc., *Syll.* ix, 283; Sampson, *TBMS*, **24**, 1940, 299.

*Loose smut of wheat (*Triticum*). Widely distributed throughout Britain but of relatively minor importance except in a few of the newer varieties (e.g. Vilmorin). Usually less than 1 % of the ears in a crop affected; occasionally 15 % or more.

Batts (*TBMS*, **38**, 1955, 465–75) traced how infection of the new grain occurs and showed that the mycelium normally enters the ovary wall and not, as had been thought, via the style. He also (*AAB*, **43**, 1955, 533–7) distinguished three physiologic races of the smut in trial plots at Cambridge, and (ibid. **44**, 1956, 437–52) experimented with various ways of controlling the smut in wheat and barley by hot-water treatment.

*Loose smut of barley. At one time less common in England and Wales than the covered smut and for many years was unimportant. After the 1939–45 War it became prevalent in introduced Scandinavian varieties. In Scotland more common than covered smut (Foister & Thompson, *Scot. Agric.* **36**, 1957, 220–1) and is conspicuous annually in north Scotland in stiff-strawed varieties (Gray, *Plant Path.* **3**, 1954, 59–62).

Ustilago olida (Riess) Cifferi, *Flora Italica Cryptogama. Ustilaginales*, 1938, 296; syn. *Tilletia olida* (Riess) Wint., in Rabenh., *Krypt. Fl.* II, i, 1884, 107; Sacc., *Syll.* vii, 482.

*Smut on *Brachypodium*. Single British record, Dorking, Surrey, 1953 (*TBMS*, **37**, 1954, 183).

Ustilago perennans Rostr. = *U. avenae.*

Ustilago scillae Cif., see *U. vaillantii.*

Ustilago striiformis (Westend.) Niessl, *Hedwigia*, **15**, 1876, 1; Sacc., *Syll.* vii, 484 (as *Tilletia striiformis* (Westend.) Magn.); Sampson, *TBMS*, **24**, 1940, 299; Ainsworth & Sampson, *Brit. Smut Fungi*, 68.

*Stripe smut of grasses. Has been found in Britain on *Agropyron repens, Arrhenatherum elatius, Dactylis glomerata, Deschampsia caespitosa, Festuca ovina, F. rubra, Holcus lanatus, H. mollis, Lolium perenne, Phleum pratense, Poa pratensis* and *P. annua*; and affecting *P. trivialis* on bowling greens, Devon. (See also *Ustilago macrospora*.)

Ustilago tragopogonis-pratensis (Pers.) Rouss., *Flor. Calvados*, 1806, 47; Sacc., *Syll.* vii, 477 (as *U. tragopogi* (Pers.) Schröt.); Sampson, *TBMS*, **24**, 1940, 299; Ainsworth & Sampson, *Brit. Smut Fungi*, 73.

*Smut of salsify (*Tragopogon*). Rare. East midlands, 1927.

Ustilago tritici (Pers.) Rostr. = *U. nuda.*

Ustilago vaillantii Tul., *Ann. Sci. nat.* 3 Ser. **7**, 1847, 90; Sacc., *Syll.* vii, 465; Sampson, *TBMS*, **24**, 1940, 300; Ainsworth & Sampson, *Brit. Smut Fungi*, 59.

*Anther smut of *Chionodoxa*. Rare. On *C. luciliae*, Kew, 1893; Oxted, Surrey, 1923; on *C. sardensis*, Cambridge, 1941 (*Report*, viii, 81); Forth.

*Anther smut of *Muscari*. Rare. On *M. botryoides*, Aberystwyth, 1938; Bucks, 1946; Jersey, 1953; also on *M. cyaneoviolaceum*.

*Scilla smut. Occasional. Som (W. G. Smith, *Gdnrs' Chron.* **15**, 1894, 463) on *Scilla bifolia* and *S. taurica alba*; on *S. verna*, Mounts Bay, Corn, 1938, Clyde, Forth.

The life history of this smut was described by Ivy Massee (*J. econ. Biol.* **9**, 1914, 9–14). Ciferri (*Flora Italica Cryptogama. Ustilaginales*, 1938, 353) distinguished this form as *Ustilago scillae* Cif. (*Ann. mycol., Berl.*, **29**, 1931, 24).

Ustilago violacea (Pers.) Fuckel, *Symb. Myc.* 1869, 39; Sacc., *Syll.* vii, 474; Sampson, *TBMS*, **24**, 1940, 300; Ainsworth & Sampson, *Brit. Smut Fungi*, 70.

*Anther smut of carnation (*Dianthus*). Fairly common in Berks, Middx and neighbouring counties. Occasional elsewhere but not known in Scotland on this host.

The disease was described by White (*Gdnrs' Chron.* **100**, 1936, 254; *Rep. exp. Res. Sta. Cheshunt for 1938*, 55–6). (See also *J. R. hort. Soc.* **32**, 1907, lxxxiii.) Baker (*Ann. Bot., Lond.*, **11**, 1947, 333–48) carried out laboratory and garden experiments to demonstrate the method of flower infection in *Melandrium* (*Lychnis*) spp.

Ustilago zeae (Bechm.) Unger = *U. maydis.*

Ustulina deusta (Fr.) Petrak, *Ann. mycol., Berl.*, **19**, 1921, 279; syn. *U. vulgaris* Tul., *Sel. Fung. Carp.* ii, 23; Sacc., *Syll.* i, 351; Bisby & Mason, *TBMS*, **24**, 1940, 156.

This root parasite of broad-leaved trees in temperate regions occurs constantly on beech (*Fagus*) in Britain, and rarely on other deciduous trees. It was thoroughly investigated by W. H. Wilkins. Having first surveyed all the previous literature on the genus *Ustulina* and provided a list of hosts (*TBMS*, **18**, 1934, 321–46), he dealt with the parasitism of *U. deusta*, especially on lime, and concluded (ibid. **22**, 1938, 47–93; **23**, 1939, 65–85) that conidia were the most probable agent of infection.

*Butt rot of beech (*Fagus*). Frequent, especially where the trees are growing under unsuitable conditions, as on thin chalky soils (Day, *Quart. J. For.* **40**, 1946, 72). (See also Wilkins, *TBMS*, **26**, 1943, 169–70.)

*On elm (*Ulmus*). Princes Risborough, Bucks (Wilkins, *TBMS*, **23**, 1939, 171–85).

*On lime (*Tilia*). Rare on this resistant tree. Oxford, Woodstock (Wilkins, *TBMS*, **20**, 1936, 133–56).

Ustulina vulgaris Tul. = *U. deusta.*

Valsa ambiens (Fr.) Fr., *Summ. Veg. Scand.* 412; Sacc., *Syll.* i, 131; Bisby & Mason, *TBMS*, **24**, 1940, 191; stat. conid. *Cytospora ambiens* Sacc., *Mich.* i, 519; *Syll.* iii, 268; Grove, *Coelomycetes*, i, 256.

*Die-back of apple (*Malus*). A weak parasite in Worcs (Ogilvie, *J. Pomol.* **11**, 1933, 205–13).

Valsa leucostoma Fr., *Summ. Veg. Scand.* 411; Sacc., *Syll.* i, 139; Bisby & Mason, *TBMS*, **24**, 1940, 142; stat. conid. *Cytospora leucostoma* (Pers.) Sacc., *Mich.* ii, 264; Sacc., *Syll.* iii, 254; Grove, *Coelomycetes*, i, 277.

*Not infrequent on plum and cherry, and formerly regarded as causing die-back. Belgrave (*AAB*, **2**, 1915, 183–94) studied this disease on plum, as also did Cayley (ibid. **10**, 1923, 253–75) on plum, and Wormald (*J. S.-E. agric. Coll. Wye*, **21**, 1912, 367–80) on cherry, but it was probably mainly the one now known as bacterial canker caused by *Pseudomonas mors-prunorum* Wormald.

Valsa rhodophila Berk. & Br., *Ann. Mag. nat. Hist.* 3 Ser. **3**, 1859, 367; Sacc., *Syll.* i, 136; Bisby & Mason, *TBMS*, **24**, 1940, 143; stat. conid. *Cytospora rhodophila* Sacc., *Syll.* iii, 253; Grove, *Coelomycetes*, i, 280.

*On branches of *Rosa*. Harborne, nr. Birmingham.

Venturia cerasi Aderh., *Landw. Jb.* **29**, 1900, 541; stat. conid. *Fusicladium cerasi* (Rabenh.) Sacc., *Syll.* iv, 346; Wakefield & Bisby, *TBMS*, **25**, 1941, 86.

The perfect stage has not been found in Britain.

*Cherry scab. Occasional. Kent, Sussex, East Anglia, Ches, Dee, Clyde. It becomes epidemic some years in Kent (M. H. Moore, *Rep. E. Malling Res. Sta. for 1943*, 54–6).

Venturia chlorospora (Ces.) Karst., *Myc. fenn.* ii, 1873, 189; Sacc., *Syll.* i, 586; Bisby & Mason, *TBMS*, **24**, 1940, 172; stat. conid. *Fusicladium saliciperdum* (Allesch. & Tub.) Lind, *Ann. mycol., Berl.*, **3**, 1905, 430; Sacc., *Syll.* xxii, 1376; Wakefield & Bisby, *TBMS*, **25**, 1941, 86.

The perfect stage has not yet been seen in Britain.

*Willow scab. Cambs, Norfolk, Som, Lanarkshire, Argyll, Angus, Moray, and doubtless elsewhere. Occurs especially on *Salix fragilis* var. *decipiens*. Alcock (*Trans. R. Scot. arb. Soc.* **38**, 1924, 128–30) described it on *S. alba* var. *vitellina* in Scotland.

See *Physalospora miyabeana* for the confusion between scab and black canker.

Venturia inaequalis (Cooke) Wint. em. Aderh., *Hedwigia*, **36**, 1897, 81; Sacc., *Syll.* i, 587; Bisby & Mason, *TBMS*, **24**, 1940, 172; stat. conid. *Fusicladium dendriticum* (Wallr.) Fuckel, *Symb. Myc.* 357; Sacc., *Syll.* iv, 345; Wakefield & Bisby, *TBMS*, **25**, 1941, 86.

*Apple scab. Of annual occurrence in all districts but varying in intensity with the season. Usually severe after a wet May.

Scab was doing much harm as long ago as 1862 (*Gdnrs' Chron.* **22**, 689). Salmon (*J. Bd Agric.* **15**, 1908, 182–95) gave a good general account of it, and since then it has received as much attention from research workers as any fungus disease occurring in Britain. About 150 scientific papers have been published about it and many of these are mentioned below. For useful summaries of progress made from time to time the following papers by R. W. Marsh should be consulted: *Rep. Long Ashton Res. Sta. for 1933*, 88–95; *for 1954*, 153–61, which covers work done at Long Ashton from 1944 to 1954. (See also M. H. Moore, *Rep. E. Malling Res. Sta. for 1948*, 153–6; *for 1956*, 161–7.)

VARIETAL SUSCEPTIBILITY. Articles dealing particularly with varietal susceptibility include those by Salmon and Ware (*J. Pomol.* **4**, 1925, 230–9), Johnstone (*J. Pomol.* **9**, 1931, 30–52, 195–227) and, as regards fruit, by Cook (*Gdnrs' Chron.* **44**, 1943, 125).

OCCURRENCE ON STORED FRUIT. Special reference to this aspect of the disease was made by Wormald (*J. Minist. Agric.* **41**, 1934, 551–6; *Gdnrs' Chron.* **107**, 1940, 257), Cheal (*TBMS*, **20**, 1936, 310–11) and Lugeon (*Gdnrs' Chron.* **107**, 1940, 315).

OVERWINTERING. The ascigerous stage was first reported and studied on the leaves by Salmon and Ware (*Gdnrs' Chron.* **75**, 1924, 190; *J. Minist. Agric.* **31**, 1924, 546–54) and on the fruit by McKay (*TBMS*, **31**, 1948, 285). M. H. Moore (*Rep. E. Malling Res. Sta. for 1938*, 267) gave dates for the discharge of ascospores from overwintered leaves in Kent over a series of years, and with the aid of spore traps Hirst *et al.* (*Plant Path.* **4**, 1955, 91–6) were able to trace the origin of scab outbreaks in the Wisbech area to ascospore discharge.

Overwintering on the bud scales was discussed by Salmon and Ware (*Gdnrs' Chron.* **89**, 1931, 437–8), by McKay (*Sci. Proc. R. Dublin Soc.* **21**, 1938, 623–40) and by Dillon-Weston, Storey and Ives (*Gdnrs' Chron.* **132**, 1952, 194); and twig infection by Salmon (ibid. **40**, 1906, 21–3), M. H. Moore (*J. Pomol.* **8**, 1930, 229–47; *Fruit Gr.* **73**, 1932, 591–3) and Marsh and Walker (*J. Pomol.* **10**, 1932, 71–90).

BIOLOGY (MISC.). Wiltshire (*AAB*, **1**, 1915, 335–50) followed the histology of infection. Gilliver (ibid. **34**, 1947, 136–43) examined the effect of plant extracts on germination of the conidia, and Fothergill and Ashcroft (*J. gen. Microbiol.* **12**, 1955, 387–95) studied the nutritional requirements of the fungus in culture. Kirkham (*J. gen. Microbiol.* **16**, 1957, 360–73) investigated the relationships between cultural characters and pathogenicity.

EFFECT OF MANURING. M. H. Moore (*J. Pomol.* **14**, 1936, 77–96) and Muskett and Colhoun (*AAB*, **25**, 1938, 50–67) paid special attention to this subject, and Moore and Bennett (ibid. **39**, 1952, 588–92) showed that moderate annual dressings of ammonium and potassium sulphate over an eleven-years period had no consistent effect on the incidence of scab.

ASSESSMENT OF DAMAGE. Croxall, Gwynne and Jenkins devised methods for the rapid assessment of the amount of scab on leaves (*Plant Path.* **1**, 1952, 39–41) and fruit (ibid. **1**, 1952, 89–92). (See also *AAB*, **40**, 1953, 600–3.)

CONTROL: *Early work.* Scab spraying in Britain has passed through a number of phases. No useful purpose would be served by enumerating the publications dealing with all the spraying trials carried out after the 1914–18 War, because materials and methods have radically changed. Those specially interested will find them in the *J. Minist. Agric.* (from vol. **35**, 1925), the *J. S.-E. agric. Coll. Wye* (from vol. **26**, 1929), the *J. Pomol.* (from vol. **2**, 1921) and in the *Reports of the Long Ashton and East Malling Research Stations* from 1928. They are concerned mainly with the relative merits of lime sulphur and Bordeaux mixture, and the problems of spray damage caused by different concentrations of the sprays on different varieties.

Developments in application. In the early 1930's fundamental changes were made. The low-pressure type of soft spray was abandoned in favour of a driving spray, which demanded higher pressures and greater mobility, and so inspired improvements in spraying machines, lances and nozzles. These developments were described by Turnbull and Kent among others (*J. Minist. Agric. 1933–8*; *Bull. Minist. Agric., Lond.*, no. 5, 1939, 76 pp.).

Meanwhile, Bordeaux mixture had been replaced by lime sulphur, and many trials were made to discover the safest effective dilution and the best times to spray, as well as to devise combined insecticidal and fungicidal washes (Kearns, Marsh & Martin, *Reps. Long Ashton. Res. Sta. for 1932–7*; M. H. Moore & Montgomery, *Reps. E. Malling Res. Sta. for 1933–6*; Austin, Jary and Martin, *J. S.-E. agric. Coll. Wye*, **36**, 1935, 95–9).

Developments in fungicides. It was not long before new fungicides, particularly organic materials, began to come into prominence, and most of these were compared with lime sulphur and with one another (Montgomery, M. H. Moore & Shaw, *Rep. E. Malling Res. Sta. for 1935*, 198–203; M. H. Moore, ibid. *for 1937*, 229–35; Marsh, *Rep. Long Ashton Res. Sta. for 1939*, 42–51). The earlier ones included mercurated lead arsenate (Shaw & M. H. Moore, *Rep. E. Malling Res. Sta. for 1944*, 128–30; also in *Fruit Gr.* **100**, 1945, 121–2), and copper sebacate and organic sulphurs (Kearns, Marsh & Martin, *Rep. Long Ashton Res. Sta. for 1945*, 132–40). Later, a wide variety was tested by the staff at Long Ashton (*J. Pomol.* **23**, 1947, 185–205; *J. hort. Sci.* **24**, 1948, 284–7; **28**, 1953, 196–206; *Reps. Long Ashton Res. Sta.* from 1948 onwards) and at East Malling (M. H. Moore & Kirby, *Rep. E.*

Malling Res. Sta. for 1951, 200–4). References will be found in these papers, for instance, to trials with the thiocarbamates (including thiram and ferbam), organo-mercury foliage sprays, glyoxalidines, the quinones and quinolines, and captan (S.R. 406).

For a general guide to the results see Marsh (*Agriculture, Lond.*, **58**, 1951, 22–4). M. H. Moore (*Rep. E. Malling Res. Sta. for 1951*, 139–47) devised a method of using apple stocks for intensive field screening of fungicides. Miller (*Plant Path.* **5**, 1956, 119–21) gave some figures for mercury residues on the fruits from trees sprayed with organo-mercury foliage fungicides.

Further changes in application—low-volume spraying. Concurrently with the testing of an increasing number of new fungicides a revolution was taking place in the methods of applying sprays. The changes in the 1930's referred to above had involved high pressures and large mobile machines capable of carrying out and applying 200–400 gal. per acre to give complete coverage for the control of scab. When hormone weed-killers were introduced, considerably lighter machines of a low-pressure type proved sufficient to apply the 100–120 gal. per acre at first used for killing weeds. With air-blast machines the quantity was soon still further reduced to 10–20 gal. per acre, and the development abroad of spraying insecticides from the air necessitated using small quantities of material, even down to ½ gal. per acre. Meanwhile, M. H. Moore (*Rep. E. Malling Res. Sta. for 1947*, 129–31) had shown experimentally that undiluted lime-sulphur could safely be applied to apple trees, and (ibid. *for 1952*, 132–5; *for 1953*, 210–12; see also *AAB*, **45**, 1957, 11–18) that it was possible to obtain good control of scab, mildew and red spider, without causing damage, with as little as 3 gal. lime sulphur per acre, provided high concentrations (81–100 %) were used (see also *Agriculture, Lond.*, **63**, 1957, 579–82). These factors had their repercussions on commercial fruit-tree spraying; growers changed over to low-volume, automatic spraying at about 50 gal. per acre, and research along these lines was intensified. High- and low-volume application of the now established materials was tested at varying concentrations (see *Reps. Long Ashton Res. Sta. for 1952–55*).

SPRAY TIMING AND WEATHER. The most recent development concerns the possibility of relating outbreaks of scab and timing of sprays to meteorological data. Many years ago M. H. Moore (*Rep. E. Malling Res. Sta.* 16–18 (1928–30), II Suppl. 1931, 157–76) discussed the effect of weather on scab in general terms, and work in America by Keitt and Wallace (*Res. Bull. Wis. agric. Exp. Sta.* 73, 1926, 104 pp.), recently underlined by Mills and Dewey (*Ext. Bull. Cornell agric. Exp. Sta.* no. 711, 1947, 18–25), had indicated that scab attacks are closely related to wetness of leaf over limited periods of time. This is now being put to the test in the Wisbech area and elsewhere. Storey and Ives (*Plant Path.* 5, 1956, 1–8) have pointed out how the organo-mercury compounds have replaced lime sulphur in the Wisbech area, and they believe that the best use of the eradicant properties of the mercury compounds can be made by correctly timing spray applications in the light of weather conditions.

SPRAY DAMAGE. Incidental references are made to spray damage in many publications on scab control. Special attention was paid to it by M. H. Moore (*J. Pomol.* 23, 1947, 139–48) when he investigated the effect of ferrous sulphate in diminishing spray damage. With Preston and Bennett (*Rep. E. Malling Res. Sta. for 1950*, 132–6) he also discussed the influence of rootstock on spray damage.

MISCELLANEOUS. M. H. Moore (*J. Pomol.* 22, 1946, 76–91) discussed the place of spreaders in spray programmes, and with Pearce (ibid. 22, 1946, 62–8) the variations in efficacy of routine spraying when the same programme is carried out on similar blocks of trees by different individuals.

Venturia pirina Aderh., *Landw. Jb.* 25, 1896, 875; Sacc., *Syll.* xxii, 150; Bisby & Mason, *TBMS*, 24, 1940, 172; stat. conid. *Fusicladium pirinum* (Lib.) Fuckel, *Symb. Myc.* 357; Sacc., *Syll.* iv, 346; Wakefield & Bisby, *TBMS*, 25, 1941, 86.

*Pear scab. Occurs annually in all districts, varying greatly in intensity from season to season.

Pear scab was known to Berkeley (*J. hort. Soc.* 8, 1853, 40–1;

Gdnrs' Chron. **8**, 1848, 398), but was given little scientific attention until after Salmon and Ware (ibid. **75**, 1924, 274) reported finding the ascigerous stage in Kent and Devon, and had described the fungus on the spur wood and bud scales (ibid. **91**, 1932, 446–7). Marsh (*J. Pomol.* **11**, 1933, 101–12) and Cheal and Dillon-Weston (*AAB*, **25**, 1938, 206–8) investigated twig infection, and Stanton (*TBMS*, **36**, 1953, 90–103) discussed field variation in the fungus and studied the factors involved in breeding for resistance (*AAB*, **40**, 1953, 184–91, 192–6). Kirkham (*J. gen. Microbiol.* **16**, 1957, 360–73) studied the relationship between cultural characters and pathogenicity.

Spraying trials were undertaken by Cheal (*Gdnrs' Chron.* **93**, 1933, 139), Marsh (*J. Pomol.* **11**, 1933, 101–12), M. H. Moore (*J. Minist. Agric.* **40**, 1933, 111–19; *Rep. E. Malling Res. Sta. for 1932*, 99–108) and Martin, Salmon and Ware (*J. S.-E. agric. Coll. Wye*, **34**, 1934, 145–54). Kirby and Bennett (*Rep. E. Malling Res. Sta. for 1951*, 187–8) followed the effects of phenyl mercury chloride on leaves.

Vermicularia atramentaria Berk. & Br. = *Colletotrichum atramentarium.*

Vermicularia circinans Berk. = *Colletotrichum circinans.*

Vermicularia herbarum Westend., *Exs.* no. 393; Sacc., *Syll.* iii, 226; Grove, *Coelomycetes*, ii, 239 (as *V. herbarum* forma *dianthi* Westend.).

*On carnation (*Dianthus*). Associated with rotting basal leaves and stems of carnation, Middx, 1932. In view of Duke's conclusions (*TBMS*, **13**, 1928, 156) the species is more correctly a *Colletotrichum* (*Report*, vii, 96).

Vermicularia holci Syd. = *Colletotrichum holci.*

Vermicularia trichella Grev. = *Colletotrichum trichellum.*

27-2

Verticillium albo-atrum Reinke & Berth., *Die Zersetzung der Kartoffel durch Pilze*, 1879, 75; Sacc., *Syll.* x, 547; Wakefield & Bisby, *TBMS*, **25**, 1941, 60.

With *V. dahliae* (q.v.) the cause of Verticillium wilt of a large number of plants, many of which are listed below. Isaac (*TBMS*, **32**, 1949, 138–57) put forward arguments for retaining specific rank for the microsclerotial, dark mycelium and chlamydospore strains, viz. *V. dahliae*, *V. albo-atrum* and *V. nigrescens* Pethyb. respectively. He also studied (*AAB*, **40**, 1953, 630–8) spread of the fungus in the soil by root contact.

*On potato (*Solanum*). Probably much more widespread than is generally supposed. It was said to have been widely distributed in 1918, especially in the south, though sporadic, but from then until 1940 thought to be uncommon and unimportant in Britain. (See Dale (*Ann. Bot., Lond.*, **26**, 1912, 129) and Pethybridge (*Sci. Proc. R. Dublin Soc.* **15**, 1916, 63–92).)

PASTURE AND FORAGE CROPS

*On lucerne (*Medicago*). Since its first recognition near Edinburgh in July 1950 and in Essex and Norfolk the following year (Noble, Robertson & Dowson, *Plant Path.* **2**, 1953, 31–3), this disease has been reported fairly frequently from East Anglia and south-eastern England, sometimes causing heavy loss. Also seen in Berks, Hants, Bucks, west midlands, west Wales. Isaac (*AAB*, **45**, 1957, 550–8) studied the disease, and Zaleski (*Plant Path.* **6**, 1957, 137–42) varietal susceptibility.

VEGETABLES

*On cucumber (*Cucumis*). Of regular occurrence but not serious. The primary cause may be something other than the *Verticillium* (see *Reps. exp. Res. Sta. Cheshunt for 1945–6*). The fungus from plants affected with wilt is sometimes reported as *Verticillium* sp. and sometimes as *V. dahliae*.

*On melon (*Cucumis*). Verticillium wilt is not infrequent but is usually attributed to *Verticillium* sp. *V. albo-atrum*, Essex, 1926; Sussex, 1946; Kent, 1929 (on the fruits). (See also *V. dahliae*.)

*On mint (*Mentha*). Kent, 1942.

*On tomato (*Lycopersicon*). Very common in England and Wales under glass in all districts from April to early June and in August–September. It usually disappears for a time in June–July in hot summers. Sporadic in Scotland.

The disease was investigated by Bewley (*AAB*, **9**, 1922, 116–34), and Williams (*Reps. exp. Res. Sta. Cheshunt for 1927*; *for 1942–6*) gave it much attention over a long period. Roberts (*AAB*, **30**, 1943, 327–31; **31**, 1944, 191–3) studied the factors influencing infection, and McKay (*Gdnrs' Chron.* **117**, 1945, 24) described outbreaks in newly erected glasshouses. Williams (*Rep. exp. Res. Sta. Cheshunt for 1949*, 27–8) tested the resistance of certain American varieties in England, and Selman and Pegg (*AAB*, **45**, 1957, 674–81) the growth response of seedlings to infection.

FRUIT

*On plum (*Prunus*). Kent (Keyworth, *Rep. E. Malling Res. Sta. for 1943*, 52–4).
*On quince (*Cydonia*). On quince A stocks, Thornbury, Glos, 1957.
*On strawberry (*Fragaria*), see *Verticillium dahliae*.

HOP

*Verticillium wilt of hop (*Humulus*). The disease occurs in two forms: a 'fluctuating' one that varies in intensity from year to year and is never very serious; and a 'progressive' form which steadily becomes more extensive and severe.

The fluctuating form is not uncommon in south-east England and occasional in the west midlands. It is caused by *Verticillium dahliae* or by a mild strain of *V. albo-atrum*.

The progressive form is locally very serious in the Kent Weald and east Sussex, with a few outbreaks in the south-east outside the main infected area. It was first confirmed in the west midlands at Rosemaund in 1957, but has not been seen elsewhere in that area or in other hop areas. This form is caused by a virulent strain of *V. albo-atrum*.

Verticillium wilt was first noticed in 1924 at Penshurst, Kent (Harris, *Rep. E. Malling Res. Sta. for 1925*, 13, II Suppl., 1927, 92–3), and it was of the fluctuating type. The progressive form did not begin to cause alarm until about 1938 (Harris & Furneaux, *Rep. E. Malling Res. Sta. for 1927*, 257–8; Keyworth, ibid. *for 1938*, 224–8; also in *J. S.-E. agric. Coll. Wye*, **44**, 1939, 23–9), by which time severe attacks had occurred in the Paddock Wood district of Kent.

By the end of 1938 there were twenty-eight known outbreaks in Kent and Sussex, mainly in the Weald area, and this number had increased to eighty-three in 1942, to about 140 in 1951 and to 200 in 1955 (Jary, *Agriculture, Lond.*, **62**, 1955, 30–4). Many acres of land had to be abandoned for hop growing, and by 1956 about 3000 acres or roughly one-seventh of the total hop acreage was infected. Meanwhile strenuous efforts were made to check its spread. In 1943 the Ministry of Agriculture introduced a scheme for the certification of hop gardens for freedom from wilt. Later, stringent statutory measures were taken for the compulsory notification of suspected outbreaks and the destruction of infected bines. These measures were incorporated in the Progressive Verticillium Wilt of Hops Orders of 1947, 1953 and 1957. These efforts unfortunately did little to affect improvement in the main affected area, but they were instrumental in delaying the progressive form from reaching the west midlands and other hop areas.

An extensive research programme has been in operation at the East Malling Research Station ever since the disease became prominent.

BIOLOGY. Since Keyworth (*AAB*, **29**, 1942, 346–57; *Agriculture, Lond.*, **51**, 1945, 556–9) clearly distinguished between the fluctuating and progressive types of attack, specialized studies have been made, including the pathogenicity of isolates from both types (Isaac & Keyworth, *AAB*, **38**, 1948, 243–9); the behaviour of a fluctuating attack (Keyworth, *J. hort. Sci.* **24**, 1948, 149–56); the incidence of symptoms in bines of different ages (Keyworth & Hitchcock, *Rep. E. Malling Res. Sta. for 1947*, 148–9); the influence of nutrition on infection (Keyworth & Hewitt, *J. hort. Sci.* **24**, 1948, 219–27); and the significance of toxic metabolites (Talboys, *TBMS*, **40**, 1957, 415–27). An interim summary of research up to 1946 was given by Keyworth (*Brew. Tr. Rev.* **61**, 1947, 100–3).

Talboys and Wilson (*Rep. E. Malling Res. Sta. for 1953*, 158–61) devised a method for determining the pathogenicity of strains of *V. albo-atrum*.

RESISTANT VARIETIES. Attempts were made to find resistant varieties. Keyworth (*J. Pomol.* **23**, 1947, 99–103) described some early Wye selections, of which the two most promising ones (Keyworth's Early and Mid-Season) were given extended trials for brewing and other qualities (Keyworth, *Rep. E. Malling Res. Sta. for 1946*, 157–9; Salmon, *J. Inst. Brew.* **46**, 1949, 234–6). Though useful they were not acceptable as a substitute for Fuggle. Further studies were made by Keyworth, Hitchcock and Goode (*Rep. E. Malling Res. Sta. for 1952*, 112–19), and the position reached in 1953 was outlined by Harris (ibid. *for 1953*, 208–9). Whitbread's Golding variety has shown satisfactory tolerance (Wilson, ibid. *for 1956*, 128–30).

Keyworth (*Nature, Lond.*, **171**, 1953, 656–7; *AAB*, **40**, 1953, 344–61) also investigated the nature of resistance and concluded that it was due to the reaction of the roots and not of the stems. By graft testing Talboys and Wilson (*Rep. E. Malling Res. Sta. for 1955*, 126–30) showed that this effect of the root does not result from translocation of resistance factors from it to the stem.

ORNAMENTALS

*Antirrhinum wilt. Occasional in southern areas since 1924. Reported also from Clyde and Tweed. The species is not usually determined, but both *V. albo-atrum* and *V. dahliae* have been implicated. In experiments under controlled conditions Isaac (*AAB*, **44**, 1956, 105–12) showed that of the five species tested *V. albo-atrum* was most virulent, with *V. dahliae* next. *V. nigrescens* (q.v.) was more mildly pathogenic, while *V. nubilum* (q.v.) and *V. tricorpus* (q.v.) did not attack antirrhinum in a well-balanced soil. Isaac (ibid. **45**, 1957, 512–15) also studied the effect of nitrogen supply on the disease.

*On *Campanula isophylla*, Berks, 1939.

*Chrysanthemum wilt. Has become prominent in many southern

and eastern districts since about 1934; occasional in northern England and in Scotland. Studied by Oyler (*Reps. exp. Res. Sta. Cheshunt for 1937–9*). *Verticillium dahliae* (q.v.) is often implicated in the south-east.

*On *Dahlia*, Kent, 1950; Westmorland, 1951.

*On *Romneya coulteri* (perennial poppy), Burghclere, Hants, 1948.

*On rose, Middx, 1938; Surrey, 1954.

*On sweet pea (*Lathyrus*), Ches, 1931; Som, 1946.

Verticillium cinerescens Wr., *Arb. Biol. Reichsanst. Land. Forstw.* **17**, 1929, 296; Wakefield & Bisby, *TBMS*, **25**, 1941, 61.

*Carnation wilt. Common and serious under glass in Middx, Berks, Bucks and other southern districts; less frequent elsewhere in England. Listed in Clyde, Moray, Dee.

Wickens (*AAB*, **22**, 1935, 630–83) clearly distinguished several diseases previously confused under the names stem rot and wilt, and White, who had studied wilt for some years and attributed it to *Fusarium* (*J. Pomol.* **7**, 1929, 302–23; *Reps. exp. Res. Sta. Cheshunt for 1934–6*) subsequently confirmed his results (*J. Pomol.* **14**, 1936, 216–26). Dowson (*AAB*, **16**, 1929, 261–80) had also attributed wilt solely to *Fusarium*. Brown (*Gdnrs' Chron.* **98**, 1935, 267; *Sci. Hort.* **6**, 1938, 93–6) published semi-popular accounts of Wickens's findings. These were that the most important of the diseases he studied was wilt caused by *Verticillium cinerescens*. The others were Fusarium wilt (*Fusarium dianthi*), stem rot (*F. culmorum* and *Fusarium* spp.) and die-back (*F. culmorum*).

White (*Rep. exp. Res. Sta. Cheshunt for 1939*, 33) tested the reactions of species of *Dianthus* other than *D. caryophyllus* to the *Verticillium*. It is not a good *Verticillium* because it has fasciculate conidiophores, a character which excludes it equally from the genus *Phialophora* to which van Beyma (*Antonie van Leeuwenhoek*, **6**, 1940, 34–47) transferred it.

Glasscock (*J. R. hort. Soc.* **81**, 1956, 313–16) described a method of testing cuttings for the presence of this and other parasites.

Verticillium dahliae Kleb., *Myc. Zbl.* **3**, 1913, 66; Sacc., *Syll.* xxv, 706; Wakefield & Bisby, *TBMS*, **25**, 1941, 61.

With *V. albo-atrum* (q.v.) the cause of Verticillium wilt in many plants, including those listed below.

*Runner bean (*Phaseolus*). Kent, 1951.
*Lucerne (*Medicago*). Cambridge (Isaac, *AAB*, **45**, 1957, 550).
*Sainfoin (*Onobrychis*). In trial plots, Cambridge, 1940 (Isaac, *AAB*, **33**, 1946, 28–34).
*Brussels sprouts (*Brassica*). In and near Evesham, Worcs, 1953 and 1956. The disease was described by Isaac (*AAB*, **45**, 1957, 276–83), who considered the fungus to be a distinct physiologic strain restricted to this host.
*Celery (*Apium*). Middx, 1925 (*Report*, vi, 36).
*Cucumber (*Cucumis*). Sometimes specifically identified as the cause of wilt.
*On mint (*Mentha*). Not infrequent, Middx, Kent, Surrey, Sussex, Lancs. Also on peppermint, Surrey, 1944 and 1951; Kent, 1953.
*Sage (*Salvia*). On a 2-acre field, Surrey, 1954; on *Salvia* sp., Herts, 1948.
*Tomato (*Lycopersicon*). Frequent as cause of wilt. (See also *Verticillium albo-atrum*.)
*On blackberry and laxtonberry (*Rubus*), Kent, 1931; *Verticillium* sp. on blackberry, Dorset, 1947; on Japanese wineberry, Beds, 1951.
*Cherry (*Prunus*). Rare. On Morello, Kent, 1931 (*Report*, vii, 77).
*Currant (*Ribes*). Rare. Killing red currant bushes, Yorks, 1940; on black currant, Kent, 1932.
*Melon (*Cucumis*). Sussex, 1953. (See also *V. albo-atrum*.)
*Peach (*Prunus*). Yorks, 1937.
*Quince (*Cydonia*). In layer rows and in rootstocks on which pears are grafted, Kent, 1932 (Wormald & Harris, *Gdnrs' Chron.* **93**, 1933, 192–3); Hants, 1936.
*Blue stripe wilt of raspberry (*Rubus*). Occasional in England and Wales, Kent, Essex, Glam, Yorks. Distributed in Scotland. Described by Harris (*J. Pomol.* **4**, 1925, 221–9; *Rep. E. Malling Res. Sta.* **14–15**, II Suppl. 1928, 128).

*Strawberry (*Fragaria*). Fairly common in East Anglia and Kent, but not always specifically identified, though Keyworth and Bennett (*J. hort. Sci.* **26**, 1951, 304–16) showed that both this species and *Verticillium albo-atrum* may be implicated. Also Worcs, Herts, Hants, Sussex, Ayrshire, Argyll.

*Vine (*Vitis*). Yorks, 1938; Worcs, 1940.

*Hop (*Humulus*), see *Verticillium albo-atrum*.

Anthemis. Middx, 1951.

Antirrhinum, see *Verticillium albo-atrum*.

Chrysanthemum. Wilt in this host is usually attributed to *Verticillium albo-atrum*, but this species is frequently involved in the south-east.

Daphne mezereum. Yorks, 1936; Northumb, 1949.

*Geranium (*Pelargonium*). Jersey, 1955.

Helenium. Yorks, 1938; Salop, 1957.

Helichrysum. Sussex, 1955–6.

*Lupin. Kent, Suffolk, Bucks, Notts.

Phlox. Oxon, Kent, Worcs, Salop. Also Som (*Verticillium* sp.).

*Iceland poppy, Suffolk, 1946; Oriental poppy, Cambs, 1949; poppy, Aberdeen, 1953.

*Privet (*Ligustrum*). Yorks, 1940.

*Rose. Purley, Surrey, 1956–7; Kent, 1948; Berks, 1950.

*Sumach (*Rhus cotinus*). Suffolk, 1954.

*Sweet pea (*Lathyrus*). Sussex, 1950; Surrey.

Verticillium malthousei Ware, *Ann. Bot., Lond.*, n.s. **47**, 1933, 781; Wakefield & Bisby, *TBMS*, **25**, 1941, 61.

*On mushroom (*Agaricus*). Not infrequent in south and south-east England. Also Northants, Essex, Yorks, Forth.

The fungus was first observed by Malthouse (*Trans. Edinb. Fld Nat. micr. Soc.* **4**, 1901, 182), and was carefully studied by Ware (*Ann. Bot., Lond.*, **47**, 1933, 763–85) when he found it at Canterbury, Kent, in 1929. Atkins (*Verticillium on Mushrooms*, Midl. Group Public., Stamford, 1945, 55 pp.) published a popular account of it. (See also La Touche (*Nature, Lond.*, **160**, 1947, 679; **163**, 1949, 69).)

Verticillium nigrescens Pethybr., *TBMS*, **6**, 1919, 117; Sacc., *Syll.* xxv, 706; Wakefield & Bisby, *TBMS*, **25**, 1941, 61.

*Originally described by Pethybridge as a saprophyte on potato, but Isaac (*TBMS*, **32**, 1949, 138–57) obtained strains that were parasitic on potato, tomato, antirrhinum (see also Isaac, *AAB*, **44**, 1956, 105–12), etc. It was found in Jersey in 1955 associated with a corky rot of potato tubers. (See also *Verticillium alboatrum*.)

Verticillium nubilum Pethybr., *TBMS*, **6**, 1919, 117; Sacc., *Syll.* xxv, 706; Wakefield & Bisby, *TBMS*, **25**, 1941, 61.

*On potato (*Solanum*) and tomato (*Lycopersicon*). Originally regarded as non-pathogenic to potato; Isaac (*TBMS*, **36**, 1953, 180–95) obtained strains that proved pathogenic to both potato and tomato.

Verticillium psalliotae Treschow, *Dansk bot. Ark.* **11**, 1941, 5.

*On mushroom (*Agaricus*). Occasional. Kent, Sussex, Derby, Staffs, Yorks. First reported by Atkins (*TBMS*, **31**, 1947, 126–7).

Verticillium tricorpus Isaac, *TBMS*, **36**, 1953, 194.

*On tomato (*Lycopersicon*). Isolated by Isaac (loc. cit.) from wilted tomato plants and shown to be pathogenic. Cambs, Hants, Norfolk, Herts.

Verticillium vilmorinii (Guég.) Westerd. & van Luijk, *Meded. phytopath. Lab. Scholten*, **8**, 1924, 50; Sacc., *Syll.* xxii, 1303; Wakefield & Bisby, *TBMS*, **25**, 1941, 61.

*Wilt of Michaelmas daisy (*Aster*). Common in many parts of Britain. Also on *Aster amellus*, Bucks, 1939. Investigated by Wiltshire (*Rep. Long Ashton Res. Sta. for 1920*, 84–5; *for 1921*, 74–6) and, as *Cephalosporium asteris* n.sp., by Dowson (*TBMS*, **7**, 1922, 283–6; *J. R. hort. Soc.* **48**, 1923, 38–57). Preston (*Gdnrs' Chron.* **103**, 1938, 338) showed that it can be prevented by taking top cuttings.

Verticillium sp.

*On cereals. Causing yellow striping and death of barley plants, Edinburgh, 1955; yellow striping and dwarfing of wheat, Kincardine, Ross-shire, Aberdeen, 1955. Subsequently identified with the disease ascribed in Japan to *Cephalosporium gramineum* Nisikado & Itaka.

*Wilt of rhubarb (*Rheum*). Kent (Wormald, *Rep. E. Malling Res. Sta.* **14** and **15**, II Suppl. 1928, 85 and 118). Also under glass in Ireland (McKay, *Gdnrs' Chron.* **117**, 1945, 24).

*Wilt of *Coleus*. Som, 1938.

*Wilt of *Dahlia*. Forth, Clyde, Cambs.

*Wilt of *Digitalis*. Worthing, Sussex.

*Wilt of *Dimorphotheca*. Corn, 1949, 1953. Mundkur (*Phytopathology*, **20**, 1930, 129) showed that a wilt in this host occurring in Iowa, U.S.A., was caused by *Verticillium albo-atrum*.

*Wilt of *Linaria*. Lincs, 1948.

*Wilt of *Lonicera*. In *L. nitida*, Cheshunt, Herts, 1937.

*Mignonette (*Reseda*). Glos, 1952.

*Wilt of *Schizanthus*. Portsmouth, Hants, 1948; Devon, 1951.

*Wilt of *Sidalcea*. Scotland, 1950.

*Wilt of *Solanum capsicastrum*. Bucks, 1947.

*Wilt of statice (*Limonium*). On *L. suworowii*, Dur, 1956.

*Attacking large trees of elm (*Ulmus*). (See Peace, *Forestry*, **6**, 1932, 125.)

Vialaea insculpta (Fr. em. Oud.) Sacc., *Bull. Soc. mycol. Fr.* **12**, 1896, 67; Sacc., *Syll.* xiv, 620; syn. *Boydia insculpta* (Oud.) Grove, *J. Bot., Lond.*, **59**, 1921, 13.

*On *Ilex*. On twigs of *I. aquifolium* and its variety *hendersonii*, Ayrshire, 1918 (*TBMS*, **6**, 1919, 151); Kew Gardens, 1919; Devon, 1950.

Volutella buxi (DC. ex Fr.) Berk. & Br., *Ann. Mag. nat. Hist.* **5**, 1850, 465; Sacc., *Syll.* iv, 685; Wakefield & Bisby, *TBMS*, **25**, 1941, 62.

*On the foliage and twigs of box (*Buxus*). The status of the fungus in this country is unknown, but it is usually regarded as a wound parasite, and may be responsible for some of the frequent deaths of twigs and branches of clipped box.

Volvaria sp.
*Invading mushroom (*Agaricus*) beds, Norfolk, 1945.

Wojnowicia graminis (McAlp.) Sacc., *Syll.* xviii, 367; Grove, *Coelomycetes*, ii, 87 (as *Woinowicia hirta* Sacc.).
*On wheat. Usually associated with foot-rot parasites, and not often recorded. Parasitism uncertain. First recognized in 1934 in Hants. Also Herts, Berks, Glam.

Xylaria pedunculata (Dickson ex Berk.) Fr., see *X. vaporaria.*

Xylaria tulasnei Nitschke, see *X. vaporaria.*

Xylaria vaporaria Berk. in Currey, *Trans. Linn. Soc.* **24**, 1863, 157; Sacc., *Syll.* i, 341; Bisby & Mason, *TBMS*, **24**, 1940, 157.
*Invades mushroom (*Agaricus*) beds. Known in England since 1862 and occurs most years in southern counties. Perhaps declining with improved sanitation (Wood, *Gdnrs' Chron.* **109**, 1941, 131). This species may be synonymous with *Xylaria pedunculata* (Dickson ex Berk.) Fr., *Summ. Veg. Scand.* 382; Sacc., *Syll.* i, 332, and the mushroom-bed fungus is perhaps to be more closely identified with *X. pedunculata* var. *pusilla* Tul., which Nitschke raised to specific rank as *X. tulasnei* Nitschke, *Pyrenomycetes germanici*, 8; Sacc., *Syll.* i, 334.

Notes on *X. vaporaria* are given by Berkeley (*Gdnrs' Chron.* **23**, 1863, 363; **31**, 1871, 610), Cooke (ibid. **14**, 1893, 299), Green (ibid. **87**, 1930, 516), Salmon and Ware (*J. S.-E. agric. Coll. Wye*, **33**, 1934, 17) and by Wood (*Gdnrs' Chron.* **97**, 1935, 213).
*Causing disease in roots of rhododendron (W. G. Smith, *Gdnrs' Chron.* **23**, 1885, 241).

Zaghouania phillyreae Pat., *Bull. Soc. mycol. Fr.* **17**, 1901, 187; Grove, *Brit. Rust Fungi*, 332; Wilson & Bisby, *TBMS*, **37**, 1954, 80; syn. *Aecidium phillyreae* DC., *Flor. Fr.* vi, 96; Sacc., *Syll.* vii, 807.

*On *Phillyrea*. On leaves and shoots of *P. latifolia*, Pevensey, Sussex, 1907 (Massee, *Dis. Cult. Pl. Trees*, 337); on var. *media*, Chichester, 1869 (aecidia) and 1874 (uredospores).

Zopfia rhizophila Rabenh., *Fung. Eur.* no. 1734; Sacc., *Syll.* i, 55; Bisby & Mason, *TBMS*, **24**, 1940, 134.

*On asparagus. Occasionally found on the roots, but doubtfully parasitic. Worcs, Kent, Surrey, Yorks. First seen 1925 (*Report*, vi, 40) and described by Salmon and Ware (*Gdnrs' Chron.* **87**, 1930, 275).

Zythia fragariae Laib. = stat. conid. of *Gnomonia fructicola*.